BIOLOGY AND ECOLOGY OF BIOLUMINESCENT MARINE FISHES

Biology and Ecology of Marine Life Series

BIOLOGY AND ECOLOGY OF BIOLUMINESCENT MARINE FISHES

Ramasamy Santhanam, PhD

AAP | APPLE ACADEMIC PRESS

First edition published 2024

Apple Academic Press Inc.
1265 Goldenrod Circle, NE,
Palm Bay, FL 32905 USA

760 Laurentian Drive, Unit 19,
Burlington, ON L7N 0A4, CANADA

CRC Press
2385 NW Executive Center Drive,
Suite 320, Boca Raton FL 33431

4 Park Square, Milton Park,
Abingdon, Oxon, OX14 4RN UK

Library and Archives Canada Cataloguing in Publication

Title: Biology and ecology of bioluminescent marine fishes / Ramasamy Santhanam, PhD.
Names: Santhanam, Ramasamy, 1946- author.
Series: Santhanam, Ramasamy, 1946- Biology and ecology of marine life book series.
Description: First edition. | Series statement: Biology and ecology of marine life series | Includes bibliographical references and index.
Identifiers: Canadiana (print) 2023022573X | Canadiana (ebook) 20230225756 | ISBN 9781774913154 (hardcover) | ISBN 9781774913161 (softcover) | ISBN 9781003388623 (ebook)
Subjects: LCSH: Bioluminescence. | LCSH: Marine fishes.
Classification: LCC QH641 .S26 2024 | DDC 572/.435817—dc23

Library of Congress Cataloging-in-Publication Data

Names: Santhanam, Ramasamy, 1946- author.

Title: Biology and ecology of bioluminescent marine fishes / Ramasamy Santhanam, PhD.

Description: First edition. | Palm Bay, FL : Apple Academic Press, 2024. | Series: Biology and ecology of marine life | Includes bibliographical references and index. | Summary: "This enlightening new volume details over 200 species of bioluminescent marine fishes along with their biotechnological and therapeutic applications. It delves into the chemistry, diversity, and functions of these fishes and also discusses their biology and ecology aspects. Bioluminescence, which is a type of chemiluminescence and is a "cold living light," is seen in a wide variety of organisms from bacteria to fish. The chemicals and circumstances involved in marine bioluminescence are studied by biologists and engineers to make use of this phenomenon for biotechnological, commercial, and therapeutic applications. Bioluminescence imaging technology has provided valuable means for the monitoring of different biological processes for immunology, oncology, virology, and neuroscience. A new emerging biological technique that combines luminescence with optogenetics involves the use of light to control cells, particularly neurons, in living tissue. Bioluminescence may also be as a potential form of green energy. The author states that more than 75% of deep-sea creatures (700 marine genera within 16 phyla) have been estimated to possess this phenomenon of bioluminescence. While marine plants do not display bioluminescence, several organisms such as the dinoflagellate, Noctiluca, cnidarian, jellyfish, ostracod crustacean, Vargula, molluscan squids and lantern fish are well known for their luminescence. Bioluminescence is employed by marine organisms mainly to hunt their prey, to defend against predators and in reproduction. This scientifically comprehensive and well-illustrated book will serve as an essential standard reference for students and teachers as well as marine biologists, fisheries scientists, zoologists, and environmentalists"-- Provided by publisher.

Identifiers: LCCN 2023016589 (print) | LCCN 2023016590 (ebook) | ISBN 9781774913154 (hardcover) | ISBN 9781774913161 (paperback) | ISBN 9781003388623 (ebook)

Subjects: LCSH: Marine fishes--Physiology. | Marine fishes--Ecology. | Bioluminescence. | Chemiluminescence.

Classification: LCC QL620 .S26 2024 (print) | LCC QL620 (ebook) | DDC 597.177--dc23/eng/20230518

LC record available at https://lccn.loc.gov/2023016589

LC ebook record available at https://lccn.loc.gov/2023016590

ISBN: 978-1-77491-315-4 (hbk)
ISBN: 978-1-77491-316-1 (pbk)
ISBN: 978-1-00338-862-3 (ebk)

BIOLOGY AND ECOLOGY OF MARINE LIFE BOOK SERIES

SERIES AUTHOR

Ramasamy Santhanam, PhD
Ex-Dean, Fisheries College and Research Institute,
Tamil Nadu Veterinary and Animal Sciences University,
Thoothukkudi-628 008, India
Email: rsanthaanamin@yahoo.co.in

Books in the Series:

- Biology and Culture of Portunid Crabs of the World Seas
- Biology and Ecology of Edible Marine Bivalve Molluscs
- Biology and Ecology of Edible Marine Gastropod Molluscs
- Biology and Ecology of Venomous Marine Snails
- Biology and Ecology of Venomous Stingrays
- Biology and Ecology of Toxic Pufferfish
- Biology and Ecology of Bioluminescent Marine Fishes
- Biology and Ecology of the Venomous Catfishes

ABOUT THE AUTHOR

Ramasamy Santhanam, PhD
Ex-Dean, Fisheries College and Research Institute,
Tamil Nadu Veterinary and Animal Sciences University, Thoothukudi, India

Ramasamy Santhanam, PhD, is the Ex-Dean of the Fisheries College and Research Institute (FC & RI), Tamil Nadu Veterinary and Animal Sciences University, India. He has 37 years of teaching and research experience in marine/fisheries sciences. He has served as a fisheries expert for various government and non-government organizations in India and abroad. Dr. Santhanam has so far published 33 books on various aspects of marine life. He has also served as a member of the American Fisheries Society, United States; World Aquaculture Society, United States; Global Fisheries Ecosystem Management Network (GFEMN), United States; and the IUCN's Commission on Ecosystem Management, Switzerland.

CONTENTS

ABBREVIATIONS

AEG	accessory escal gland
DPL	discrete projected luminescence
FMNH2	flavin mononucleotide
LO	light organ
LOS	light organ system
PAC	posterior AC
RCHO	reacts with an aldehyde
rRNA	ribosomal RNA
SL	standard length

PREFACE

Bioluminescence, which is not the same thing as fluorescence or phosphorescence, is a type of chemiluminescence and is a "cold living light." It is seen in a wide variety of organisms, from bacteria to fish. While some bioluminescent organisms, like fireflies and fungi, are found on land, there are almost no bioluminescent organisms in freshwater environments. Interestingly, most bioluminescent organisms are found in the seas and oceans at all depths. It is more common in tropical waters, than in higher latitudes.

More than 75% of deep-sea creatures (700 marine genera within 16 phyla) have been estimated to possess this phenomenon. Among the known 20,000 known species of marine fishes, bioluminescence has been recorded in about 1500 species of fish. It is also reported that 90% of fishes observed below 500 m depth are bioluminescent. While marine plants do not display bioluminescence, several organisms such as the dinoflagellate, *Noctiluca,* cnidarian, jellyfish, ostracod crustacean, *Vargula,* molluscan squids, and lantern fish are well known for their luminescence. Bioluminescence is employed by marine organisms mainly to hunt their prey, defend against predators, and in reproduction.

The chemicals and circumstances involved in marine bioluminescence are being studied by biologists and engineers to make use of this phenomenon for biotechnological, commercial, and therapeutic applications. For instance, bioluminescent trees may help to light highways and city streets which could considerably reduce the need for electricity. Further, bioluminescent crops could emit light when there is a need for water or nutrients; or ready for harvest. When this is possible, it could reduce the cost for agribusiness and farmers.

Though several titles are available presently on the Biology and Ecology of Marine Life, a comprehensive book on the "Bioluminescent Marine Fishes" is the need of the hour. Keeping this in consideration, the present title, which deals with about 200 species of bioluminescent marine fishes; and biotechnological and therapeutic applications of bioluminescent marine fishes is being brought out. It is hoped that this well-illustrated book will certainly serve as a standard reference for students and teachers of Marine Biology, Fisheries Science, and Environmental Science. Further, it would also attract the shelves of libraries in universities and research establishments.

The author is immensely thankful to Dr. K. Venkataramanujam, former Dean of Fisheries College and Research Institute, Tamil Nadu Veterinary and Animal Sciences University, for his valued comments for the improvement of the manuscript. Thanks are also due to my international friends who permitted me to share their images. The secretarial assistance rendered by Mrs. Albin Panimalar Ramesh is also gratefully acknowledged.

—**Author**

CHAPTER 1

INTRODUCTION

Bioluminescence, an amazing biological phenomenon, is the heatless light generated chemically by living organisms in contrast to 'biofluorescence,' which is only a process, in which blue light hits the surface of an animal and is reemitted as a different color. Bioluminescence is extremely common in all the seas and oceans and occurs at all depths. Bioluminescent animals can be found throughout the ocean. It is estimated that 76% of pelagic organisms (those that live in the open water) and 30–40% of the benthic organisms (those that live close to the seafloor) have the ability to emit light (Martini and Francis, 2020).

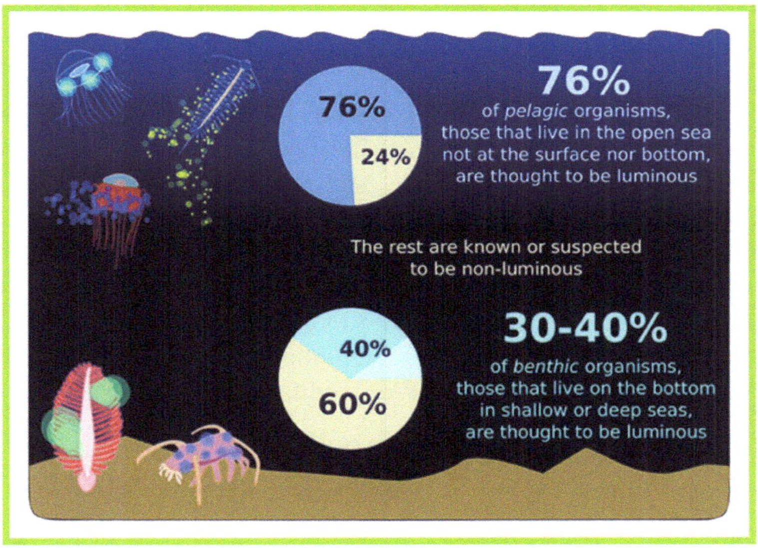

Source: Reprinted from Martini and Francis, 2020. https://creativecommons.org/licenses/by/4.0/

Bioluminescence in the Oceans

Biology and Ecology of Bioluminescent Marine Fishes. Ramasamy Santhanam (Author)

The phenomenon, bioluminescence is especially abundant in the deep-sea (e.g., lanternfishes and anglerfishes) and is less common on coral reefs (e.g., cardinalfishes and pineconefishes) and other localities near the shore (e.g., ponyfishes and croakers) (Davis et al., 2016). Interestingly bioluminescence is rare on land, and it is nearly non-existent in freshwater. At present, the only known terrestrial animals capable of bioluminescence are arthropods (e.g., fireflies and millipedes) whereas in marine environments, the majority of luminescent taxa are invertebrate animals such as ctenophores, annelids, crustaceans and mollusks. Vertebrates lack bioluminescent light organs, with the exception of teleosts and some shark species. It is estimated that about 80% of the marine animals possess bioluminescence. Almost all marine bioluminescence is blue in color, for two reasons, viz. (i) blue-green light (wavelength around 470 nm) gets transmitted furthest in water; and (ii) most organisms are sensitive only to blue light. The color of light emitted by marine fishes; however, may range from nearly violet to green-yellow (and very occasionally red (Johnson, https://oceanexplorer.noaa.gov/explorations/15biolum/background/biolum/biolum.html).

1.1 BIOLUMINESCENCE IN MARINE FISH

While about 70% of the deep-sea bony fishes are luminous, only a small fraction of deep-sea cartilaginous fishes (about 6%) is endowed with this capability. It is worthy of mention here that in chimaeras (Holocephali), this phenomenon is totally absent (Claes and Mallefet, 2009). Paitio et al. (2016) reported that a total of about 1500 species of fish (i.e., about 8% of the approximately 20,000 known species) representing 201 genera (bony fishes, 190; and cartilaginous fishes, 11) and 45 families (bony fishes, 43; and cartilaginous fishes, 2) display bioluminescence. The luminous marine fishes representing the different orders and families are given below.

1.2 EMISSION MAXIMA OF LUMINESCENT MARINE FISH

While the terrestrial animals are mainly yellow-green luminescent, marine species of pelagic environment are mostly blue-emitting and of benthic environment are green to red light emitting (Herring, 1983). The emission maxima of marine species may therefore have a range of 443–689 nm. The range of values of emission maxima (λmax (nm) of different groups of marine animals and the color of their light are given in the following table.

Luminous Marine Fishes

Order	Family	Genera / Species
Cartilaginous Fishes Squaliformes	Dalatiidae	All genera
	Etmopteridae	All genera
	Somniosidae	*Zameus squamulosus*
Bony Fishes Anguilliformes	Congridae	*Lumiconger arafura*
Aulopiformes	Chlorophthalmidae	*Chlorophthalmus*
	Evermannellidae	*Cocorella atrata, Odontostomops normalops*
	Paralepididae	*Lestidium, Lestroplepis*
	Scopelarchidae	*Benthalbella, Scopelarchoides*
Batrachoidiformes	Batrachoididae	*Porichthys*
Beryciformes	Anomalopidae	All genera
	Monocentridae	All genera
	Trachichthyidae	*Aulotrachiththys*
Clupeiformes	Engraulidae	*Coilia dussumieri*
Gadiformes	Macrouridae	*Cetonurus, Coelorinchus, Haplomacrurusd, Hymenocephalus, Kumba, Lepidorynchus, Lucigadus, Malacocephalus, Nezumia, Odontomacrurus, Sphagemacrurus, Spicomacrurus, Trachonurus, Ventrifossa*
	Merlucciidae	*Steindachneria argentea*
	Moridae	*Gadella, Physiculus, Salilota, Tripterophycis*
Lophiiformes	Centrophrynidae	All genera
	Ceratiidae	All genera
	Diceratiidae	All genera
	Gigantactinidae	*Gigantactis*
	Himantolophidae	All genera
	Linophrynidae	All genera
	Melanocetidae	All genera
	Oneirodidae	All genera
	Thaumatichthyidae	All genera
	Ogcocephalidae	*Dibranchus atlanticus*
Myctophiformes	Myctophidae	All genera

(Continued Table)

Order	Family	Genera / Species
	Neoscopelidae	*Neoscopelus*
Osmeriformes	Alepocephalidae	*Microphotolepis, Photostylus, Rouleina, Xenodermichthys*
	Microstomidae	*Nansenia*
	Opisthoproctidae	*Opisthoproctus, Rhynchohyalus, Winteria*
	Platytroctidae	All genera
Perciformes	Acropomatidae	*Acropoma, Synagrops*
	Apogonidae	*Apogon, Archamia, Jaydia, Rhabdamia, Siphamia*
	Chiasmodontidae	*Pseudoscopelus*
	Epigonidae	*Epigonus, Florenciella, Rosenblattia*
	Howellidae	*Howella*
	Leiognathidae	All genera
	Pempheridae	*Pempheris, Parapriacanthus*
	Scianidae	*Collichthys, Larimichthys, Sonorolux*
Saccopharyngiformes	Eurypharyngidae	All genera
	Saccopharyngidae	All genera
Stomiiformes	Gonostomatidae	All genera
	Phosichthyidae	All genera
	Sternoptychidae	All genera
	Stomiidae	All genera

Source: Reprinted from Paitio et al. (2016). https://creativecommons.org/licenses/by/3.0/

Emission Maxima (λmax) Values and Color of Light of Marine Organisms

Phyla/class	λmax (nm) range	Color
Protozoa	443–458	Blue
Cnidaria	444–488	Blue
Ctenophora	478–496	Blue
Annelida	565	Green
Crustacea	444–492	Blue
Mollusca	449–514	Blue
Tunicata	471, 493	Blue
Pisces	477 – 689	Blue to Red

Herring (1983).

1.3 TYPES OF LUMINESCENCE IN MARINE FISH

Marine fish bioluminescence may be either directly or indirectly, i.e., intrinsic/intracellular or extrinsic/extracellular. While the first method is achieved by making endogenous proteins to carry out the light-producing reaction, the latter is by housing bioluminescent (symbiotic) bacteria or luminous, digested crustacean substances in the light organ; or by secreting luminous fluids. Intrinsic luminescence is caused by chemicals produced by the organism itself. Intracellular bioluminescent organisms which are self-luminescent have specialized luminous cells, i.e., photocytes which are grouped inside special organs called photophores. On the other hand, in the extracellular, bacterially luminescent fishes, symbiotic luminous bacteria are housed in extensions of their gastrointestinal tract, which are elaborate and specialized organs (Tanet et al., 2020). It is reported that 52% of the luminescent fishes produce intracellular luminescence and 48% (30% bacterial and others 18%) extracellular luminescence (Davis et al., 2016). While the intracellular, luminescent fishes can be found everywhere from coastal habitats down to the depths of the ocean, majority of bacterially luminescent fishes are found as deep-sea fishes, e.g., the rattails (Gadiformes: Macrouridae). The major types of luminescence observed in marine fishes are given in the following table.

Intrinsic and Bacterial (extrinsic) Luminescence in Marine Fishes

Order	Type of Luminescence
Telest fishes	
Saccopharyngiformes	Intrinsic
Anguilliformes	Intrinsic
Stomiiformes	Intrinsic
Myctophiformes	Intrinsic
Batrachoidiformes	Intrinsic
Aulopiformes	Intrinsic and bacterial
Lophiformes	Intrinsic and bacterial
Perciformes	Intrinsic and bacterial
Gadiformes	Bacterial
Beryciformes	Bacterial
Cartilaginous fishes	Intrinsic

Davis et al. (2016).

1.4 FUNCTIONS OF BIOLUMINESCENCE IN MARINE FISHES

The most important function of bioluminescence of marine fishes is to illuminate the surroundings for the purpose of camouflage, schooling and for recognition of movement of predators in the water. In the dark of the ocean, bioluminescence can help fish survival in at least three critical ways:

i) In several deep-sea fish, such as anglerfish and viperfish, bioluminescence can serve as an aid in attracting prey by the use of glowing lures. The dangling appendage which arises from the snout of the anglerfish has a light organ at the end.
ii) It can help in reproduction by attracting a mate by means of species-specific spatial or temporal patterns of light emission.
iii) It can serve as a defense against predators.

1.5 POTENTIAL OF FISH BIOLUMINESCENCE IN BIOMEDICAL AND BIOTECHNOLOGICAL APPLICATIONS

Brain activity studies: Fish luminescence can be combined with optogenetics, a new biological technique that deals with the use of light from fish to control neurons in living tissue. This technique may serve as a powerful new tool for studying brain activity (Anon., https://www.sciencedaily.com/releases/2016/10/161027164348.htm).

Cancer immunotherapies: The enzyme luciferase from marine organisms, when introduced into cancer cells relating to chronic myelogenous leukemia, acute myelogenous leukemia, and Burkitt lymphoma, has been reported to leak out as the cancer cells die, leaving a visible glow. This assay has been reported to accurately recognize the death of a single cancer cell in 30 minutes (Dybas, 2019).

Hygiene control and pollutant monitoring: The ATP-bioluminescence assay relating to the firefly bioluminescence system is presently used in applications relating to hygienic control and pollutant monitoring in marine environment. Similar applications could also be made possible in the future with bioluminescent marine fishes if and when adequate research is attempted (Syed and Anderson, 2021).

CHAPTER 2

BIOLUMINESCENCE: CHEMISTRY, DIVERSITY, AND FUNCTIONS

Bioluminescent light is normally generated in specialized light organs through oxidation of a light emitting luciferin in the presence of oxygen and the enzyme luciferase. It has been established that the glandular tissue of luminescent fish secretes a chemical substance called as luciferin, which is an indole derivative consisting of tryptamine, arginine and isoleucine. Under the influence of the enzyme luciferase, this substance is converted into oxy-luciferin and emits blue or blue-green light. In this chemical reaction, after the release of light, the luciferin becomes inactive until it is replaced by the fish. It is also interesting to note that when the ingredients luciferin and luciferase are mixed together, either in a cell or in the water itself, one may get light. Most of the bioluminescent species produce blue- green light which has an emission maximum (λmax) wavelength around 475 nm. Due to the inclusion of chemical substance, this phenomenon may also be termed as "chemical luminescence."

$$\text{Luciferin (substrate)} + O_2 \xrightarrow{\text{Luciferase (Enzyme)}} \text{Oxyluciferin} + CO_2 + h\nu \text{ (light)}$$

To harness this biochemical reaction in a bioluminescent system, luminous organisms employ modules to perform several physiological functions viz. (i) production/dietary acquisition of a photoprotein/luciferase, (ii) production/dietary acquisition of a luciferin, (iii) maintenance of a light organ producing bioluminescence or house luminous bacterial symbionts, (iv) control of light production, (v) catabolism and/or recycling of bioluminescent molecules, (vi) storage of bioluminescent substrates (in some cases), and (vii) modification of light emission using biochemical or physical mechanisms such as green fluorescent proteins, filters and chromatophores (Lau and Oakley, 2021).

Biology and Ecology of Bioluminescent Marine Fishes. Ramasamy Santhanam (Author)

2.1 LUCIFERINS

The first luciferin (from the Latin lucifer, "light-bearer"), was isolated by Green and McElroy in 1956 (Green and Mcelroy, 1956). Luciferins of the fish may be of two types, viz. coelenterazine (coelenterate –type luciferin) and cypridine type luciferin. This coelenterazine, an imidazolopyrazine derivative is produced by the luminescent fishes and is primarily found in the marine invertebrates, viz. coelenterates has also been reported in marine luminous organisms such as radiolarians, arthropods and mollusks (Lee, info/LeeBasicBiolum.html). Vargulin or Cypridina-type luciferin (cypridinidinluciferin) is originally found in the crustacean, ostracod ("seed shrimp") *Vargula* spp. and *Cypridina* spp. It is reported that fishes like the midshipman fish *Porichthys* are extrinsically luminescent due to their feeding with these ostracods (Haneda, 1980). Further, there is also a clear dietary link, with fish losing their ability to luminesce until they are fed with luciferin-bearing food. It has been shown that these ostracods synthesize vargulin molecule from the amino acids tryptophan, isoleucine, and arginine (Anon., https://en.wikipedia.org/wiki/Vargulin]. The bacterial luciferin which is synthesized by the symbiotic bacteria of luminescent fishes is a reduced riboflavin phosphate. This bacterial luciferin is oxidized in association with a long-chain aldehyde, oxygen, and a luciferase in the chemical reaction of producing light (Anon., https://biolum.eemb.ucsb.edu/chem/detail2.html).

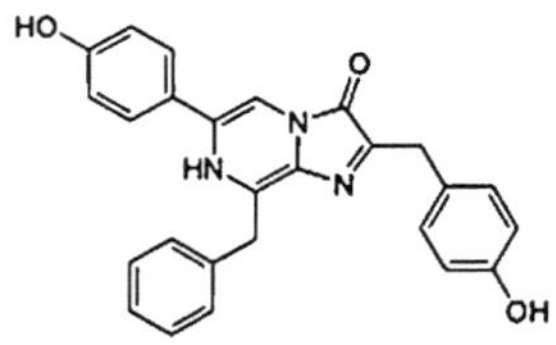

Coelenterazine

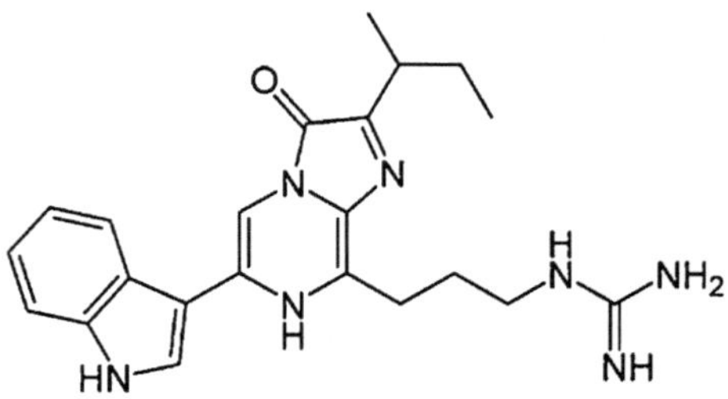

Vargulin (cypridinluciferin)

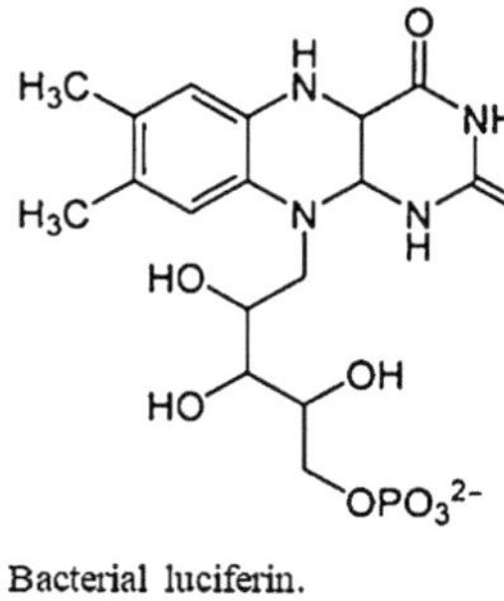

Bacterial luciferin.

2.2 LUMINOUS ORGANS OR PHOTOPHORES IN MARINE FISHES

Marine fishes are known to produce characteristic light through their special light-emitting cells called luminous organs or photophores. These organs, which are special gland cells of the epidermis, are commonly found in fishes living in deep-sea where the sunlight stops entering. The distribution of photophores on the body type and adaptive value may vary in different species of fish. Luminescent sharks possess numerous, small, external light organs (photophores) with intrinsic light production. In bony fishes, bioluminescence reaches its maximum in terms of complexity and diversity of light organs with biological and ecological functions. All the species of families like Myctophidae, Leiognathidae, Anomalopidae, and Platytroctidae and several families of anglerfishes are bioluminescent. In all these families, the light organs' patterns or shapes are generally species-specific and can even be used for taxonomic purposes. Similarly, all the species of the order Stomiiformes produce their own light. On the other hand, only some genera (with very few luminescent species in each genus) of fishes are bioluminescent in families of the order Perciformes or with more luminescent species as in the family Macrouridae. However, only a single luminous species is known in the order Clupeiformes (*Coilia dussumieri*) and in Anguiliformes (*Lumiconger arafura*); and in the families Merluciidae (*Steindachneria argentea*) and Ogcocephalidae (*Dibranchus atlanticus*).

2.3 TYPES OF PHOTOPHORES

The photophores of luminescent fishes may range from simple clusters of cells to complex organs surrounded by reflectors, lenses, color filters, and muscles. In many fishes, the photophores appear as luminous spots on the body. In lanternfishes possessing intrinsic luminescence, the photophores are present ventrally along the belly, laterally on the flank and head, and on the tail. While their ventral series of photophores are mainly used for camouflage while the lateral photophores are normally used for intra-specific signaling. In bacterially luminescent fishes, there are usually one or two bacterial light organs and rarely four. Due to this restriction, in these fishes, each organ serves several purposes like illuminating the prey, for schooling and for sexual communication (e.g., flashlight fishes). The bacterial photophores are either connected to the gut or are open to the sea. These photophores glow continuously and the fish switch them off by masking them with a shutter or by rotating them into a black-lined pocket. Fishes with high-density luminous bacteria (e.g., *Photoblepheron*) flash intermittently by blinking the shutter over its light organs. On the other hand, some greeneye fish of the family Chlorophthalmidae have weak bacterial, peri-anal light organs which are mainly used to illuminate the ventral surface presumably for camouflage. In *Opisthoproctus*, the light organ which is located near the anus is also used for ventral camouflage. Most complex photophores have been developed for ventral camouflage in fishes of the upper 1000 m of the ocean and these photophores help match the background when viewed obliquely as well as directly from below.

On the basis of anatomy, the photophores may be categorized into two types:

- **i) Simple Photophore:** These photophores are small in size, 0.1 to 0.34 mm in width, and each photophore consists of light-generating cells called photocytes. Simple type photophores may be provided with or without mantle of pigment. The lenses are made of a grouping of cells known as lenticular cells. The distal part of photocyte is provided with acidophilic granules. A layer of melanophres surrounds the photophore. This simple type of photophores is seen in the elasmobranchs viz. sharks. In the bony fish, Stomias, the luminous organs are found logged in gelatinous corium of the epidermis (Anon., https://www.lkouniv.ac.in/site/writereaddata/siteContent/202003251324430605shelly_Sound_Producing.pdf).
- **ii) Compound Photophore:** This type of photophores consists of additional structures like reflectors, pigmented mantle and sub-ocular organs. The sub-ocular organ is a large organ deeply embedded in dermal tissue. The photocytes of these compound photophores are arranged in the form of cords and bands. The photogenic tissues which are found in the center of the photophore consist of two types of glandular cells and are provided with nerves and blood vessels (Anon., https://www.lkouniv.ac.in/site/writereaddata/siteContent/202003251324430605shelly_Sound_Producing.pdf).

2.4 ADDITIONAL LUMINESCENT ORGANS

Deep-sea fishes of the order Lophiiformes (anglerfishes) have evolved a peculiar organ to assist in the attraction and capture of prey viz. a modified first dorsal-fin spine (illicium) bearing a fleshy bait (esca) that acts as a fishing apparatus to lure prey. The esca is a spherical, luminescent bacteria-filled organ which contains one or more small openings to the external environment. In certain deep-sea fish species, in addition to the long "fishing rod" (illicium) and bioluminescent lure, and two or three soft fleshy wart-like bioluminescent appendages (caruncles) on the back. These luminescent bacteria-containing caruncles, which are similar to the esca, are modified anterior dorsal-fin rays and are located in front of the soft dorsal fin.

2.5 CONTROL OF THE FUNCTION OF LIGHT-PRODUCING ORGANS

The function of light-producing organs in luminescent fishes is often controlled by their nervous or endocrine system (Anon., https://www.lkouniv.ac.in/site/writereaddata/siteContent/202003251324430605shelly_Sound_Producing.pdf).

i) **Nervous Control:** In luminescent fishes, the light production by their luminous organs is controlled by the nervous system, probably by the peripheral sympathetic system. The efferent nerves enter the photogenic cells and activate them.
ii) **Hormonal Control:** It has been reported that some fishes have hormonal control on the photophores. Endocrine glands like adrenal glands (also called as supra renal glands) activate the photophores besides controlling their light emission.
iii) **Mechanical Control:** In certain luminescent fishes, the muscles located beneath the photophores contract and rotate the photophores in such a way that they get concealed. In *Photoblepharon palpebratus*, the ventral part of its luminous organ has a fold of black tissue, and this fold can be drawn over the photophores and conceal the light. In some fishes, light production seems to be influenced by the movement of pigment in the chromatophores.

2.5.1 MECHANISM OF LIGHT PRODUCTION

The mechanism of light production is amazing in marine fishes, and it takes place at the special sets of muscles located around the photocytes. When these muscles contract, they pull the outer surface of the photophore downwards, thereby causing a brighter surface to be concealed. On the other hand, the relaxation of these muscles exposes the bright surface of the photophores. In some species of fish, movement of pigmented layer executes the works of concealing and rotating photophores.

2.5.2 TYPES OF LIGHT PRODUCTION

Luminescence in marine fishes can broadly be divided into two different types of light production, viz. intrinsic (intracellular or autogenic or endogenous) bioluminescence and extrinsic (extracellular or exogenous) bioluminescence.

Some fishes can make luciferin themselves, or it may be synthesized by symbiotic bacteria inside the photophore in some cases. While intrinsic luminescence has been reported in 25 families of marine fishes, extrinsic luminescence is available in 30 genera in 11 families of fishes (Nealson and Hastings, 1979).

- **i) Intracellular Luminescence:** In this type, the fishes have specialized glandular, luminous cells, i.e., intrinsic photocytes, grouped inside specialized organs called photophores (Herring, 1977). These luminous organs are developed from the epidermis. These fish species are self-luminescent and show their own luciferin-luciferase system in these light organs. Fishes possessing this type of luminous organs belong largely to the family of teleosts, i.e., Sternoptychidae (hatchetfish), Myctophidae (lanternfish), Halosauridae (Halosaurid eel), Stomiatidae (scaly dragon fishes), Brotulidae (Brotulus), Lophiidae (anglerfish), and Zoarcidae (eelpouts). Interestingly intrinsic luminescence is confined to only elasmobranch fishes (Anon., https://www.lkouniv.ac.in/site/writereaddata/siteContent/202003251324430605shelly_Sound_Producing.pdf).
- **ii) Bacterial Luminescence:** In this type, fish species host bioluminescent, symbiotic bacteria (bacteriogenic bioluminescence) within light-emitting cells called photophores or in specialized light organs. The biochemical step in bacterial luminescence is linked to the electron transport chain of oxidative phosphorylation, in which flavin mononucleotide (FMNH2) from the electron transport chain reacts with an aldehyde (RCHO) to form a complex (luciferin) that is oxidized to an acid (RCOOH) with emission of light (Anon., https://www.lkouniv.ac.in/site/writereaddata/siteContent/202003251324430605shelly_Sound_Producing.pdf). Barton (2016) reported that this exogenous bioluminescence depends more on diverticula and extensions of the digestive tract (e.g., coelom) of fishes to house symbiotic bacteria

derived from the environment. Fishes with such extrinsic bioluminescence are not able to control their light emission directly, and they use mechanical structures to cover the light organs and produce impressive blink patterns during the night.

iii) **Extra Cellular Luminescence:** In some fishes, light may be generated by luminous secretion from the glandular tissues. These extracellular luminescent organs are found in a very limited species of fishes. For example, fishes like rattails emit light by secreting extracellular slime. These rattails possess special glands near their anus, which secretes slime of sufficient luminosity (Haneda, 1980; Anon., https://www.lkouniv.ac.in/site/writereaddata/siteContent/202003251324430605shelly_Sound_Producing.pdf).

iv) **Extrinsic Luminescence by Luminous Ingested Crustaceans:** Extrinsic bioluminescence of some species of fishes may also be due to the inclusion of luminous, digested crustacean substances in the light organ (Haneda, 1980). It is worthy of mention here that regardless of whether the luminescence is made in cells, made by bacteria, made by ingested crustaceans, or it is vomited, the light is created by the same basic reaction. In this reaction, a small organic molecule called the substrate (luciferin) is oxidized. This reaction requires energy and is helped along by an enzyme (luciferase).

2.5.3 LOCATIONS OF PHOTOPHORES IN LUMINESCENT MARINE FISHES

In luminescent sharks, the light organs are predominantly present in the ventral region of their body. In the dalatiid species, lower densities of light organs may be present even in the dorsal areas. In luminescent, teleost fishes, the light organs (photophores) may be located on or in different parts of their body. In deep-sea fishes, the photophores are present along their body, under the eyes, and often on barbels or antennae. In luminescent, bathypelagic fishes, the photophores are located ventrally and they display ventral bioluminescence. The latter ventral light (counterillumination) helps these deep-sea fishes from their predators swimming above. Similar light ventral light which is a counter or disruptive illumination helps the inshore fishes decreases predation from below, especially during day time (Pauly, 1977). McMillan et al. (2019) reported on the location of photophores in selected species of luminescent marine species as given in the following table.

Location of Photophores in Selected Luminescent Marine Species

Family/ Fishes	Location of photophores
Diplophidae (diplophids)	2 rows of photophores along lower body; and photophores on isthmus (throat)
Gonostomatidae (bristlemouths)	Rows of photophores on lower, long thin body
Sternoptychidae (marine hatchetfishes)	Prominent row of photophores low on sides of deep body; and groups of photophores in cluster-like organs
Phosichthyidae (lightfishes)	2 rows of photophores on ventral body from head to about anal fin origin; and 1 row of photophores extends back to tail
Stomiidae (barbeled dragonfishes)	Postorbital photophore present; rows of small photophores running along lower body, one row lateral, another ventral
Neoscopelidae (blackchins)	Photophores on body in horizontal rows and on tongue in *Neoscopelus*
Myctophidae (lanternfishes)	All species with groups of photophores on head and body
Himantolophidae (prickly anglerfishes, footballfishes)	Large bioluminescent esca with tentacles at tip of illicium
Linophrynidae (leftvents)	Illicium short, less than head length with large terminal bioluminescent organ (esca)

Source: McMillan et al. (2019).

2.5.4 ORIGINS OF THE LIGHT EMISSIONS IN MARINE FISHES

Generally, the origins of the light emissions do not vary much within families and it may be either intrinsic or extrinsic (bacterial). However, exceptions exist in the Apogonidae where only the species of the genus *Siphamia* are known to have bacterial symbionts, whereas the remainder of the luminescent species employ intrinsic bioluminescence with *Cypridina* luciferin. Similarly, a rare case has been observed in anglerfishes of the families Lynophrinidae and Centrophryinidae which have an intrinsic bioluminescent barbel in addition to the symbiotic luminous lure. The latter is the modification of the first ray of the dorsal fin and projects from the dorsum of the anglerfishes.

2.5.5 MARINE FISHES WITH INTRACELLULAR LUMINESCENCE

In the intrinsic bioluminescence, the light is emitted by specialized cells called photocytes which are sometimes grouped into complicated lensed structures called photophores. These photophores look very much like eyes, except that light comes out instead of going in. This bioluminescent system produces light through the oxygenation of a substrate, called luciferin (lat. lucifer, the light bringer), and an enzyme, luciferase (photoprotein). This reaction is typically described as a luciferase catalyzed production of an excited intermediate from oxygen and luciferin that emits light when returning to its ground state. The families of marine fishes possessing intrinsic bioluminescence are listed below.

Families of Intracellular Luminescent Fishes

Order	Family/Subfamily	Genera
Elasmobranchs Squalformes	Dalatiidae	*Dalatias, Euprotomicroides, Euprotomicrus, Heteroscymnoides, Isistius, Mollisquama, Squaliolus,*
	Etmopteridae	*Aculeola, Centroscyllium, Etmopterus, Miroscyllium, Trigonognathus*
	Somniosidae	*Somniosus, Zameus*
Teleosts Acropomatiformes	Epigonidae	*Epigonus, Florenciella, Rosenblattia*
	Howellidae	*Howella*
	Pempheridae	*Parapriacanthus, Pempheris*
Alepocephaliformes	Alepocephalidae	*Bathyprion, Mirognathus, Microphotolepis, Photostylus, Rouleina, Xenodermichthys*
	Platytroctidae	*Holtbyrnia, Maulisia, Persparsia, Sagamichthys, Searsia, Searsioides*
Aulopiformes	Evermannellidae	*Coccorella*
	Paralepidida	*Lestidium, Lestrolepis*
	Scopelarchidae	*Benthalbela, Scopelarchoides*
Batrachoidiformes	Batrachoididae	*Porichthys*
Clupeiformes	Engralulidae	*Coilia*
Lophiformes	Ogcocephalidae	*Dibranchus*
Myctophiformes	Myctophidae Subfamily: Myctophinae	*Benthosema, Centrobranchus, Ctenoscopelus, Dasyscopelus, Diogenichthys, Electrona, Gonichthys, Hygophum, Krefftichthys, Loweina, Metelectrona, Myctophum, Protomyctophum, Symbolophorus, Tarletonbeania*

(Continued Table)

Order	Family/Subfamily	Genera
	Subfamily: Diaphinae	*Diaphus, Idiolychnus, Lobianchia*
	Subfamily: Gymnoscopelinae	*Gymnoscopelus, Hintonia, Lampanyctodes, Lampichthys, Notoscopelus, Scopelopsis*
	Subfamily: Lampanyctinae	*Bolinichthys, Ceratoscopelus, Lampadena, Lampanyctus, Lepidophanes, Parvilux, Stenobrachius, Taaningichthys, Triphoturus*
	Subfamily: Notolychninae	*Notolychnus*
	Neoscopelidae	*Neoscopelus*
Notacanthiformes	Halosauridae	*Halosauropsis*
Saccopharyngiformes	Eurypharyngidae	*Eurypharynx*
	Saccopharyngidae	*Saccopharynx*
Scombriformes	Chiasmodontidae	*Pseudoscopelus*
Stomiiformes	Gonostomatidae	*Cyclothone, Gonostoma, Sigmops, Triplophos*
	Phosichthyidae	*Ichthyococcus, Phosichthys. Polymetme*
	Sternoptychidae	*Argyripnus, Argyropelecus, Maurolicus. Polyipnus, Sternoptyx, Valenciennellus*
	Stomiidae	*Aristostomias. Chirostomias, Malacosteus, Pachystomias, Astronesthes, Borostomias, Chauliodus, Echiostoma, Eustomias, Grammatostomias, Idiacanthus, Melanostomias, Opostomias, Photonectes, Photostomias, Rhadinesthes, Stomias, Tactostoma, Thysanactis*
Eupercaria incertae sedis (Uncertain)	Sciaenidae	*Collichthys, Larimichthys, Sonorolux*

2.6 MARINE FISHES WITH BACTERIAL LUMINESCENCE

Fishes with extrinsic luminescence lack their own luciferine luciferase, light emitting system. For example, in the anglerfish (order Lophiiformes), the luminous organ consists of a modified luminescent fin ray (illicium and esca) which acts as a lure for other fish. In these anglerfish, the luminescence comes from symbiotic bacteria, which are thought to be acquired from seawater and are dwelling in and around the esca.

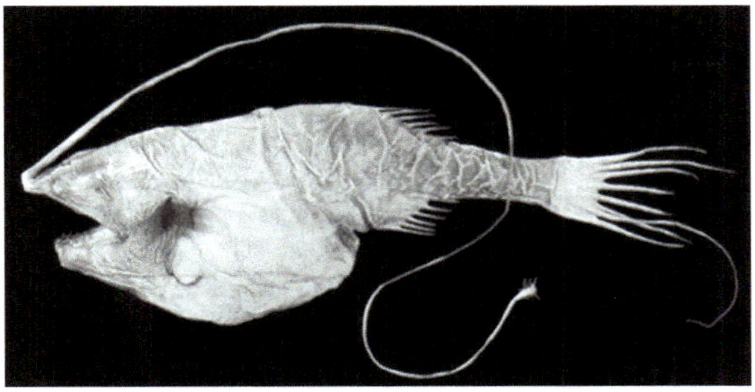

Image courtesy: Wikwand

Anglerfish showing a fishing rod (illicium) and esca
(with bacteria containing escal bulb).

Source: Reprinted from Martini and Francis, 2020. https://creativecommons.org/licenses/by/4.0/

Luminous lure of anglerfish

In other extrinsic luminescent fishes, these luminous bacterial species are harbored in specialized light organs, which may be either appendage of the alimentary tract (e.g., Family: Apogonidae) or they show their light organs under the eyes (subocular) (e.g., Family: Anomalopidae), in a modified dorsal fin ray (e.g., Family: Ceratioidei) or light organs at the anterior angle of the lower jaw (e.g., Family: Monocentridae) (Hellinger et al., 2017). The marine fish families and genera possessing extrinsic bioluminescence are listed below.

Families/Genera of Marine Fishes with Bacterial Luminescence

Order	Family	Species
Acanthuriformes	Leiognathidae	*Aurigequula, Deveximentum, Equulites, Eubleekeria, Gazza, Karalla, Leiognathus, Nuchequula, Photolateralis, Photopectoralis, Secutor*
Acropomatiformes	Acropomatidae	*Acropoma*
Anguilliformes	Congridae	*Lumiconger*
Argentiniformes	Microstomatidae	*Nansenia*
	Opisthoproctidae	*Dolichopteryx, Monacoa, Opisthoproctus, Rhynchohyalus, Winteria*
Aulopiformes	Chlorophthalmidae	*Chlorophthalmus*
	Synodontidae	*Harpadon*
Gadiformes	Bathygadidae	*Gadomus*
	Macrouridae	*Cetonurus, Coelorinchus, Haplomacrurus, Hymenocephalus, Hymenogadus, Kumba, Kuronezumia, Lepidorhynchus, Lucigadus, Malacocephalus, Mesovagus, Nezumia, Odontomacrurus, Sphagemacrurus, Spicomacrurus, Trochonurus, Ventrifossa*
	Merlucciidae	*Steindachneria*
	Moridae	*Gadella, Physiculus, Salilota, Tripterophycis*
Kurtiformes	Apogonidae	*Jaidia, Rhabdamia, Taeniamia, Siphamia*
Lophiformes	Centrophrynidae	*Centrophryne*
	Ceratiidae	*Ceratias, Cryptopsaras*
	Diceratiidae	*Bufoceratias, Diceratias*
	Gigantactinidae	*Gigantactis*
	Himantolophidae	*Himantolophus*
	Linophrynidae	*Haplophryne, Photoorynus*
	Lophiidae	*Lophiodes, Lophiomus, Lophius*
	Melanocetidae	*Melanocetus*
	Oneirodidae	*Bertella, Chaenophryne, Danaphryne, Leptacanthichthys, Lophodolos, Microlophichthys, Oneirodes, Phyllorhinichthys, Spiniphryne, Tyrannophryne*
	Thaumatichthyidae	*Lasiognathus, Thaumatichthys*
Trachichthyiformes	Anomalopidae	*Anomalops, Kryptophanaron, Parmops, Photoblepharon, Protoblepharon, Phthanophaneron*
	Monocentridae	*Cleiodopus, Monocentris*
	Trachichthyidae	*Aulotrachichthys, Paratrachichthys*

Haneda (1980).

2.7 FISHES WITH LUMINESCENT SYMBIONTS

Within the light organs of the luminescent fishes, the bacteria have been reported to emit light continuously, and in all cases, the control of light emission involves some specific host mechanisms, such as shutter, the rotation of the organ, or chromatophores (e.g., *Monocentris japonica*). As the bacterial symbionts are extracellular, they are maintained in canals or tubules which communicate with the exterior either directly (monocentrid, anomalopid, ceratioid, and macrourid fishes) or via the intestinal tract. The bacteria are characteristically densely packed within the tubules (109 to 1010 cells per ml of organ fluid); (e.g., leiognathids and monocentrids). In all known cases, the organs are extensively vascularized, presumably for the supply of nutrients and oxygen to the bacteria (Nealson' and Hastings, 1979).

Among the marine fishes with luminescent symbiotic bacteria, herbivorous fish have been reported to have a slightly higher proportion of culturable luminous bacteria compared to the carnivorous fish. Further, higher incidence of luminescent bacteria has been observed in pelagic than in reef-associated fish. Furthermore, filter-feeder-fish guts contain more luminous bacteria compared to other feeding types (e.g., predator). Among the luminescent bacterial genera, the genus *Photobacterium* represents 78% of all present genera and *Photobacterium phosphoreum* clade is the most abundant *Photobacterium* lineage. It is also believed that the luminous species *Photobacterium kishitanii* constitutes over 26% of the Vibrionales community, which is the dominant clade (Tanet et al., 2020). The tropical and temperate teleost fishes, which occur in shallow and warmer waters have been reported to host higher temperature tolerating *Photobacterium leiognathi and Photobacterium fischeri*. On the other hand, the deep-sea and midwater fishes harbor psychrotrophic *Photobacterium phosphoreum* species as their symbionts (Nealson and Hastings, 1979). The families of fishes with luminescent symbionts are listed below.

2.8 MARINE FISHES WITH EXTRA CELLULAR LUMINESCENCE (LUMINOUS SECRETION)

In the extrinsic bioluminescent fishes, light may also be generated by luminous secretion from the glandular tissues. These extra cellular luminescent organs are found in a very limited species of fishes. For example, rattails and searsids have been reported to emit light by secreting extra cellular slime. Rattails possess special glands near their anus, which secretes slime of sufficient luminosity (Jha, https://www.yourarticlelibrary.com/fish/anatomy-and-physiology/luminous-organs-or-photophore-of-the-fishes-with-diagram/88411).

Families of Extracellular Luminescent Fishes and the Associated Species of Luminescent Symbionts

Family	Species of bacteria	Light organ opening
Macrouridae	*Photobacterium phosphoreum*	External
Chlorophthalmidae	-do-	External
Merlucciidae	-do-	Rectum
Moridae	-do-	Rectum
Opisthoproctidae	-do-	Rectum
Trachichthyidae	-do-	Rectum
Monocentridae	*Photobacterium fischeri*	External
Leiognathidae	*Photobacterium leiognathi*	Oesophagus
Apogonidae	-do-	Intestine
Ceratiidae	Not culturable	External
Anomalopidae	-do-	External
Acropomatidae	Not identified	Rectum

Source: Nealson and Hastings (1979); Haneda (1980); Barnes (2003).

2.9 MARINE FISHES WITH EXTRINSIC LUMINESCENCE (LUMINOUS INGESTED CRUSTACEANS)

In these fishes, the extracellular luminescence is mainly due to the accumulation of digested crustacean substances which could be due to the consumption of luminescent ostracods containing cypridina-type luciferin, vargulin (Haneda, 1980). It has also been reported that the chemical luminescence of these fishes is due to their glandular tissue, which secretes a chemical substance called luciferin, an indole derivative consisting of tryptamine, arginine, and isoleucine. The latter is converted into oxy-luciferin under the influence of the enzyme luciferase and blue or blue-green light is emitted thereby. The fishes of the genera *Apogon* and *Parapriacanthus* are known to possess luminous glands containing crude form of luciferin and luciferase (Jha, https://www.yourarticlelibrary.com/fish/anatomy-and-physiology/luminous-organs-or-photophore-of-the-fishes-with-diagram/88411).

2.10 BIOLOGICAL SIGNIFICANCE OF LUMINOUS ORGANS AND THE FUNCTION OF LUMINESCENCE

The functions of the light in luminescent marine fishes may be brought under three main categories viz. attracting prey, assisting in escaping or diverting predation, and communication. While the fish use their illumination in sex determination within species and for preying, some fish use their bioluminescence for counter-illumination (camouflage). Normally, if these fish swim at the surface during the day, their silhouette (outline) against the sun would be visible to their predators swimming beneath them. However, some fish can produce light from their bellies to disrupt the silhouette and hide them from would-be predators. Haddfock et al. (2010) reported that camouflage is found in many midwater sharks and bony fishes which are able to control their light intensity to match the variations of the environment. The luminous organs are useful in a variety of ways in marine fishes especially in deep-sea fishes (Anon., https://www.lkouniv.ac.in/site/writereaddata/siteContent/202003251324430605shelly_Sound_Producing.pdf). The biological significance of these luminous organs are given below.

i) **Illumination of Surroundings:** Some luminescent fishes use their luminous organs to illuminate their surroundings, especially during dimness. This would help them search for their prey in the dark waters. Some species, like stomiatoid fishes have been reported to emit a beam of light from their luminous cheek organ to catch ,the small animals like plankton. The cheek organs of *Anamalops* spp. produce light like a torch. Angler fish (Lophiiformes: Ceratiidae) possess an illuminated fishing lure-like structure that may attract would-be predators, which are then converted into prey. Flashlight (anomalopid) and pinecone (monocentrid) fishes use their luminescence to illuminate their surroundings and for intraspecies communication.

ii) **Defensive Device:** Many luminescent fishes are known to produce sudden flash of light from their luminous organs, which helps largely in diverting the attention of their predators. Further, such emission of light facilitates the escape of fish by puzzling their enemies. For example, alepocephalid fishes have been reported to produce a glowing spark, which is believed to confuses the predator for a spur of moment, and help the fish to escape from such predators. However, some fishes use their luminous organs to illuminate their ventral surface which makes them inconspicuous against the lighted background above. For example, both lanternfishes and bristlemouths use their bioluminescence for camouflage. In pony fish (Leiognathidae), the light organ which is located deep within the body as an outpocketing of the esophagus presents a uniform ventral glow which may act as camouflage for avoiding predation.

Source: Reprinted from Martini and Francis, 2020. https://creativecommons.org/licenses/by/4.0/

Counter-illumination in bioluminescent fish.

iii) Warning Signal: A number of luminescent fishes use their luminous organ to warn their predators. For example, the midshipman *Porichthys* flashes light as a toxic sign when it is attacked by a predatory fish and thereby it avoids the danger.

iv) Recognizing Own Species: Each species has its own unique arrangement and distribution of photophores on its body, and this helps this species to recognize the species of the same type, especially for its schooling behavior. As the light organs may be different both in male and female, these organs may also be helpful in recognizing their mates for courtship and breeding. For example, the male lanternfish has one or many photophores above and t the female possess it below the caudal peduncle. In some other species, the size of luminous organ may be different in both sexes. For example in melanostommiatid fishes, the postorbital luminous organ is fairly larger in the male and smaller in the female. Lanternfishes have evolved a suite of light organs that act as a beacon for communication for the opposite sex.

CHAPTER 3

BIOLUMINESCENT MARINE FISHES: ECOLOGY AND BIOLOGY

3.1 MARINE FISHES WITH INTRACELLULAR BIOLUMINESCENCE

3.1.1 CARTILAGINOUS FISHES (ELASMOBRANCHII: SQUALIFORMES)

The luminescent sharks of all three families (Dalatiida, Etmopteridae, and Somniosidae) of cartilaginous fishes have been reported to possess only intrinsic type of bioluminescence and light emission associated symbiotic bacteria is absent in these cartilaginous fishes.

3.1.1.1 KITEFIN SHARKS (DALATIIDAE)

All Dalatiidae are essentially solitary pelagic (except *Dalatias licha,* which is benthopelagic) predators, which undergo daily vertical migrations. They are found distributed worldwide in open ocean or on the bottom in mostly temperate to tropical seas (Ebert and Fowler, 2014). All the seven genera of this family viz. *Dalatias, Euprotomicroides, Euprotomicrus, Heteroscymnoides, Isistius, Mollisquama,* and *Squaliolus* are luminous. In these dalatiid species, the luminous organs viz. photophores are arranged following a density gradient from dorsal to ventral region, where their density may be about 60 units per square millimeter. These sharks use their luminescence only for camouflage by counterillumination. The latter helps these sharks to avoid their predators (Claes and Mallefet, 2015).

Biology and Ecology of Bioluminescent Marine Fishes. Ramasamy Santhanam (Author)

3.1.1.1.1 Photophores of Dalatiid Sharks

The photophores of dalatiid sharks are present in the stratified squamous epidermis of the skin and they lack reflectors or any kind of specialized structure. In the luminescent sharks of the genera *Dalatias, Euprotomicroides, Euprotomicrus* and *Isistius* of this family, the photophores Pavement-like squamation pattern These photophores are extremely small (100 μm) and are in several thousands. Each photophore contains only a single photogenic cell (= photocyte) plus a pigmented sheath and a lens made up of a group of small cells overlying the apical surface of the photogenic cell. The photocyte contains photogenic granules and a large nucleus which is placed at the apical side. The lens cells contain a homogeneous cytoplasm. (Claes and Mallefet, 2009).

1. Dalatias licha

Common name(s): Kitefin shark or seal shark.

Global distribution: It is found distributed sporadically around the world oceans except eastern Pacific and northern Indian Oceans.

Ecology: It is an offshore, deep-water, benthopelagic species living at a depth of 200–600 m.

Biology: This is the only species in this genus and is the largest luminous vertebrate on record.

It has a slender body with a very short, blunt snout, large eyes, and thick lips. It reaches a maximum length of 1.4 m. Its diet ranges from bony fishes, sharks and rays, to cephalopods, crustaceans, polychaete worms, and siphonophores.

Luminous organs and bioluminescence: The mean photophore diameter of this species is 83.9 μm (Claes and Mallefet, 2009). A blue glow was observed on the ventral surface of this species when the animals are kept in a fully dark environment. Further, it is also known to emit a faint blue glow from the lateral and dorsal areas and at the two dorsal fins (Mallefet et al., 2021).

2. Euprotomicroides zantedeschia

Common name(s): Taillight shark.

Global distribution: Southern Atlantic Ocean and Pacific Ocean.

Ecology: It is a solitary, pelagic species inhabiting deep oceanic waters at depths of 458–641 m.

Biology: It is a small shark with a laterally compressed body, a bulbous snout and large mouth and oval eyes. Its pectoral fins are paddle-like and are used for propulsion.

Luminous organs and bioluminescence: It has a pouch-like gland on its abdomen that emits clouds of luminescent blue fluid. On the belly in front of the cloaca is a pouch-like groove devoid of denticles and lined with a luminescent tissue formed into numerous, tightly packed papillae (nipple-like structures). The entrance to the pouch is a slit lined with folds of skin. In life, the pouch emits a glowing blue fluid of unknown function (Wikipedia).

3. *Euprotomicrus bispinatus*

Common name(s): Pygmy shark or dwarf pelagic shark.

Global distribution: It has circumglobal distribution from subtropical to temperate waters. It has been recorded in Southeast Atlantic, Indian Ocean, and North and South Pacific.

Ecology: This solitary, bathypelagic species has a depth range of 0–1800 m.

Biology: It is a small dogfish with a cigar-shaped body with a distinct dark collar marking around its gill region. It has a conical snout, large round eyes, very small dorsal fins, and a paddle-shaped caudal fin with almost symmetrical upper and lower lobes. It feeds on squid, bony fishes, and crustaceans. The body is black with white fin edges. It attains a maximum length of 31 cm.

Luminous organs and bioluminescence: It has a luminous belly. Spontaneous bioluminescence has been observed with a greenish glow (Claes and Mallefet, 2009). Anon. (https://www.naturepl.com/search/preview/close-up-of-the-skin-of-a-bioluminescing-pygmy-shark-euprotomicrus/0_01534178.html) reported glowing blue photophores (light organs) on its skin. In this species, it's very numerous minute light organs form an uniform network over the entire venter and lower sides of the body and head. Further, the upper fringe of the eyes, the basal parts of the paired fins, and the posterior tip of the caudal axis also bear similar type of photophores. Each photophore is more compact and it contains a single photogenic cell (Hubbs et al., 1967).

4. *Heteroscymnoides marleyi*

Common name(s): Longnose pygmy shark.

Global distribution: It is widespread in the Atlantic, Pacific, and Indian Oceans.

Ecology: It is a solitary, pelagic species and is often seen at or near the surface at night. It is believed to descend to a depth of about 1500 m.

Biology: This diminutive shark has a slender, dark brown body with a very long, bulbous snout. Two spineless dorsal fins of equal size are seen. It eats fish, squid, and crustaceans and attains a maximum length of 37 cm.

Luminous organs and bioluminescence: Its underside is covered by minute light-producing photophores, which are surrounded by small flat denticles (Anon., http://www.biodiversityexplorer.info/chondrichthyes/elasmobranchii/squalomorphii/heteroscymnoides_marleyi.htm).

5. *Isistius brasiliensis*

Common name(s): Cookiecutter shark or cigar shark.

Global distribution: It occurs in warm, oceanic waters worldwide, particularly near islands.

Ecology: This deep-sea shark is a solitary, pelagic species.

Biology: It has a long, cylindrical body with a short, blunt snout and large eyes. Two tiny spineless dorsal fins, and a large caudal fin are present. The body coloration is dark brown. It feeds on a wide variety of marine mammals and fishes and attains a maximum length of 56 cm.

Luminous organs and bioluminescence: Its thousands of light-emitting photophores cover the underside except for a dark "collar" around its throat and gill slits. These ventral photophores emit downwelling light. Claes and Mallefet (2009) reported that this species is known for its spontaneous bioluminescence and it a bluish glow with a peak of emission at 455 nm. Anon. (https://en.wikipedia.org/wiki/Cookiecutter_shark) reported that the intrinsic green luminescence of this shark species is the strongest known of any shark, and this light persisted for three hours after it has been taken out of water. This shark uses its bioluminescence to deceitfully attract its prey.

6. *Mollisquama mississippiensis*

Common name(s): American pocket shark.

Global distribution: Pacific Ocean; eastern Gulf of Mexico.

Ecology: This solitary pelagic, deep-sea shark inhabits at a depth of 3,038 m.

Biology: Its cylindrical body tapers towards a slender caudal peduncle which is followed by a bilobed homocercal caudal fin with rounded apexes. Its wide head is considerably deeper than the rest of the body with a broadly-rounded bulbous snout with a subterminal mouth. There are two low dorsal fins which are positioned above the pelvic fin. The body coloration is coal grey with brown undertones and fins are dark. It has an average length of 14 cm.

Luminous organs and bioluminescence: Along its ventral-abdominal surface, there is a series of 16 photophore aggregations which extend from the posterior of the lower jaw to the anterior of pelvic fin origin. There are also additional photophores which are irregularly distributed along many other areas of the body (Anon., https://www.gbif.org/species/10051334/treatments). Anon. (https://news.tulane.edu/pr/researchers-identify-new-species-pocket-shark) reported that this species possesses two small glands (one on each side near the gills) that produce luminous fluid to draw its prey close.

7. Squaliolus aliae

Common name(s): Smalleye pygmy shark.

Global distribution: Western Pacific Ocean: Japan, the Philippines, and Australia.

Ecology: It is a solitary, pelagic species inhabiting depths of 150–2,000 m.

Biology: Body of this species is shaped like a cigar with a bulbous, pointed snout. The eyes are relatively small. While its first dorsal fin is tiny, second dorsal fin is long and low. Pectoral fins are short with rounded margins, and pelvic fins are long and low. Anal fin is absent. Caudal fin is broad and triangular. The body coloration is dark brown to black, and it is becoming light towards the fin margins. feeds on small squid, krill, shrimp, and bony fishes. It attains a maximum length of 22 cm.

Luminous organs and bioluminescence: In this species, bioluminescent photophores occur on its underside. These photophores may serve to camouflage its silhouette from predators by matching the downwelling light, a phenomenon known as "counter-illumination" (https://en.wikipedia.org/wiki/Smalleye_pygmy_shark).

8. Squaliolus laticaudus

Common name(s): Spined pygmy Shark.

Global distribution: Atlantic Ocean, Indian Ocean, and Pacific Ocean.

Ecology: This solitary pelagic species inhabits a depth range of 200–500 m; it is a diel vertical migrator, spending the daytime at 500 m deep and migrating towards a depth of 200 m at night.

Biology: It is one of the world's smallest sharks with a maximum length of 22 cm for males and 28 cm for females. It has an elongated, spindle-shaped body with a long, bulbous, moderately pointed snout. The eyes are large with straight upper rim of orbit. It has a long but low second dorsal fin, and a symmetrical caudal fin. The body coloration is dark brown to black. It feeds on small bony fish and squid. This species attains a maximum length of 28 cm.

Luminous organs and bioluminescence: Its ventral surface has numerous photophores which are responsible for its intrinsic bioluminescence. This shark is believed to use its photophores to match ambient light conditions, which break up its silhouette and help the shark to predators swimming below (https://en.wikipedia.org/wiki/Spined_pygmy_shark).

Credits: 1: CSIRO National Fish Collection. https://creativecommons.org/licenses/by/3.0/deed.en; 2, 3 & 4: Image courtesy: Mark A. Grace – NOAA Federal; 5: Public domain.

1. *Dalatias licha*; 2. *Euprotomicroides zantedeschia*; 3. *Euprotomicrus bispinatus*; 4. *Mollisquama mississippiensis*; 5. *Isistius brasiliensis*.

3.1.1.2 LANTERN SHARKS (ETMOPTERIDAE)

The photophores of this family are more complex than those of Dalatiidae. as they show aggregations of photophores on the ventral side, but also on the flanks, the tail and even on the upper eyelid (e.g., *Centroscyllium* and *Trigonognatus*). In some species of lantern sharks, the photophores can also be inconspicuous because they are not arranged in specific markings. Further, the luminous pattern of lantern sharks is believed to be species-specific. These photophores are composed of a black circular pigmented sheath encompassing a group of 6–13 photogenic cells in general, a lens formed by 2 to 3 lens cells, and an iris-like structure formed by chromatophores. The etmopterid circular photophores may also fuse together to produce larger linear photophores (300 μm in length by 100 μm broad) with a higher number of lens cell. Counterillumination is considered to be one of the main functions of luminescence of these lanternsharks. Utilizing the glow from their photophores, these lanternsharks can mask their shadow and match the sunlight filtering above. This process helps these sharks to be away from their predators (Ebert and Fowler, 2014).

Ebert et al. (2021) reported that the photophores of these lanternsharks are small pigment-lined structures distributed on the belly, flanks and fins. Each photophore contains up to 12 photocytes and one or two lens cells to focus the light emitted. There may be as many as 500000 photophores on a single shark which is more than any other luminescent animal known so far. While bony fishes control their light organs with nerves, these lanternsharks use their hormones to switch the photophores on and off. These hormones are believed to stimulate pigmented cells or melanophores to cover or uncover the photocytes (Ebert et al. 2021).

1. Aculeola nigra

Common name(s): Hooktooth dogfish.

Global distribution: Southeast Pacific: Peru to central Chile.

Ecology: This solitary pelagic and demersal species inhabits a depth range of 110–560 m.

Biology: The body of this species is stocky. Its head is broad and flattened, and is wider than deep. Its short snout is broadly rounded or subangular, flattened and truncated. Mouth is fairly long and broadly arcuate. The body coloration is blackish above and below. It attains a maximum length of 67 cm.

Luminous organs and bioluminescence: In this species, the intrinsic photophores are more dense on the ventral surface than the dorsal surface. But there are no conspicuous black photomarks on underside of head and abdomen, flanks, tail and caudal fin (Claes and Mallefet, 2009; Anon., https://shark-references.com/species/view/Aculeola-nigra).

2. Centroscyllium fabricii

Common name(s): Black dogfish.

Global distribution: Atlantic Ocean.

Ecology: This solitary pelagic species is Common over the outer continental shelf and continental slope at depths of 180–2,250 m.

Biology: 60–75 cm (24–30 in) long. It has a stocky, dark brown body that is darker below than above, Its two dorsal fins are preceded by stout spines, and the anal fin is absent. It is an opportunistic predator and scavenger that mainly consumes bony fishes, crustaceans, and cephalopods.

The body of this fish is moderately stout and compressed with a short caudal peduncle and a long abdomen. Snout is moderately long with an arched mouth. The second dorsal fin is much larger than the first dorsal fin. Both dorsal fins have prominent, grooved spines. Anal fin is absent. The body coloration is dark brown that is darker below than above. It is an opportunistic predator and scavenger that mainly consumes bony fishes, crustaceans, and cephalopods. It attains a maximum length of 75 cm.

Luminous organs and bioluminescence: It bears scattered, minute bioluminescent organs which look like deeply pigmented dots similar to those seen in other luminescent squaloid species (https://en.wikipedia.org/wiki/Black_dogfish). Claes and Mallefet (2009) reported that this species intrinsically bioluminescent.

3. Etmopterus carteri

Common name(s): Cylindrical lanternshark or Carter Gilbert's lanternshark.

Global distribution: Western Central Atlantic: Colombia.

Ecology: This solitary pelagic and oceanic species inhabiting depths of 285–355 m.

Biology: It is a very small black shark with a short and bluntly rounded snout. The head and body are almost cylindrical anteriorly and are tapering unevenly to a slender caudal peduncle. It has moderately large eyes. Its dermal denticles are very small and are needle to spine-like covering the whole body except along lips, gill slit margins, around cloacal opening, and dorsal surface of claspers. Pectoral fins are lobate and pelvic fins are small. It attains a maximum length of 21 cm.

Luminous organs and bioluminescence: Though this species is believed to be luminescent, it lacks any concentrated photophores (Compagno, 2005).

4. Etmopterus granulosus

Common name(s): Southern lanternshark.

Global distribution: Southeast Pacific.

Ecology: This solitary pelagic species is often found on outermost continental shelves and upper slopes at depths of 220–1460 m.

Biology: It is a large, heavy-bodied lanternshark with a big head. Its stocky body is with conspicuous lines of denticles on the dorsal side and conspicuous black markings on underside of body and tail. The body coloration is dark brown or black and is darker below. It feeds on bony fish, squid, shrimp and crabs and attains a maximum length of 60 cm.

Luminous organs and bioluminescence: It exhibits a more complex pattern of light emission with its dorsal photophores, flank markings, and brighter pectoral fin and claspers. The mean photphore diameter of this species has been reported to be 132 μm While the light emitted by these organs is used for intraspecific communications, its ventrally emitted light is used for counterillumination (Mallefet et al., 2021; Anon., https://www.fishbase.in/summary/679).

5. Etmopterus lucifer

Common name(s): Blackbelly lanternshark or lucifer shark.

Global distribution: It is found distributed around the world in tropical to temperate seas; Western Pacific: Japan to the Philippines.

Ecology: This solitary, pelagic species is found on the outer continental and insular shelves and upper slopes at depths of 150–1,350 m.

Biology: Body coloration of this species is dark grey or brown dorsally, black ventrally and is with a distinct black streak above and behind pelvic fins. The body is small and slender with a moderately elongate caudal peduncle. Flank denticles are found arranged in regular rows. It feeds mainly on squid, small bony fishes, and shrimps. It attains a maximum length of 47 cm.

Luminous organs and bioluminescence: Its bioluminescent belly presents a more complex pattern of light emission with flank marks, and dorsal, and rostral patterns. The mean photphore diameter of this species has been reported to be 122 μm (Mallefet et al., 2021). The function of its light emission is very much similar to that of *Etmopterus granulosus.*

6. Etmopterus molleri

Common name(s): Slendertail lantern shark or Moller's lantern shark.

Global distribution: Western Indian Ocean.

Ecology: This solitary, pelagic species is more common in upper continental slopes at depths of 250 and 860 m.

Biology: Body coloration of this species is light brown above and black below. It has a pale stripe on the side and a black streak above and behind pelvic fins. The body is small and slender, with a moderately elongate caudal peduncle. Its flank denticles are found mostly arranged in regular rows. Its belly is luminous and it attains a maximum length of 46 cm.

Luminous organs and bioluminescence: The light emission from its dorsal photophores, flank markings, and brighter pectoral fin and claspers is believed to be used for intraspecific communications while its ventrally emitted light is used for counterillumination (Mallefet et al., 2021).

Claes and Mallefet (2015) reported that the photophore density of this species ranged from 3,862 photophores cm^2 in the ventral luminous zones to 10,085 photophores cm^2 in the pectoral luminous zones. Its photophores were observed to be small (~100 μm in diameter) as hemispheric pigmented cups containing several photocytes and are surmounted by iris-like pigmented structures and lens cells. Photocytes of these photophores harbored apical vesicles that appeared autofluorescent. Further, the zones adjacent to pectoral and pelvic fins of this species showed higher photophore density than the ventral counterilluminating continuum, with the exception of the elongated flank photogenic markings.

7. *Etmopterus pusillus*

Common name(s): Smooth lanternshark or slender lanternshark.

Global distribution: Atlantic and Pacific Oceans.

Ecology: This solitary, pelagic species inhabits depths of 274–1,000 m.

Biology: This species has a slender, dark brown body with an indistinct black band on the sides over the pelvic fins. Head is large with a pointed snout and large oval eyes. Its nostrils have short anterior skin flaps. Its first dorsal fin bears a stout spine in front and second dorsal fin is much larger than the first dorsal fin and has a longer spine.Pelvic fins are low and angular, and there is no anal fin. Caudal fin is short and broad. This shark feeds on smaller squid, fishes, and fish eggs and reaches 50 cm in length.

Luminous organs and bioluminescence: Spontaneous bioluminescence has been observed in this species which emits a “whitish” glow (Claes and Mallefet, 2009).

8. Etmopterus spinax

Common name(s): Velvet belly lanternshark or velvet belly.

Global distribution: Northeastern Atlantic Ocean.

Ecology: It is a deep-sea shark spending most of its time at mesopelagic depths of 20–2490 m.

Biology: It is a small shark of about 60 cm long, with black underside and brown coloration on the rest of its body. The body is fairly stout, with a moderately long snout and tail, and very small gill slits. It is believed to harbor defensive mineralized spines in front of its dorsal fins. Adults feed on squid and shrimp while young velvet bellies feed mainly on krill and small bony fish.

Luminous organs and bioluminescence: Its ventral surface is covered by thousands of tiny (<200 μm in diameter) photophores (Claes et al., 2013). These photophores are found concentrated more in the belly than in the flanks of this species. Stimulated by melatonin or prolactin, these photophores produce a long lasting blue-green luminescence (Claes and Mallefet, 2009). Anon. (https://www.central.edu/writing-anthology/2019/04/30/bioluminescence-in-lanternsharks-and-its-impact-on-communication-via-identification/) reported that a similarity exists between all lanternshark species which is the photophore concentration on their bellies. This similarity explains counterilluminating camouflage as one of their main functions. Utilizing the glow from these ventral photophores, these lanternsharks mask their shadow and match the sunlight filtering above. Further, the bioluminescence of this species is intended to attract it to mate, lure food, and to hide from predators.

Renwart et al. (2013) reported that the luminescence of this species is intrinsic and this is due to its three potential alimentary sources of luciferin obtained from the euphausiid *Meganyctiphanes norvegica* (tetrapyrrole luciferin), the shrimp *Pasiphaea multidentata* (coelenterazine) and the teleost *Maurolicus muelleri* (coelenterazine).

Claes and Mallefet (2011) reported that the luminescence of this shark was under a dual control of hormones and neurotransmitters (or neuromodulators). According to these authors, the light switch on in the photophores of this species is induced by two hormones viz. (i) the methyl testosterone which is produced by the pineal gland and (ii) the prolactin which is produced by the pars distalis of the pituitary gland.

9. Etmopterus splendidus

Common name(s): Splendid lanternshark.

Global distribution: Western Pacific: Japan, Taiwan, and Java (Indonesia).

Ecology: This solitary species is probably demersal on the outer continental shelves and upper slopes up to a depth of 730 m.

Biology: This species is characterized by the following features: black flank marking on side of tail base with short truncated posterior lobe; and additional oval dark marking near base of caudal fin. It feeds mainly on squids and attains a maximum length of 30 cm.

Luminous organs and bioluminescence: In all the species of *Etmopterus*, the light organs appear as dark markings. This species has the most complex photophore pattern with maximum ventral density area. These photophore patterns have been reported to maximize downward emission of luminescence, making the shark seem brighter to match their backgrounds (https://www.sharkwater.com/shark-database/sharks/splendid-lanternshark/).

10. Miroscyllium sheikoi (= *Centroscyllium sheikoi; Etmopterus sheikoi*)

Common name(s): Rasptooth dogfish.

Global distribution: Northwest Pacific near Japan.

Ecology: This solitary, pelagic species is found distributed on the upper slopes of the submarine ridge between 340 and 370 m.

Biology: It is dark brown dorsally and black ventrally. It has a long, flat snout. Mouth is short. It attains a maximum length of 43 cm.

Luminous organs and bioluminescence: There are black photophores on the caudal peduncle and caudal fin (https://planetsharkdivers.com/rasptooth-dogfish/). These photophores are responsible for its intrinsic bioluminescence.

11. Trigonognathus kabeyai

Common name(s): Viper dogfish or viper shark.

Global distribution: Pacific Ocean: off southern Japan, Bonin Islands, and off northern Taitung County; and Northwestern Hawaiian Islands.

Ecology: This solitary, pelagic species inhabits the upper continental slopes and seamounts. It may migrate vertically, shifting between bottom waters of 270–360 m depth during the day and upper waters less than 150 m at night.

Biology: This slender, black shark is recognized by its narrow, triangular jaws and well-spaced, fang-like teeth. It also has two spined dorsal fins. Its dermal denticles possess faceted crowns. It feeds mainly on bony fishes and attains a maximum length of 54 cm.

Luminous organs and bioluminescence: Almost the entire ventral surface of its body and head has many minute luminous organs or photophores. Each of these photophores is rounded in structure with a black margin and is about 100 microns in dia. Its ventral surface is covered with photophores in high density and is black in color. Other body areas are found with scattered photophores in very low density (Mochizuki and Ohe, 1990).

1

2

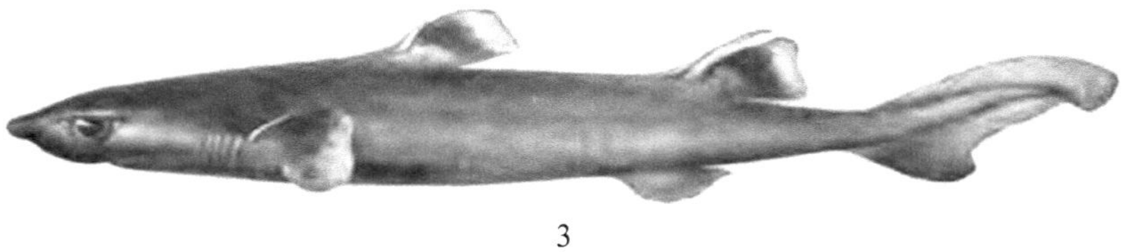

3

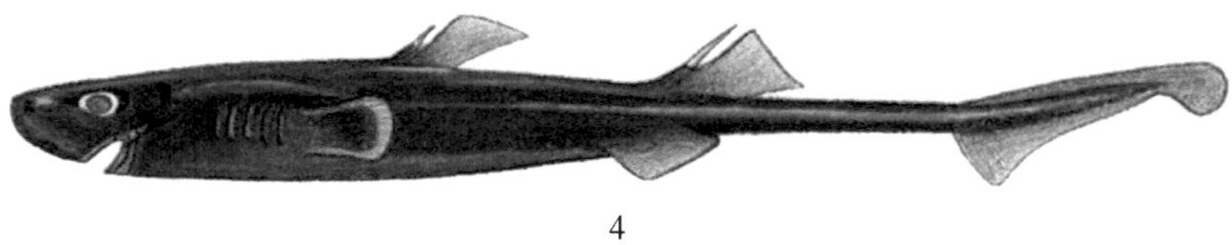

4

5

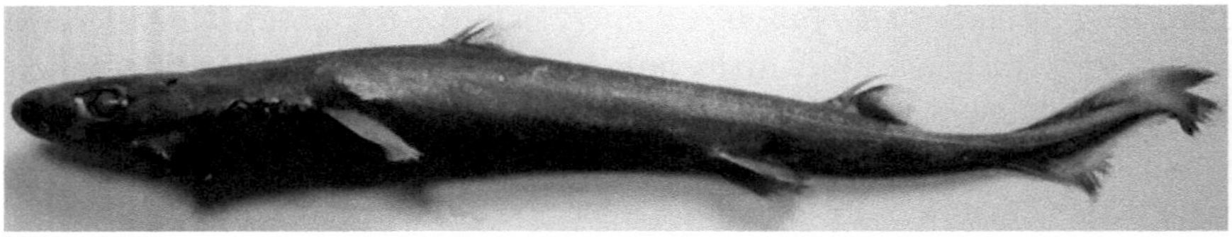

6

7

8

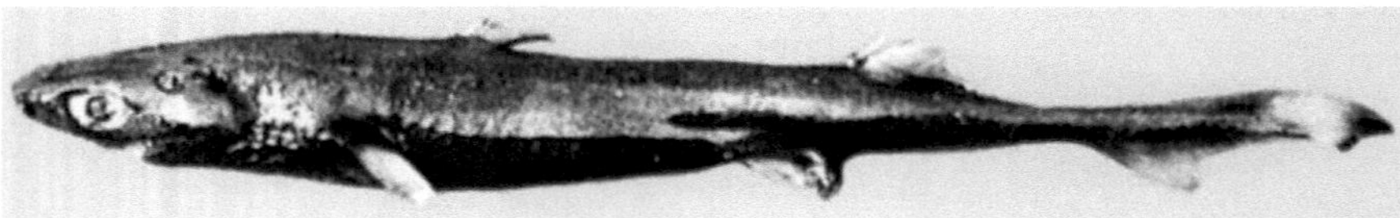

9

10

Credits: 1: Image courtesy: Nicolas Straube; 2 & 4: Tambja.https://creativecommons.org/licenses/by-sa/3.0/deed.en; 3, 5, 6, 8. Public domain; 7 & 10: *Image courtesy Mark A. Grace, NOAA Federal; 9*: Image courtesy Kwang-Tsao Shao, Taiwan Fish Database.

1. *Aculeola nigra;* 2. *Centroscyllium fabricii;* 3. *Centroscyllium ritteri;* 4. *Etmopterus carteri;* 5. *Etmopterus granulosus;* 6. *Etmopterus pusilllus;* 7. *Etmopterus sheikoi;* 8. *Etmopterus spinax;* 9. *Etmopterus splendidus;* 10. *Trigonognathus kabeyai.*

3.1.1.3 SLEEPER SHARKS (SOMNIOSIDAE)

1. *Somniosus rostratus*

Common name(s): Little sleeper shark.

Global distribution: Northeast Atlantic, western Mediterranean, and western Pacific.

Ecology: This bathydemersal species is found distributed above sandy mud substrate on the outer continental shelves and on the upper slopes at depths between 200 and 1,000 m.

Biology: In this species, its head is short and snout is short and broadly rounded. The eyes are small and rounded. There are no fin spines on both dorsal fins. There is no anal fin. The body coloration is uniformly blackish without transverse dark bands, small light spots, and blotches. It feeds mainly on cephalopods and it reaches a maximum length of 143 cm.

Luminous organs and bioluminescence: In this species, photophores occupy incongruous positions for light-giving organs along the lateral line and on the anterior dorsum. This species has been reported to exhibit splendor of the spectral colors (Hubbs et al., 1967).

2. *Zameus squamulosus*

Common name(s): Velvet dogfish.

Global distribution: Atlantic, Indian and Pacific tropical oceans.

Ecology: This demersal or pelagic species is found distributed near continental slopes and seamounts in depths of 550–2,000 m.

Biology: It is a black bioluminescent shark with small dorsal fin spines and a moderately long snout. Caudal fin has a strong subterminal notch and a short lower lobe. It attains a maximum length of 84 cm.

Luminous organs and bioluminescence: It has a new bioluminescent-associated squamation in its rostral area. Its bioluminescence-associated placoid scale is highly translucent or possesses specific physical characteristics to allow efficient light transmission (Mallefet et al., 2021).

Anon. (https://en.wikipedia.org/wiki/Velvet_dogfish) reported that its photophores located on its ventral epidermis exhibit intrinsic bioluminescence which is used for ventral counter-illumination.

In a fully dark environment, this species has been reported to emit a dim blue-green light from the ventral side of the body (Duchatelet et al., 2021).

1. *Somniosus rostratus;* 2. *Zameus squamulosus.*

3.1.2 BONY FISHES (TELEOSTS)

3.1.2.1 ORDER: ACROPOMATIFORMES

3.1.2.1.1 Deepsea Cardinalfishes (Epigonidae)

1. Epigonus angustifrons

Common name(s): Deepwater cardinalfish.

Global distribution: Southern part of the Indian Ocean.

Ecology: This bathydemersal species occurs at depths of 150–880 m. It is also a mesobenthic-pelagic species living mainly above the bottom.

Biology: It has a slender body with a maximum length of 36 cm. Other diagnostic characters of this species include: anal fin is with 2 spines and 9 segmented rays; first dorsal has 7–8 fin spines; and 14 or more gill rakers on lower limb of first arch.

Luminous organs and bioluminescence: Its luminous system consists of luminous pyloric caeca and luminous window. Its luminous organ contains luminous substances of ingested substances and is open to the outside (Haneda, 1980).

2. Epigonus macrops

Common name(s): Luminous deepsea cardinalfish.

Global distribution: Indo-Pacific and Western Atlantic.

Ecology: This bathydemersal species inhabits both marine and brackishwaters at a depth range of 550–1300 m. It is common above the bottom in the continental slope.

Biology: This species has a maximum length of 21.2 cm and has the following diagnostic features: anal fin is with 2 spines and 9 segmented rays; first dorsal-fin has 7–8 spines; 17–21 gill rakers on first arch; and 7–8 pyloric ceca.

Luminous organs and bioluminescence: In this species, the light organ is present in one of the pyloric caeca.

3. *Florenciella lugubris*

Common name(s): Deepwater cardinalfish.

Global distribution: It is native to the Indian Ocean.

Ecology: This bathydemersal species occurs at depths from 150 to 800 m.

Biology: It grows to a maximum length of 10 cm. No other information is available.

Luminous organs and bioluminescence: It has bioluminescent oesophageal diverticula which is able to eject a pulse of luminescent material from beneath the operculum (Herring, 1992).

4. *Rosenblattia robusta*

Common name(s): Stout cardinalfish.

Global distribution: Circumglobal in southern temperate waters.

Ecology: This bathypelagic species occurs at depths between 700 and 2000 m.

Biology: Adults of this species are uniformly dark with a maximum length of 10 cm only.

Luminous organs and bioluminescence: In this species, its ventral reflective strip is bioluminescent (Herring, 1992).

1

2

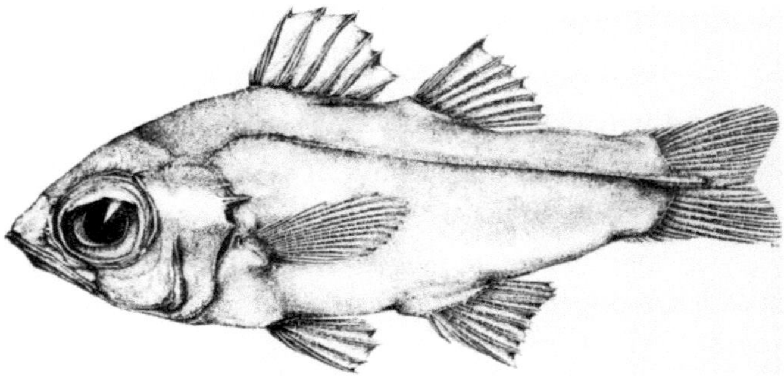

3

Credits: 1: Mindat.org, https://www.mindat.org/taxon-2395048.html; 2 & 3: National Museum of Natural History (EOL), https://creativecommons.org/licenses/by-nc-sa/3.0/

1. *Epigonus angustifrons*; 2. *Florenciella lugubris*; 3. *Rosenblattia* sp.

3.1.2.1.2 Oceanic Basslets (Howellidae)

1. Howella brodiei

Image courtesy: https://inaturalist.ca/taxa/88968-Howella. https://creativecommons.org/licenses/by-sa/3.0/

Common name(s): Pelagic basslet.

Global distribution: It is native to the Indo-Pacific.

Ecology: This oceanodromous species occurs in the bathypelagic at depths of 100 to 1,829 m. It is often staying in deeper waters (below 305 m) during the day and is moving to shallower waters during the night.

Biology: It is a dark brown colored fish with a lighter head. It attains a maximum length of 90 cm.

Luminous organs and bioluminescence: The pyloric caeca and posterior intestine of this species are bioluminescent, and are associated with internal reflective tissues. Unlike other apogonid species, it does not depend upon Vargula luciferin (Herring, 1992). Herring (1983) reported that the wavelength of emission maxima of this species was at 495 nm.

3.1.2.1.3 Bullseyes (Pempheridae)

***Pempheris* spp.**

Common name(s): Sweepers.

Global distribution: These species are native to the Atlantic, Indian, and Pacific Oceans.

Ecology: These reef-associated species inhabit depths of 4–30 m and they take shelter in the reef by day, often in aggregations.

Biology: These fishes have a compressed body with large eyes, short snout, strongly oblique mouth, protrusible upper jaw, and numerous long gill rakers. They are also characterized by a short-based, pointed dorsal fin and a very long, low anal fin. Mostly these fishes are copper, bronze, or silvery in color. They feed mainly on zooplankton at night.

Luminous organs and bioluminescence: Pempheris spp.

As in the case of *Parapriacanthus,* the kleptoprotein-based autogenic bioluminescence is seen in the species of this genus also (Lau and Oakley, 2021). In these fishes, both anterior and posterior light organs are seen.

In *Pempheris ornata*, an upwardly curving mass of small finger-like projections form the anterior light organ which is communicating posteriorly with pyloric caeca. Its posterior light organ is found to be linear with a slight bifurcation around anus. On the other hand, in *Pempheris ypsilychnus,* both the anterior and posterior light organs are V-shaped. Its anterior and posterior light organs communicate with the pyloric caeca and rectum, respectively (Mooi and Jubb, 1996).

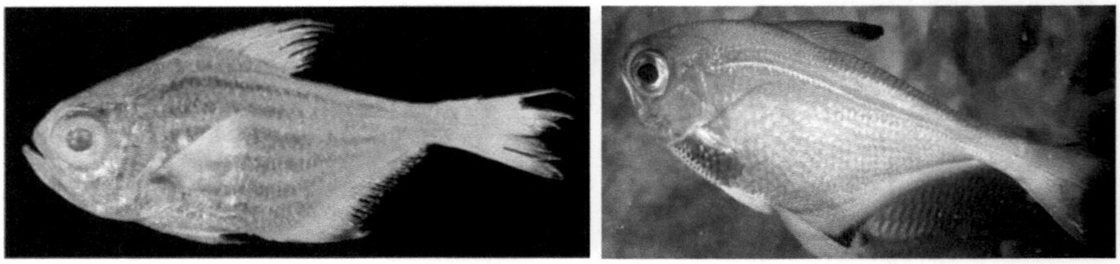

1 2

Credits: 1: Museums Victoria, https://fishesofaustralia.net.au/home/species/2487. https://creativecommons.org/licenses/by/4.0/; 2: Roberto Pillon. http://www.fishbase.de/Photos/ThumbnailsSummary.php?ID=10350. https://creativecommons.org/licenses/by/3.0/deed.en

1. *Pempheris ornata*; 2. *Pempheris schwenkii.*

3.1.2.2 ORDER: ALEPOCEPHALIFORMES

The fishes of this order have the unique characteristic feature of possessing three types of photophores or light organs viz. regular photphores, secondary photophores or simple photophores and mental light organs which are peculiar formation of photophores located near the tip of the lower jaw (Sazonov et al., 1996).

3.1.2.2.1 Slickheads (Alepocephalidae)

1. Bathyprion danae

Common name(s): Fangtooth smooth-head.

Global distribution: Eastern Atlantic, North Atlantic, and Western Pacific.

Ecology: These bathypelagic fishes inhabit a depth range of 100–3200 m.

Biology: The body of this species is very slender and elongated. Head is large; snout is pointed and long; and eye is much smaller than snout. It attains a maximum length of 38 cm.

2. Mirognathus normani

Common name(s): Beaked slickhead or Norman's smooth-head.

Global distribution: Northeast Atlantic.

Ecology: This benthopelagic; species inhabits in depths of 1500–3000 m.

Biology: It is a slender dark greyish to black fish becoming a paler brownish behind the origin of the pelvic fins. It grows to a maximum length of 22 cm.

3. *Microphotolepis schmidti* (= *Alepocephalus schmidti*)

Common name(s): Schmidt's slickhead.

Global distribution: Western Central Pacific; Indian Ocean.

Ecology: It is a deep-water, bathydemersal species.

Biology: It attains a maximum length of: 7.2 cm.

4. *Photostylus pycnopterus*

Common name(s): Starry slickhead or starry smooth-head.

Global distribution: It has got circumglobal distribution and is found in all oceans.

Ecology: This bathypelagic species inhabits depths of 1000–2000 m.

Biology: The body of this fish is fairly elongated. Head is small with steep forehead. The eye is small, and is less than snout. Its upper jaw is reaching to about the hind margin of eye. The body coloration is brownish. It grows to a total length of 13 cm.

5. *Rouleina* spp.

Common name(s): Black stickhead.

Global distribution: This species has circumglobal distribution and is common in the Atlantic, most Indian Ocean submarine ridges, Subantarctic, North Pacific and tropical western and eastern Pacific.

Ecology: This bathypelagic species inhabits depths of 450–2300 m.

Biology: It attains a maximum length of 48.0 cm.

6. *Xenodermichthys copei*

Common name(s): Bluntsnout slickhead.

Global distribution: Atlantic, Pacific, and Indian Oceans.

Ecology: This bathypelagic species inhabits depths of 100–2650 m.

Biology: This species has a short head, a broadly rounded snout, a very large eye, and short jaws which do not extend beyond the middle of the eye. It feeds mostly on crustaceans and small cephalopods and its body coloration is bluish-black. It attains a maximum length of 31 cm.

Luminous organs and bioluminescence: Alepocephalid fishes.

In the family Alepocephalidae, regular and secondary photophores are seen only in the genera.

a. Regular photophores of fishes of the family Alepocephalidae

In *Microphotolepis schmidti* (a single species of this genus), the regular photophores are rounded structures (up to 0.8 mm in diameter) and are serially arranged on the body and head and also under the scales when the latter are developed. Most of these photophores are found located near the ventral parts of the body and head, but they are also numerous on the body sides and near the back (Sazonov, 1995). In this species, adults also have small, secondary photophores which are situated almost over each scale pocket above the scales.

b. Secondary photophores ("simple" photophores) of fishes of the family Alepocephalidae

In three alepocephalid genera related to *Microphotolepis* viz. Photostylus (in one species, *Photostylus pycnopterus), Xenodermichthys* (in two species) and *Rouleina* (in 7 luminescent species from a total of 9 known). The head photophores have been reported to be the homologs of regulär photophores of *Microphotolepis,* but their size and functions are more or less the same as those of other secondary photophores. In the species of *Microphotolepis,* the secondary photophores (ca. 0.05–0.15 mm in dia.) are present over almost everybody scale (except those in and along the lateral line and on the belly*).* On the other hand, in the species of *Persparsia,* there are a few small secondary photophores all along the lateral line and along the bases of the vertical fins, on the belly, and along the lower caudal peduncle. In the species of *Xenodermichthys*, the whole scaleless body is covered with about 500 small, nodular, secondary photophores of about 0.5 mm in dia. and they form rather regular and vertically oblique rows. A small number of photophores of similar shape and size are situated on the head. In several species of *Rouleina,* the secondary photophores are similar to those in *Xenodermichthys,* but in *Rouleina maderensis* and *Rouleina livid*a, these photophores seem to be rudimentary and are less numerous than in other species of this genus. Further, these photophores are restricted to the areas near the lower body profile. The species of the genus *Photostylus* have unique and the most unusual photophores among all fishes. In the fishes of this genus, the secondary photophores are situated on stalks and poorly ely set on the head and body. Each secondary photophore is covered with epidermis (Sazonov et al., 1996).

c. Mental light organs of fishes of the family Alepocephalidae

Two monotypic genera of the family Alepocephalidae viz. *Bathyprion (Bathyprion. danae)* and *Mirognathus (Mirognathus normani)* showed peculiar mental light organs near the tip of the lower jaw. This organ consists of modified fibers of the anterior portion of the luminescent musculus intermandibularis and transparent tissues in front of it. These transparent tissues are believed to serve as the lenses. The skin which covers the head behind this organ and the epithelium of the tongue are darkly pigmented. In both the genera, the eyes are considerably reduced in size (Saznov et al., 1996; Finn et al., 2020).

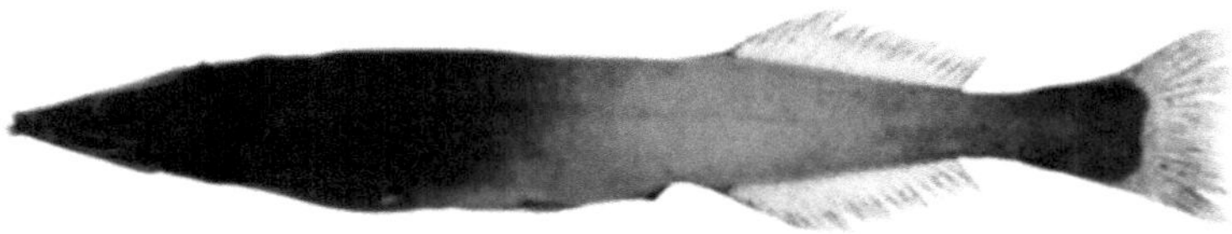

1

2

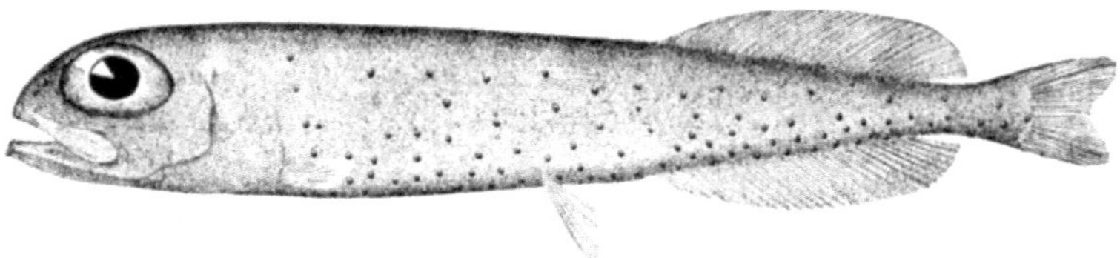

3

Credits: 1. Museums Victoria, https://fishesofaustralia.net.au/home/species/3593; 2. Museums Victoria, https://fishesofaustralia.net.au/home/genus/133. Museums Victoria, https://fishesofaustralia.net.au/home/genus/133; 3. Public domain.

1. *Mirognathus normani*; 2. *Rouleina sp.*; 3. *Xenodermichthys copei*.

3.1.2.2.2 Tubeshoulders (Platytroctidae)

The fishes of the family Platytroctidae are known for their best-developed photophores and the absence of gas bladder. They are unique in having a shoulder organ and an extensive subcutaneous canal system. The regular photophores of the family Platytroctidae are fairly large, variably shaped and developed; and are serially arranged structures resembling true photophores of other luminescent fishes (e.g., Myctophidae, Stomiiformes) are found in 6 genera of the family Platytroctidae and one genus of the family Alepocephalidae. In the family Platytroctidae, regulär photophores are developed in the genera *Persparsia* (a single species, *Persparsi kopua), Holtbyrnia* (in all species except for *Holtbyrnia anomala*), *Sagamichthys* (all 3 species), Searsia (a single species, *Searsia koefoedi), Searsioides* (in both of two known species), and *Maulisia* (in 2 of 5 species). It is reported that all other platytroctid genera except for *Mentodus* and *Normichthys* have no photophores primarily, whereas in the two latter genera, the photophores have been secondarily lost. The photophores of the family Platytroctidae are relatively variable in shape and the position of each photophore is rather constant. Generally, the photophores of the fishes of Platytroctidae consist of luminescent tissue embedded in the surrounding skin, a lens, and a reflector (only in some photophores on the head); and the whole luminous organ has a black lining (Sazonov 1986).

1. Holtbyrnia cyanocephala

Common name(s): Slickhead.

Global distribution: Eastern Atlantic.

Ecology: It is a benthopelagic species living at depths of 150–1500 m.

Biology: The body of this species is fusiform and is more or less compressed laterally. There are two well-developed supramaxillae and the gill openings wide. Head is usually scaleless and swimbladder absent. Colortion of head and body is uniformly dark (blackish, brown, or dark grey). It attains a maximum length of 22.1 cm.

2. Holtbyrnia innesi

Image courtesy: FAO.

Common name(s): Teardrop tubeshoulder.

Global distribution: Eastern and Western Atlantic.

Ecology: This mesopelagic, bathypelagic and benthopelagic species is also seen exceptionally in shallow water depths of 100–120 m.

Biology: The diagnostic features of this species include modified lateral line scales, 22–23 gill rakers, and medium gill filaments. It attains a maximum length of 24 cm.

3. *Holtbyrnia intermedia*

Common name(s): Not known.

Global distribution: Southeast Pacific: Chile.

Ecology: It is a bathypelagic species.

Biology: It is characterized by its modified lateral line scales. It reaches a maximum length of 18.4 cm.

4. *Holtbyrnia latifrons*

Common name(s): Streaklight tubeshoulder.

Global distribution: Eastern Pacific.

Ecology: It is a bathypelagic species living at depths of 225–1400 m. It is also frequently seen in mesopelagic zones and near coasts.

Biology: This dark-colored fish has the characteristic of two white tissue patches behind the eye. Light organ is seen at tip of lower jaw and its long and well-developed gill filaments are not united at their bases. It attains a maximum length of 25 cm.

5. *Holtbyrnia macrops*

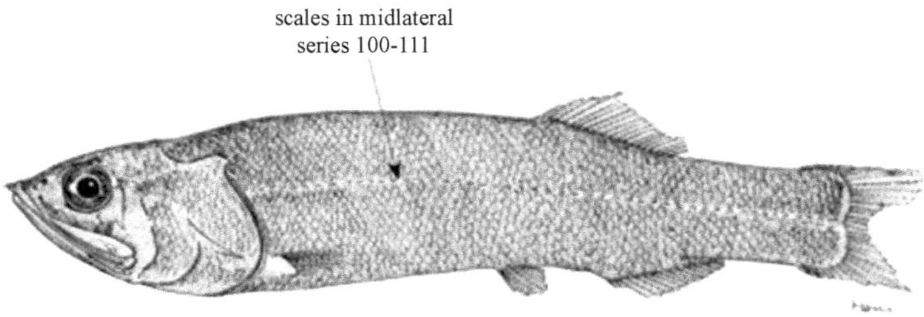

Image courtesy: FAO.

Common name(s): Bigeye searsid.

Global distribution: Eastern Atlantic.

Ecology: This bathypelagic to benthopelagic species is seen at depths of 300–1,000 m in the daytime and 100–120 m at night.

Biology: The diagnostic features of this species include modified lateral line Scales, 27–29 gillrakers, and short gill filaments. It reaches a maximum length of 20 cm.

6. *Holtbyrnia ophiocephala*

Common name(s): Not available.

Global distribution: Eastern Central Atlantic.

Ecology: It is bathypelagic species living at depths around 960 m.

Biology: It attains a maximum length of 21.7 cm.

7. *Maulisia argipalla*

Common name(s): Palegold searsid.

Global distribution: Eastern Pacific and Eastern Atlantic.

Ecology: This meso- and benthopelagic species inhabits at depths of 850–1620 m.

Biology: The body of this species is compressed. Head is large and is almost one-third of body length. Snout is pointed, and premaxillary tusks are pointing straight forward. It attains a maximum length of 19.5 cm.

8. *Maulisia microlepis*

Image courtesy: Museums Victoria, https://fishesofaustralia.net.au/home/genus/874; https://creativecommons.org/licenses/by/3.0/au/

Common name(s): Smallscale searsid.

Global distribution: Eastern and Western Atlantic and Eastern Indian Ocean.

Ecology: This bathypelagic to benthopelagic species is found at depths of 500–2000 m.

Biology: It has the following diagnostic features: Presence of about 50 modified lateral-line scales present; 112–125 body scales along midline; anal fin origin is opposite 9th dorsal ray; pyloric caeca are with 4 primary branches; and 8–9 terminal diverticula. It attains a maximum length of 25.5 cm.

9. Persparsia kopua

Credit: Museums Victoria, https://fishesofaustralia.net.au/home/species/1852; https://creativecommons.org/licenses/by/3.0/au/

Common name(s): Spangled tubeshoulder.

Global distribution: Atlantic, Indian and Pacific Oceans.

Ecology: This pelagic-oceanic species inhabits depths of 700–1500 m.

Biology: It has an elongate, compressed body with short snout, large eyes and single rows of fine needle-like teeth in both jaws. The body is covered with small cycloid scales and has violet-black coloration. Head however is pale. It attains a maximum length of 14 cm.

10. Sagamichthys abei

Common name(s): Shining tubeshoulder.

Global distribution: Western and Eastern Pacific.

Ecology: This mesopelagic and bathypelagic species is seen at depths of 300–1500 m. Juvenile fishes migrate to within 200 m of the surface at night.

Biology: It has the following diagnostic features: modified lateral line scales; and presence of reversely imbricated scales on opercle, and on head to posterior margin of nasal sac. Above the lateral line also, the scales are reversely imbricated from shoulder organ anteriorly. The body scales are found normally imbricated below lateral line. It feeds extensively on small crustaceans and it attains a maximum length of 33 cm.

11. Sagamichthys gracilis

Common name(s): Not known.

Global distribution: Western Indian Ocean and Western Central Pacific.

Ecology: This bathypelagic species lives at depths of 400–1800 m.

Biology: It has the following diagnostic features: modified lateral line scales; and presence of reversely imbricated scales on opercle and on head to before eye present and on cheek. It attains a maximum length of 20.3 cm.

12. Sagamichthys schnakenbecki

Image courtesy: Jan Y. Poulsen.

Common name(s): Schnakenbeck's searsid.

Global distribution: Eastern Atlantic:

Ecology: It is bathypelagic and benthopelagic, living at depths of 365–1200 m.

Biology: It has a long and slender body r, with a short snout. It has 27–32 gill rakers, 7 or 8 branchiostegal rays and 9–11 pyloric caeca. Dorsal fin is far back and the anal fin is further back. The body coloration is dark. It attains a maximum length of 27 cm.

13. Searsioides calvala

Common name(s): Not known.

Global distribution: Western Central Pacific.

Ecology: This bathypelagic species is found at depths of 1000–1400 m.

Biology: It has the greatest body depth (5.8–6.0 in SL). There is a single infracaudal organ, and the cleithral symphysis is with a forward-directed spine. Dorsal fin origin is ahead of that of anal by about a fourth of its length. Mouth is large. A row of 8–10 teeth is present on the external lateral face of the dentary. A small supraorbital bone is present. Pectoral base and adjacent area are without scales. There are 29–34 gill rakers on 1st arch and 120 scales in midlateral series. It attains a maximum length of 9.3 cm.

14. Searsioides multispinus

Common name(s): Not assigned.

Global distribution: Western Central Pacific.

Ecology: This bathypelagic species occurs at depths of 600–1000 m.

Biology: It has the greatest body depth which is 4.6–5.9 in SL. Anal fin origin is nearly opposite that of dorsal, at about 4th dorsal ray. Anus is opposite dorsal origin. There are 40–45 gill rakers on 1st arch and 76–81 scales in midlateral series. It attains a maximum length of 11.8 cm.

Searsioides sp.

Image courtesy: Bulletin of the Scripps Institution of Oceanography (Open access).

Luminous organs and bioluminescence: Fishes of the family Platytroctidae.

***Holtbyrnia* spp.:** In *Holtbyrnia cyanocephala*, the photophores are well developed in adults. Interventral and posterior gular oorgans and anal organ are present (Orrell and Hartel, 2016).

***Maulisia* spp.:** In two closely related species of *Maulisia* viz. *Maulisia mauli* and *Maulisia argipalla* these photophores are poorly developed. Further, the former species has a pineal window, whereas in the latter, it is closed (overlying skin is densely pigmented) (Sazonov et al., 1993).

***Persparsia* spp. and *Sagamichthys* spp.:** In these genera, most photophores located on the head and ventrolateral parts of the body are either rounded or semicircular in shape, but they face downward usually. Further, the photophores situated near the midline of the belly are more variable in shape, i.e., from rounded to looking like transverse bars, chevrons, or tear-drops. While most of these photophores face downward, some (gular and interventral) face forward and downward. Juveniles of the species of *Persparsia* and *Sagamichthys* have small photophores within the eye-ball (in the iris) (Saznov et al., 1996; Finn et al., 2020).

***Persparsia* spp.:** The photogenic tissue is red in the species of this genus. Most photophores found on the head and ventrolateral parts of the body are rounded or semicircular in shape, and they usually face downward. Photophores found near the midline of the belly are more variable in shape: from rounded to transverse bars, chevrons, or tear-drops; most of them face downward, and some (gular and interventral) face forward and downward (Sazonov 1986).

Sagamichthys abei: In this species, its photophores are more elongated, resembling a narrow ellipse or a short transverse bar (Matsui and Rosenblatt, 1987). Allen and Horn (2006) reported that this species has two types of light organs viz. ventrally placed photophores and a shoulder organ or sac with an external tube through which luminous particles are expelled.

Sagamichthys schnakenbecki: Poulsen (2018) reported that photogenic tissues were observed in this species at the pectoral-fin base in early juveniles; as a pair of large globule-like tissues inside the caudal peduncle of early juveniles and at the pelvic girdle of late juveniles and early adults; and as photogenic tissue as pigment over the cleithral bone in adults. Further, a unique skin-slit structure with certain bioluminescent structures has also been reported in this species.

Image courtesy: Jan Y. Poulsen.

Luminescent tissues in *Sagamichthys schnakenbecki* (ventral view).

3.1.2.3 ORDER: AULOPIFORMES

3.1.2.3.1 Sabretooth Fish (Evermannellidae)

1. Coccorella atrata

Source: Public domain

Common name(s): Sabertooth fish.

Global distribution: Indo-West Pacific, SE Pacific.

Ecology: This bathypelagic species inhabits depths of 300–2626 m.

Biology: The body of this species is moderately elongated and deep, and strongly compressed. Both body and head are naked without scales. The head and body are dark brown. It attains a maximum length of 4.2 cm.

2. *Coccorella atlantica*

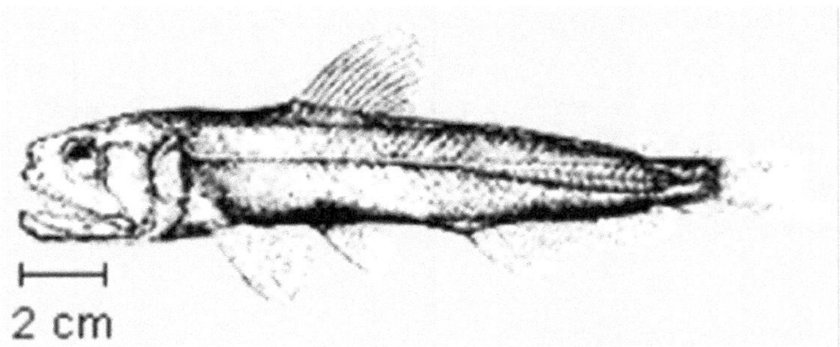

Image courtesy: André SG.

Common name(s): Atlantic sabretooth.

Global distribution: Atlantic, Indian and Pacific.

Ecology: This mesopelagic and bathypelagic species is oceanodromous and inhabits depths of 50–1000 m.

Biology: The body and fins of this species are dark brown in alcohol. It feeds mainly on fish and sepia and attains a maximum length of 18.5 cm.

Luminous organs and bioluminescence: Evermannellid fishes: These fishes possess ventral luminescent organs derived from hepatopancreatic tissue. The luminescent quality of these organs is due to their endogenously luminescent cells containing distinct granular inclusions (Barton, 2016).

3.1.2.3.2 Barracudinas or Naked Barracudinas [Paralepididae (= Lestiidae)]

1. Lestidium atlanticum

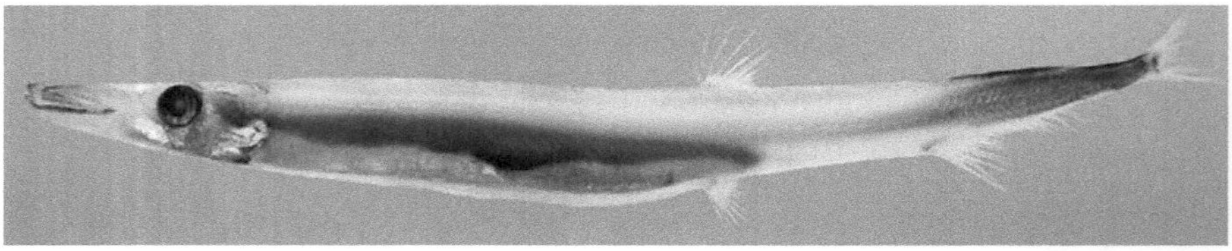

Source: Public domain

Common name(s): Atlantic barracudina.

Global distribution: It is widely distributed in tropical and subtropical waters of the three major oceans.

Ecology: This bathypelagic species has a depth range of 50–1270 m.

Biology: The body of this species is naked without scale, except for the lateral line, which is covered by a single row of scales. The body coloration is pale with a greyish dorsal band. Adults possess a heavily pigmented caudal peduncle except for the light band along the lateral line. This species attains a maximum length of 25 cm.

2. Lestidium longilucifer

Common name(s): Long light-organ barracudina.

Global distribution: Indo-West Pacific: Northwestern Australia and Southern Taiwan.

Ecology: This mesopelagic and coastal species occurs at depths of 12–1000 m.

Biology: The head of this species is moderately long. The snout is long and blunt. The mouth is terminal and its gape is extending to under anterior half of eye posteriorly. It is a slightly translucent fish with greyish back and blackish top of head and lower lips. Body above anal-fin base to the caudal fin base is silvery white; upper abdominal cavity is black; abdomen is silvery-white below, caudal fin is slightly blackish; and the other fins are clear. It grows to a maximum length of 29 cm.

3. Lestidium orientale

Common name(s): Oriental barracudina.

Global distribution: Indo-west Pacific Ocean.

Ecology: This bathypelagic species lives at a depth range of 50–1270 m.

Biology: The body of this species is stout and compressed. Head is relatively stout and blunt. Mouth is terminal and moderately large. Its gape extends to the anterior margin of the eye. The lower jaw is slightly upturned at the tip, and the eye is large. The body is translucent and it grows to a maximum length of 17 cm.

4. Lestrolepis japonica

Image credit: Wikipedia

Common name(s): Japanese barracudina.

Global distribution: Western Pacific.

Ecology: It is a mesopelagic and oceanic species.

Biology: The body of this species is elongated and compressed. Head to caudal is gradually narrowing. Mouth is large and snout is piercing. Lower jaw is slightly protruding. It grows to a maximum length of 19 cm.

Luminous organs and bioluminescence: Naked barracudinas.

While the species of *Lestidium* have single abdominal bands of luminous tissue, *Lestrolepis* possess double abdominal bands. Further, some species have been reported to possess ventral photophores. The ventral light organs of these fishes have been derived from hepatopancreatic and are "self-luminous" (i.e., endogenous luminescence without the presence e of bacterial symbionts), bilobed, yellow organs. The luminescent quality of these organs is further supported by their endogenously luminescent cells containing distinct granular inclusions. The luminescent species both *Lestidium* and *Lestrolepis* inhabit the mesopelagic zone where their ventral counterilluminating bioluminescence is especially beneficial (Barton, 2016).

Ho et al. (2020) reported that *Lestidium longilucifer* and its congeners *Lestidium orientale* and *Lestidium atlanticum* form a group possessing a luminescent duct extending forward to in front of the eyes. Further, in all these species, light organs are absent on the lateral and ventral surfaces of the body.

Barton (2016) reported that *Lestrolepis* spp. possess an antorbital light organ (a pigmented spot lying in front of the eye), and the luminescent duct is divided into two branches. This light organ is believed to serve light-matching purposes.

3.1.2.3.3 Pearleyes (Scopelarchidae)

1. Benthalbela infans

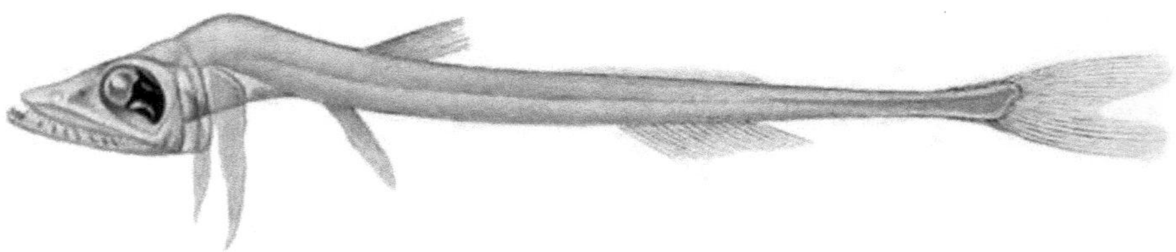

Source: Public domain

Common name(s): Zugmayer's pearleye.

Global distribution: Tropical and subtropical waters of Atlantic, Indian and Pacific Oceans.

Ecology: This bathypelagic, oceanodromous species lives at depths of 0–4740 m.

Biology: In alcohol, the body of this species is uniformly dull brown and it is slightly darker on the edges of scale pockets. A dense black stripe is seen along the belly. Adults feed mainly on midwater fishes and they reach a maximum length of 13.8 cm.

Luminous organs: This self-bioluminescent species has four relatively large luminescent organs along its ventral surface of the trunk. The light-emitting cells in these organs are believed to be derived from its skeletal muscle (Anctil, 2018; Finn, 2020).

2. Scopelarchoides kreffti

Image courtesy: Hiroyuki Motomura.

Common name(s): Twin-striped pearleye, silvery pearleye or silvery pearleye.

Global distribution: Southwest Atlantic and Southwest Pacific.

Ecology: This bathypelagic species lives at depths up to 2000 m.

Biology: This species is characterized by the presence of a black stripe along the side above the lateral line from behind the head almost to the caudal fin. Further, the bases of the first dorsal and tail fins are also blackish. It has large upward-directed eyes and large scales along the lateral line. It grows to a maximum length of 19 cm.

Luminous organs: This self-bioluminescent species has two ventral luminescent organs. The photogenic tissues of its luminous organs is believed to have been derived from its muscle cells (Suntsov et al., 2008)

3.1.2.4 ORDER: BATRACHOIDIFORMES

3.1.2.4.1 Toadfishes (Batrachoididae)

1. Porichthys myriaster

Image courtesy: John Snow.

Common name(s): Specklefin midshipman.

Global distribution: Eastern Pacific.

Ecology: It is a reef-associated, demersal species living at depths up to 126 m. It is common in rocky areas and soft bottom, and in bays. It is more active during night time.

Biology: The body of this species is compressed and elongated. Head is slightly flattened above. One solid spine is seen on opercle; Body coloration is iridescent purplish grey to brown above and yellowish below. Upper body may have 8–12 saddles and pectoral and dorsal fins have small spots and bars. Anal fin is with dark to black margin in adults. It grows to a maximum length of 51 cm.

2. *Porichthys notatus*

Source: Public domain

Common name(s): Plainfin midshipman.

Global distribution: It is native to the eastern Pacific Ocean.

Ecology: It inhabits sandy and muddy bottoms from shallow water just below the tide to depths of 366 m.

Biology: It is brownish to olive to iridescent purple dorsally, and is becoming lighter on the sides and yellowish/golden on the belly. A whitish patch and black crescent are seen below the eye. Juveniles of this species may have a dark saddle mark. It has wide pectoral fins and a narrow but rounded tail fin. Its diet consists of mainly crustaceans and fish. This species reaches up to 38 cm in length.

Luminous organs and bioluminescence: The luminous fishes of this family have hundreds of ventral dermal photophores, which are composed of a pear shaped lens, photogenic tissue, a reflector, and pigment. The autogenic (intrinsic) bioluminescence of these fishes is due to its dietary acquisition of cypridina-type luciferin "vargulin" from their bioluminescent ostracod prey.

It is also reported that the photophores of these fishes do not contain enough luciferase and their bioluminescence is therefore much dimmer compared to that of ostracods (Lau and Oakley, 2021). In *Porichthys myriaster,* light organs are seen on its four lateral lines. Further, a row of light organs is also seen under each side of the chin of this species. (Anon., https://biogeodb.stri.si.edu/sftep/en/thefishes/species/4655). In *Porichthys notatus*, photophores, or light producing cells are seen on its belly (Abnon., https://fishbio.com/field-notes/fish-biology-behavoir/ghouls-goblins-glowing-fish).

3.1.2.5 ORDER: CLUPEIFORMES

3.1.2.5.1 Anchovies (Engralulidae)

1. Coilia dussumieri

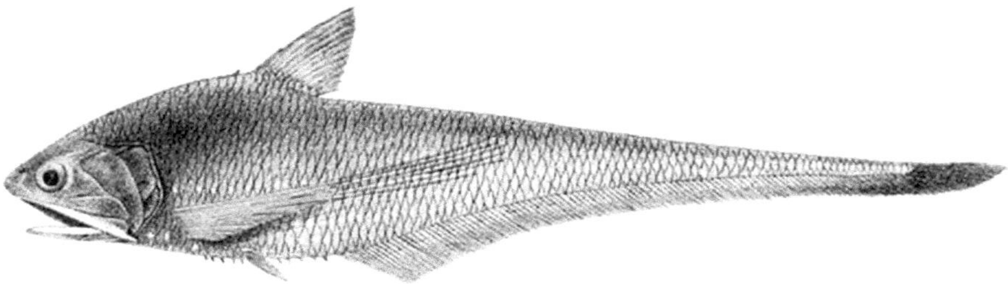

Credit: BEDO (Thailand). Wikipedia. https://creativecommons.org/licenses/by-sa/4.0/deed.en

Common name(s): Goldspotted grenadier anchovy.

Global distribution: India, Bangladesh, Myanmar, Thailand and Malaysia; western Indo-Pacific.

Ecology: It is often encountered in coastal waters and estuaries also. This pelagic, neritic and amphidromous species dwells in depths of only 0–50 m.

Biology: 20.0 cm; Body of this species is tapering and its belly is rounded before pelvic fins. It feeds on copepods, prawn fish larvae, etc., and it attains a maximum length of 20 cm.

Luminous organs and bioluminescence: The only known bioluminescent species of the family Engraulidae (Suntsov et al., 2008) in which light organs (golden or pearly spots) are present in its flanks and belly in rows below scales; and also along isthmus, edge of lower jaw, on cheek and gill cover. Suntsov et al. (2008) reported that the photophores present in six rows along the ventral and lateral surface of its body showed endogenous (intrinsic) bioluminescence without the use of symbiotic luminescent bacteria.

3.1.2.6 ORDER: KURTIFORMES

3.1.2.6.1 Cardinalfishes (Apogonidae)

1. Jaydia argyrogaster

Common name(s): Silvermouth siphonfish.

Global distribution: Western Central Pacific.

Ecology: This demersal, reef associated species inhabits at depths of 18–86 m.

Biology: It is a pale brown to silvery cardinalfish with large brown spots on the sides and clear- cut brown bands radiating from the rear of its eye. Males of this species are mouth brooders. It grows to a maximum length of 6 cm.

2. Jaydia photogaster

Common name(s): Silverbelly cardinalfish.

Global distribution: Western Pacific.

Ecology: It is a solitary species moving in small groups occasionally. It is found in lagoon patch reefs at depths of 18–52 m.

Biology: It grows to a maximum length of 6 cm

***3. Jaydia poeciloptera** (= Apogon poecilopterus; Apogon bilaciniatus)*

Common name(s): Pearly-finned cardinalfish or pearlyfin cardinalfish.

Global distribution: Indo-Pacific.

Ecology: It is a brackish water, reef associated and demersal species inhabiting depths of 19–146 m. It is a nocturnal species often seen in inshore over soft bottom and river mouths. It may also be seen in holes in the mud during the day.

Biology: It is greyish above and silvery-grey below and is with dark spots on scales forming indistinct lines along side. There is a dark blotch on its first dorsal fin; several rows of small spots on second dorsal fin; and a broad dusky margin on the caudal fin. This species grows to a maximum length of 14 cm.

*4. **Jaydia smithi*** *(= Apogon smithi; Jaydia hungi)*

Image credit: Sergey Bogorodsky.

Common name(s): Smith's cardinalfish.

Global distribution: Indian and Pacific Oceans.

Ecology: This pelagic, neritic and nocturnal species that hides among rocks and reefs during the day and emerges at night.

Biology: It has a slightly compressed, oblong body with two dorsal fins, Caudal fin is rounded or truncated. Mouth is slightly oblique and terminal. It is light tan in color and is becoming paler towards its whitish belly, with five or six faint, dark bars. It feeds on invertebrates and grows to a maximum length of 10 cm.

5. Jaydia striatodes

Common name(s): Cardinalfish.

Global distribution: Indo-West Pacific: Thailand, Hong Kong, Philippines.

Ecology: This reef-associated species inhabits at depths of 0–35 m.

Biology: Its diagnostic feature is its blackish distal half of anal fin. It grows to a maximum length of 5 cm only.

6. *Jaydia truncata*

Common name(s): Flagfin cardinalfish.

Global distribution: Indo-Pacific: Red Sea and Persian Gulf to Japan; and Australia.

Ecology: This reef-associated and nocturnal species inhabits at depths of 0–132 m and it occurs mainly in inshore waters of the continental shelf.

Biology: The body of this fish is silver grey without any vertical band. Preopercular ridge is weakly serrated and preopercular margin is with a black spot. Upper half of its first dorsal fin is black and the caudal fin is rounded. Males are mouth brooders and these fishes grow to a maximum length of 15 cm.

7. *Rhabdamia gracilis*

Image credit: Kwang-Tsao Shao, Taiwan Fish Database.

Common name(s): Luminous cardinalfish.

Global distribution: Indo-West Pacific.

Ecology: This species inhabits lagoon and coastal reefs and among coral and rocks at depths of 3–91 m.

Biology: This species is characterized by the presence of a bluish longitudinal stripe on the middle of its body; translucent white or pink, silvery on head and abdomen; caudal peduncle is with a posteriorly placed black spot; and one or both caudal fin lobes is having black tip. It grows to a maximum length of 7.4 cm.

8. *Taeniamia zosterophora* (= *Archamia zosterophora*)

Source: Image by Graham Edgar. https://creativecommons.org/licenses/by/3.0/deed.en

Common name(s): Girdled cardinalfish or blackbelted cardinalfish.

Global distribution: It is native to the western Pacific Ocean.

Ecology: It inhabits on reefs in protected bays and lagoons at depths of 1–40 m. It may form dense schools.

Biology: It is characterized by the presence of two narrow vertical red stripes next to its gills and by a distinct small black dot at the base of its tail. A black band may or may not be present below the second dorsal fin. It has a maximum length of 8 cm.

Luminous organs and bioluminescence: The fishes of the genera *Apogon, Achamia* and *Rhabdamia* possess internal light organs which ranges from pouch-like protrusions of the intestine, rectum or modified pyloric caeca (Haneda et al., 1969). These fishes show endogenous bioluminescence and produce light from their own luciferase and using Cypridina-like luciferin apparently acquired in the diet (Dunlap et al., 2011).

3.1.2.7 ORDER: LOPHIFORMES

3.1.2.7.1 Netdevils (Linophrynidae)

1. Linophryne algibarbata

Common name(s): Leftvent sea devil.

Global distribution: North Atlantic Ocean.

Ecology: This bathypelagic species inhabits at depths of 400–2,200 m.

Biology: This species grows to a maximum length of 4.1 cm.

2. Linophryne arborifera

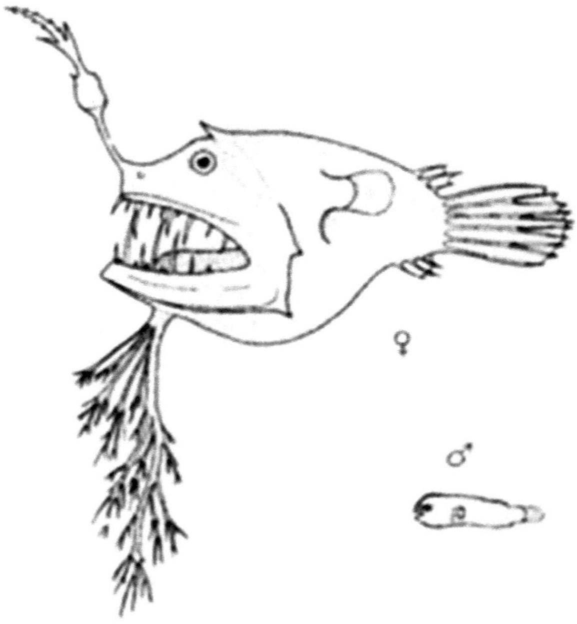

Credit: Tony Ayling. https://creativecommons.org/licenses/by-sa/1.0/deed.en

Linophryne arborifera Female and Male

Common name(s): Illuminated netdevil.

Global distribution: All tropical and subtropical oceans.

Ecology: This mesopelagic and bathypelagic species inhabits at depths below 1,000 m.

Biology: Fishes of this species have no scales and gelatinous skin. The eyes of males are very well developed and are tubular and directed somewhat forward. Nostrils of males are also large and well developed. These fishes have black colored body and, feed on fishes and crustaceans, and grow to a maximum length of 7.7 cm.

3. Linophryne bicornis

Common name(s): Leftvent.

Global distribution: Northwestern Atlantic and Southeastern Indian oceans.

Ecology: This bathypelagic species has a depth range of 620–1220 m.

Biology: It grows to a maximum length of only 3.0 cm.

4. Linophryne densiramus

Common name(s): Thickbranch angler.

Global distribution: Atlantic and Pacific Oceans.

Ecology: This bathypelagic species has a depth range of 150–1400 m.

Biology: In this species its barbel is divided nearly at base into 3 unpaired primary branches. Subdermal pigment is seen in two rows alongside of body. Its parasitic males do not possess without pointed sphenotic spines. While the metamorphosed female has a maximum total length of 9 cm, the parasitic male is only 0.9 cm.

5. Linophryne indica *(= Linophrynx corymbifera)*

Common name(s): Headlight angler.

Global distribution: Pacific Ocean.

Ecology: It occurs in the bathyal zone at depths below 1000 m.

Biology: It has a laterally compressed, oval body. Head is broad and large and its oblique mouth extends behind the eye. Teeth are large and sharp and are consisting of four longitudinal rows in the upper jaw and three in the lower jaw, as well as a single pair of vomerine teeth in the roof of the mouth. Its sphenotic spines (above the eyes) are large and there is a distinct angular spine. Coloration of head and body is black. In the body melanophores are arranged in two longitudinal bars. Fins are colorless. While the female has an average length of 5 cm, the male is only 1.8 cm.

6. Linophryne maderensis

Common name(s): Not known.

Global distribution: Eastern Central Atlantic.

Ecology: This bathypelagic and non-migratory species is seen up to a depth of 880 m.

Biology: Males are unknown in this species.

Luminous organs and bioluminescence: Linophrynid fishes.

Female fishes of *Linophryne* display a dual double bioluminescent system, with both a self-luminescent hyoid (chin) barbels and a bacterial light organ within the esca (Finn, 2020; Suntsov et al., 2008). While the esca (with a open type of gland) containing symbiotic, luminous bacteria is ectodermal in origin, the barbel light organs are believed to be derived from the mesoderm. Anon. (https://www.frogfish.ch/deepsea-anglerfish.html) reported that the filaments of the multibranched barbels hanging from the lower jaw of these fishes contain many more bioluminescent organs with complex paracrystalline photogenic granules and they are without luminous bacteria. The luminescent organs of the different species of *Liniophryne* are described below.

Linophryne algibarbata: In juvenile females of this species, the escal bulb is globular and is surrounded distally and laterally by a blunt ridge. Escal filaments or other appendages are absent. Its self-luminescent barbel is divided at base into 4 primary branches of nearly equal length and each of these branches has a varying number of secondary and tertiary branches, which are more or less prolonged into filaments. Each of these filaments is with a distal photophore.

Linophryne arborifera: The female fishes of this species have not only a luminous lure on their head, called the esca (Latin for "food"), but also have a multibranched barbel hanging from the lower jaw. These barbel filaments contain many more self-bioluminescent organs.

Linophryne bicornis: In this species, its escal bulb is without distal prolongation or filaments. In this species also, posterior and anterior appendages of the esca are absent. However, stout, tapering filament is seen on each side of the upper part of escal bulb. Each filament is with distal series of internal photophores.

Linophryne densiramus: Metamorphosed females possess the following characteristics: illicia maximum length is 36 mm; esca is without distal appendage, with two simple filaments on upper part of bulb and there is no appendage behind escal pore; barbel is about 50% of standard length and unbranched part of its stem is about 30% standard length; of its 4 primary branches, proximal branch is simple with few photophores in unpigmented tip and the others are with 2–4 terminal, slender, unpigmented filaments; and each of these filaments is with an internal series of photophores

Linophryne indica: In this species, the illicium (so-called fishing rod) on its snout is short and broadens into a larger esca at its tip. This has a number of short filaments and a long trailing appendage. There is a self-luminescent barbel on the chin with a single stem and is branching into a bunch of short branches.

Linophryne maderensis: Metamorphosed females are distinguished by the following characteristics: distal escal appendage is about 40% of standard length; presence of compressed distal escal appendage; and absence of additional escal appendage.

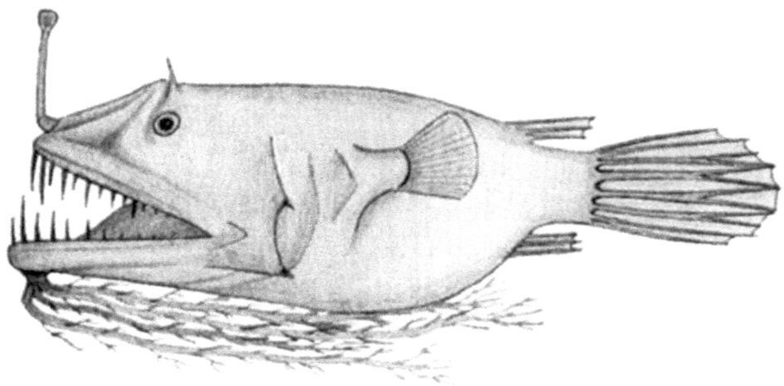

1

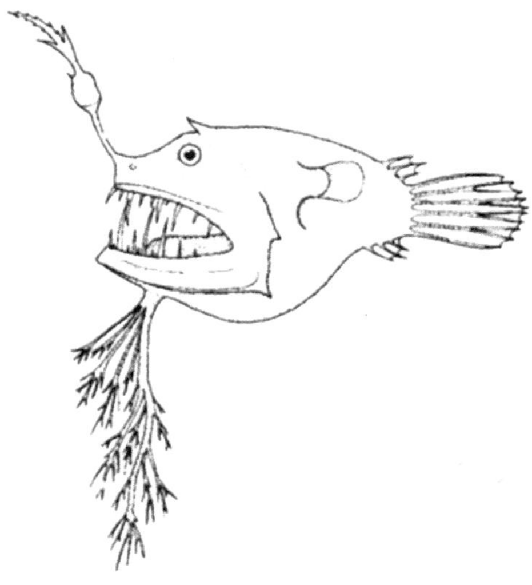

2

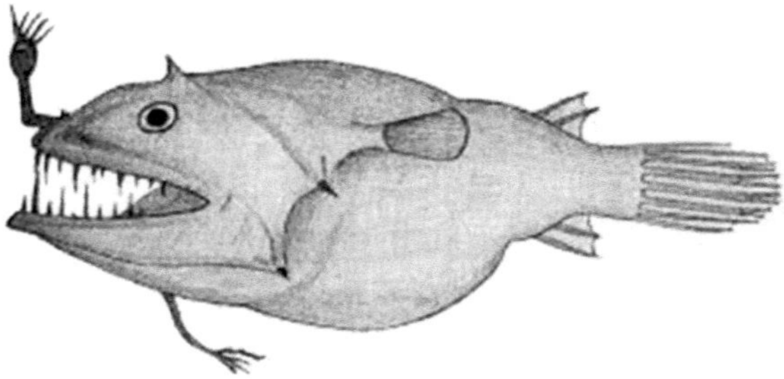

3

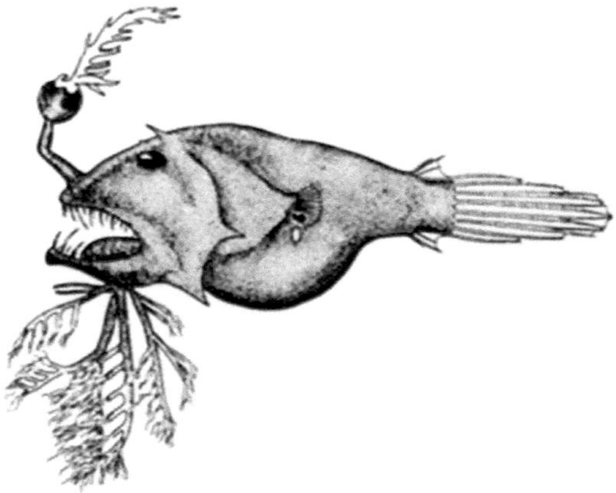

4

5

Credits: 1. Tambja. https://creativecommons.org/licenses/by-sa/3.0/deed.en; 2. Tony Ayling & Geoffrey Cox. https://creativecommons.org/licenses/by-sa/1.0/deed.en; 3. Tambja. https://creativecommons.org/licenses/by-sa/3.0/deed.en; 4. Museums Victoria, https://fishesofaustralia.net.au/home/species/3426 https://creativecommons.org/licenses/by/4.0/; 5. Ray Simpson. https://creativecommons.org/licenses/by-sa/2.5/

1. *Linophryne algibarbata;* 2. *Linophryne arborifera;* 3. *Linophryne indica;* 4. *Linophryne densiramus;* 5. *Linophryne maderensis.*

3.1.2.7.2 Batfishes (Ogcocephalidae)

1. *Dibranchus atlanticus*

Source: Reprinted from https://www.fishbase.se/photos/PicturesSummary.php?StartRow=0&ID=4956&what=species&TotRec=5. https://creativecommons.org/licenses/by/3.0/

Common name(s): Atlantic batfish.

Global distribution: Western and eastern Atlantic Ocean.

Ecology: It is a bathypelagic species seen on the seabed on muddy and sandy areas at depths of 45–1300 m.

Biology: It is a disc-shaped, species with a vertically compressed body and a flat ventral surface, Dorsal fin, which is found far back near the tail, has no spines and 5–7 soft rays, and the anal fin has no spines but is with four soft rays. Pectoral fins are very strong and are held at right angles to the body. The dorsal surface of the fish is uniformly reddish-grey, but fishes in the eastern Atlantic may have reticulated markings. The ventral surface is paler. It feeds mainly on small invertebrates and it grows to a standard length of 14.5 cm.

Luminous organs and bioluminescence: In this species, its dark skin has been found to be luminescent and spontaneous light emission was observed from the dorsal surface only (Finn, 2020; Suntsov et al., 2008). However, no measurable spontaneous light emission was observed from the esca of this species (Crane, 1968). It was experimentally observed that both the skin and the esca of this species showed a flash of light when the esca was treated with 0.5 ml sodium phosphate buffer with pH 7.0. Further, when hydrogen peroxide was added to a non-glowing ventral skin, a light response was observed (Crane, 1968). Furthermore, when 0.25 ml adrenalin hydrochloride was injected in the region of the heart of the fish, and the fish was found to luminesce. Within a few min the entire dorsal surface of the fish was also found to show a low level glow. However, neither the ventral surface nor the esca glowed under these experimental conditions (https://pdfslide.net/documents/bioluminescence-in-the-batfish-dibranchus-atlanticus.html).

3.1.2.8 ORDER: MYCTOPHIFORMES

3.1.2.8.1 Lanternfishes (Myctophidae)

The lanternfish or myctophids are the most widely distributed, populous, and diverse of all vertebrates and they form prey for larger organisms, especially in the deep sea. They are the most widely distributed worldwide in all oceans and seas from the Arctic to Antarctic (Anon., https://www.fao.org/3/x2401e/x2401e43.pdf). They form nearly 5% of the living species of ray-finned fishes and account for as much as 90% of all deep sea fish biomass with its 34 known genera and 248 spp. (Paxton, 1972). Interestingly more than 50% of these species have been reported to inhabit the eastern Pacific Ocean (Cruz-Agüero, 2016). While most of the myctophids inhabit open oceanic waters, some species live in continental slopes and continental shelves including coral reefs. It is also reported that these fishes for largely responsible for the deep scattering layer of the world's oceans with its estimated global biomass of about 600 million metric tonnes (Anon., https://www.sheppardsoftware.com/content/animals/animals/fish/flashlightfish.htm).

a. Luminous organs and bioluminescence in myctophid fishes

An important characteristic of the family Myctophidae is the presence of intrinsic (endogenous) bioluminescence in all its species excepting a single species, *Taaningichthys paurolychnus*. The serially arranged small light organs or photophores produce yellow, green, or blue light. (http://tolweb.org/Myctophidae/) which helps these fishes to provide a camouflaging effect protecting them from predators below by means of counter-illumination (Anon. https://invertebrate.w.uib.no/2016/12/14/door-14-where-the-sun-doesnt-shine-lucifer-luciferin-and-luciferase/). These light organs are round, or keyhole shaped and are found underlying modified scales on the head and ventral body surface The arrangement and type of the photophores are different in different species of Myctophidae. While all the myctophid species possess primary photophores which are arranged in distinct groups on their head and body, small secondary photophores are also present on the head, body and median fins in some species. While a variety of luminous patches are present in some species, luminous tissue of various shapes and sizes are seen on head, caudal peduncle, and/or at bases of various fins in most species. Further, in some species, eyes are found associated with photophores (Suntsov et al., 2008; Anon., https://www.fao.org/3/x2401e/x2401e43.pdf). In some species of the genera of myctophids viz. *Scopelopsis, Hintonia, Lampichthys, Bolinichthys, Lepidophanes,* and *Lampanyctus* minute secondary photophores are seen on their body in addition to the primary photophores. These minute photophores have been reported to occur either at every scale pocket (as in the species of *Scopelopsis*) or they may be restricted to a particular areas of the body (as in the case of *Hintonia* spp.).

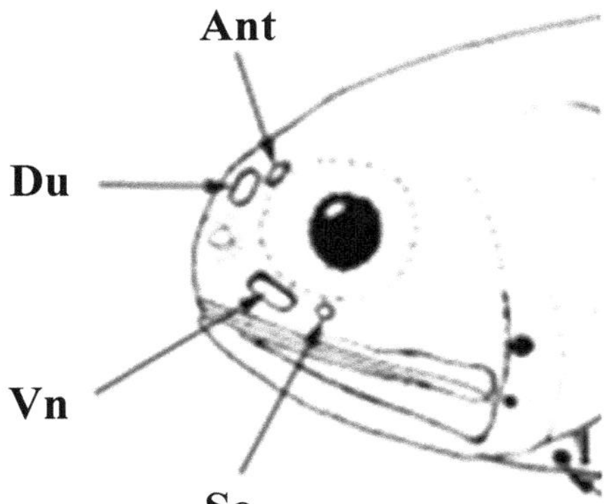

Photophores in the head of a fish of Myctophidae.

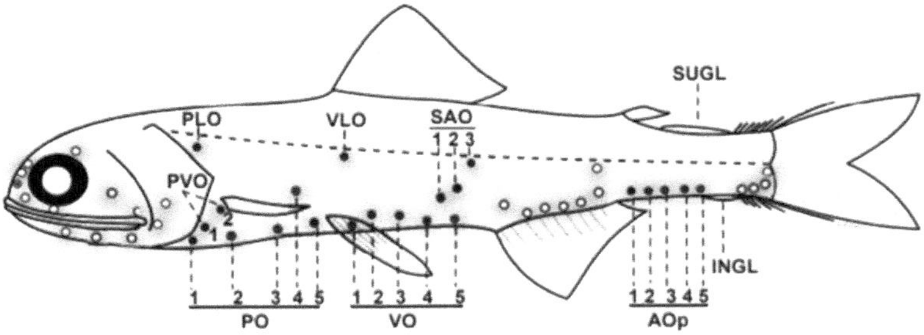

Ant = antorbitalgland; Suo = supraorbital photophore; Cl = postorbital photophore; Dn = dorsonasal gland; Vn = ventronasal gland; Br = branchio-stegal photophores; Bu = cheek photophore; So = suborbital photophore; Op = opercular photophores; PLO, suprapectoral organ; PVO, subpectoral organ; VLO, supraventral organ; SAO, supralateral organ; AO, anal organs:

(a) anterior, (p) posterior; INGL, infracaudal gland; VO, ventral organs; PO, thoracic organs; SUGL, Supracaudal gland.

Generalized photophore pattern for Myctophidae showing the location of the photophores

Image courtesy: De La Cruz-Agüero.

The nature of luminous organs viz. sexual dimorphism of caudal luminous glands; luminous tissue; and location of photophores in the different genera of the family Myctophidae are given in the following tables.

Characters of the luminous organs in the fishes of the family Myctophidae

Characters	Genera	SDCG	CG	LT
Benthosema	+	+	–	
Centrobranchus	+	+	–	
Ceratoscopelus	–	–	+	
Ctenoscopelus	+	+	–	
Diaphus	–	–	–, +	
Diogenichthys	+	+	–	
Electrona	+, –?	+	–	
Gonichthys	+	+	–	
Gymnoscopelus	–	–	+	
Hierops	+	+	–	
Hintonia	–	–	+	
Hygophum	+	+	–	
Lampadena	–	–	–	
Lampanyctodes	–	–	+	
Lampanyctus	–	–	+	
Lampichthys	–	+	+	
Lepidophanes	–	–	+	
Lobianchia	+	+	–	
Loweina	+	+,–	–	
Metelectrona	–?	+	–	
Myctophum	+	+	–	
Nasolychnus	–	–	+	
Notolychnus	+	–	–	

(Continued Table)

Characters	Genera	SDCG	CG	LT
Notoscopelus	+	+	+	
Parvilux	–	–	–	
Scopelopsis	+	+,–	+	
Stenobrachius	–	–	–	
Symbolophorus	+,	–?	+,–	
Taaningichthys	–	–	–	
Tarletonbeania	+	+,–	–	
Triphoturus	–	–	–	

SDCG: Sexual dimorphism of caudal glands; CG: Caudal glands; LT: luminous tissue.
Source: Paxton (1972).

Characters of the photophores in the fishes of the family Myctophidae

Genera Characters	EPO	VO	EVO	Pol	Prc	CP	SP
Benthosema	–	4	2	1	2	0	–
Centrobranchus	–	4	–	1	2	0	–
Ceratoscopelus	–	5	–	2	4	0	–
Ctenoscopelus	–	4	–	1	2	0	–
Diaphus	4	5	2;3	1–3	4	0	–
Diogenichthys	–	4	2	1	2	0	–
Electrona	–	4	–	0	2	0	–
Gonichthys	–	4	–	1	2	0	–
Gymnoscopelus	–	5–6	2	4	4–9	0	–
Hierops	–	4	–	0	2	0	–
Hintonia	–	5	2	3	5–6	1	LL
Hygophum	–	4	–	2	2	0	–
Lampadena	–,4	–	4- 6	1	3	0	–
Lampanyctodes	3;5	5	2;3	1	5	0	–
Lampanyctus	4	4	–,2	2	3–4	0–2 LL,	+
Lampichthys	–	5	–	3	4	3–7	–
Lepidophanes	4	5–6	2	2	3–4	0	–
Lobianchia	4	5	2;3	1–2	4	0	–
Loweina	–	2- 4	–	1	2	0	–
Metelectrona	–	4	2	0–2	2	0	–

(Continued Table)

Genera Characters	EPO	VO	EVO	Pol	Prc	CP	SP
Myctophum	–	4	–	1	2	0	–
Nasolychnus	–	5–6	2	2	4–6	0	–
Notolychnus	3;4?	4	2	2	2	0	–
Notoscopelus	–	5	–	2–3	3–4	0	–
Parvilux	4	3–6	–	2	4	0	–
Stenobrachius	4	4–5	–	1	3–4	0	–
Symbolophorus	–	4	–	1	2	0	–
Taaningichthys	–	5–8	–	1	3	0	–
Tarletonbeania	–	6	–	1	1	0	–
Triphoturus	4	5	2	2	3	0	–

EPO: Elevated thoracic organs; VO: Ventral organs; EVO: Elevated ventral organs; Pol: Number of posterolateral organs; Prc: Number of precaudal organs; CP: Check photophores; SP: Secondary photophores.

LL, Secondary photophores may be restricted to the scales immediately above and below the lateral line +; Secondary photophores are present under every scale on the body and even on the head.

Source: Paxton (1972).

a) Subfamily: Diaphinae

1. Diaphus lucidus

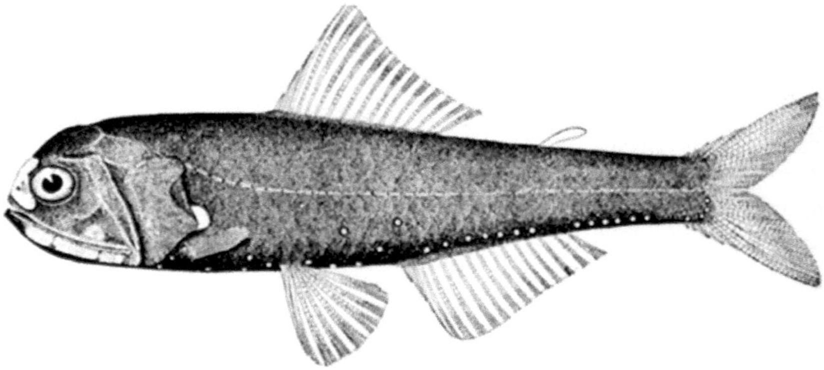

Source: Public domain

Common name(s): Spotlight lanternfish.

Global distribution: Eastern and Western Atlantic Ocean, Indian Ocean and Indo-west-central Pacific Ocean.

Ecology: This bathypelagic- mesopelagic and oceanodromous dwells up to a depth of 2999 m. It has been reported to occur between 425–750 m during the day time and between 40–550 m at night.

Biology: This species is diagnosed by the following characteristics: Seventeen dorsal fin rays; 18 anal fin rays; 11 pectoral finrays; and 17 gillrakers. It grows to a maximum length of 11. 8 cm.

Luminous organs and bioluminescence: In the *Diaphus* spp., the primary photophores are seen as spots on the side and belly of the fish (Anon., https://www.amnh.org/explore/news-blogs/research-posts/bioluminescent-fish-flashing-patterns-might-facilitate-mating). Unlike the fishes of other genera of this family, the species of this genus have evolved a unique complex system of anteriorly oriented light organs on the head. These additional light organs otherwise called "headlight" organs are sexually dimorphic. These organs are usually larger and well defined in males than in females and they are used additionally to confuse, brightly illuminate, or induce fluorescence in their prey (Martin and Davis, 2016; Anon., http://tolweb.org/Myctophidae/).

2. *Diaphus watasei*

Source: Public domain

Common name(s): Watases lanternfish.

Global distribution: Indo-West Pacific.

Ecology: This mesopelagic–benthopelagic has a depth range of 100–2005 m. It is often found over continental shelves and slopes.

Biology: It is diagnosed by the following features: absence of dorsal spines; 14–15 dorsal soft rays; absence of anal spines; and 14–15 anal soft rays. It grows to a maximum length of 17 cm.

Luminous organs and bioluminescence: In this species, the third thoracic and first and second ventral light organs are important besides the secondary photophores located on the head. Reflector of its photophore possesses monolayered iridophores with multilayered guanine crystals which enable high reflection with light interference coloration. Further, these reflectors have been reported to be helpful in matching sunlight filtering down through the depths.; In vitro luminescence reaction using purified luciferase and the substrate coelenterazine showed the wavelength of light emission maximum (λmax) at 454 nm (Paitio et al., 2020).

3. Idiolychnus urolampus

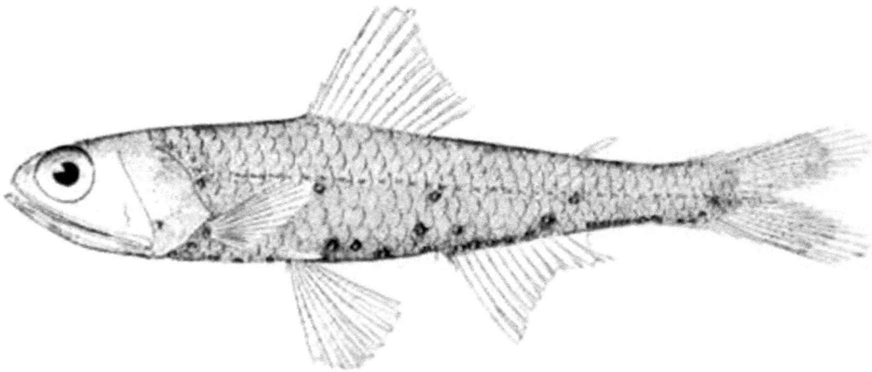

Image credit: World Register of Marine Species (WoRMS). https://www.marinespecies.org/aphia.php?p=taxdetails&id=217702#images. https://creativecommons.org/licenses/by/4.0/deed.en

Idiolychnus sp.

Common name(s): Not assigned.

Global distribution: Pacific: off Hawaii; and Indian Ocean: Madagascar.

Ecology: This pelagic-neritic species has a depth range of 124–582 m. It is also benthopelagic occurring in nearshore areas.

Biology: It grows to a maximum length of 11 cm.

Luminous organs and bioluminescence: It has a small dorsonasal luminous organ and well developed sexually dimorphic luminous caudal glands. Further, the suprapectoral, supraventral, precaudal, and anterior anal light organs are highly elevated compared to other species of this genus (Nafpaktitis and Paxton, 1978).

4. Lobianchia gemellerii

Source: Public domain

Common name(s): Cocco's lantern fish.

Global distribution: Eastern and Western Atlantic; Indian Ocean: Eastern and Western Pacific.

Ecology: This high-oceanic and mesopelagic – bathypelagic and oceanodromous species has a depth range of 25–800 m. It is often found between 300 and 800 m during the daytime and between 25 and 100 m at night.

Biology: It is diagnosed by the following features: absence of dorsal spines; 16–18 dorsal soft rays; absence of anal spins; and 13–15 anal soft rays. It grows to a maximum length of 6 cm only.

Anal organs 10–11; mature males have supracaudal gland consisting of a series of 6 large, well-defined scale-like structures flanked by 5 pairs of smaller, triangular ones; the infracaudal gland of mature females consists of 2 heart-shaped scales; flanked by smaller, triangular luminous scales.

Luminous organs and bioluminescence: In this species, the male bears a well-developed supracaudal luminous organ while the female lacks the supracaudal organ but bears a small infracaudal luminous organ (Anon., http://tolweb.org/Myctophidae/).

b) Subfamily: Gymnoscopelinae

1. Gymnoscopelus braueri

Image credit: Museums Victoria, https://fishesofaustralia.net.au/home/genus/61; https://creativecommons.org/licenses/by/4.0/

Common name(s): Brauer's lanternfish.

Global distribution: It is found distributed between the coasts of the Antarctic.

Ecology: This mesopelagic- bathypelagic and oceanodromous species dwells up to a depth of 2700 m. However, it is found in the upper 200 m at night.

Biology: It has been reported to feed mainly on the euphausiids, *Euphausia superba* and *Thysanoessa macrura;* copepods *Euchaeta antarctica,* and *Rhincalanus gigas;* and amphipods *Primno macropa,* and *Themisto gaudichaudii.* It grows to a maximum length of 13.2 cm.

Luminous organs and bioluminescence: It has a large number of photophores (light-producing organs) distributed in ventrolateral rows of its body and head (Vacquié-Garcia et al., 2012).

2. Gymnoscopelus nicholsi

Common name(s): Nichol's lanternfish.

Global distribution: Antarctica.

Ecology: This bathypelagic and oceanodromous species dwells from a depth of 300 m.

Biology: It feeds on euphausiids and euphausiid larvae, hyperiids, mysids, copepods, and amphipods. Its maximum recorded length is 19.1 cm.

Luminous organs and bioluminescence: In all the species of this genus, there are two precaudal luminous organs which are vertically or subvertically arranged. Further, its anal organ is elevated and there are also 5 (4+1) precaudal luminous organs.

3. Hintonia candens

Common name(s): Lanternfish.

Global distribution: South-western Atlantic and Southern Ocean.; Oceanic and mesopelagic.

Ecology: This is a mesopelagic – bathypelagic and oceanodromous species.

Biology: It grows to a maximum length of 13 cm.

Luminous organs and bioluminescence: In this species, minute, secondary photophores are found distributed on definite areas of the body in addition to the primary photophores.

4. Lampanyctodes hectoris

Image credit: Museums Victoria, https://fishesofaustralia.net.au/home/species/ 2811. https://creativecommons.org/licenses/by/3.0/au/

Common name(s): Hector's lanternfish.

Global distribution: Eastern Atlantic; and Eastern and Western Pacific.

Ecology: It is a shallow water species often found at a depth of about 300 m. It has been reported to hide in this depth during the day, and migrates to the surface at night to feed mainly on planktonic crustaceans.

Biology: It grows to a maximum length of 7.3 cm.

Luminous organs and bioluminescence: It has three ventral photophores which are running along its belly in a line (Anon., https://www.blakenz.org/person/myctophids/).

5. Lampichthys procerus

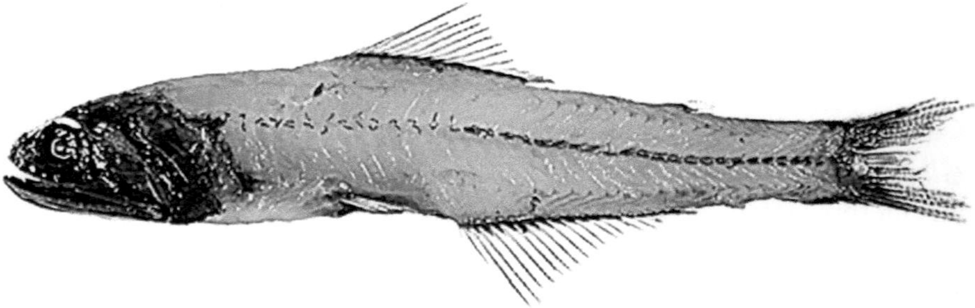

Image credit: Australian National Fish Collection, CSIRO

Common name(s): Blackhead lanternfish.

Global distribution: It has circum- global distribution in the southern hemisphere.

Ecology: This bathypelagic and oceanodromous species dwells up to a depth of 2000 m.

Biology: It grows to a maximum length of 10 cm.

Luminous organs and bioluminescence: In this species, minute, secondary photophores are found distributed on its body in addition to the primary photophores.

6. Notoscopelus bolini

Common name(s): Not designated.

Global distribution: North Atlantic Ocean and the Mediterranean Sea.

Ecology: This mesopelagic – oceanodromous species dwells up to a depth of 1300 m. It undertakes a daily vertical migration. During the day it is usually about 1,000 m beneath the surface and it rises at night to within about 125 m of the surface.

Biology: It feeds mostly on euphausiid krill and its diet is also composed of a smaller proportion of copepods and fish larvae. It attains a maximum length of 10.2 cm.

Luminous organs and bioluminescence: In this species, males do not possess luminous gland at the top of the caudal peduncle but it has large patches of bioluminescent tissue on its cheek and above the eye (Anon., https://en.wikipedia.org/wiki/Notoscopelus_bolini).

7. ***Notoscopelus caudispinosus*** *(= Scopelus caudipsinosus)*

Image credit: Jeff Dubosc

Common name(s): Lobisomem.

Global distribution: Eastern and Western Atlantic:

Ecology: This bathypelagic and oceanodromous species has a depth range of 0–360 m. It is also known to occur deeper than 1,000 m during the day and it is nyctoepipelagic at the surface and down to 175 m.

Biology: It has a laterally compressed head and body. The mouth is at the tip of its rounded snout and it extends backwards to just behind the eye. Teeth are small. It grows to a maximum length of 14 cm.

Luminous organs and bioluminescence: There are groups of small light organs on its head and sides of the body. In adult males, there is a row of eight to nine luminous scale-like segments above the caudal peduncle but these males lack luminous organs above the eye and on the cheek (Anon., https://en.wikipedia.org/wiki/Notoscopelus_caudispinosus).

8. ***Notoscopelus elongatus*** **(=** *Scopelus elongatus***)**

Common name(s): Not designated.

Global distribution: It is endemic to the Mediterranean Sea.

Ecology: It is found in deep water habitats at a depth range of 45–1000 m. It is known to rise to near the surface to feed at night and going down to great depths (375–1,000 m) by day and rise to remain at 45–150 m in night.

Biology: It grows to a maximum length of 14.2 cm.

Luminous organs and bioluminescence: In this species, there is a definite arrangement of photophores (light-emitting organs) on its head and body. At a length of 5.2 cm, the males have been reported to develop luminous, glandular patches on its dorsal surface of the caudal peduncle (Anon., https://en.wikipedia.org/wiki/Notoscopelus_elongatus). Anon. (http://

species-identification.org/species.php?species_group=fnam&id=2100&menuentry=soorten) reported that the males of this species possess only supra-caudal gland which is consisting of about 9 scale-like luminous segments.

9. Notoscopelus kroyeri

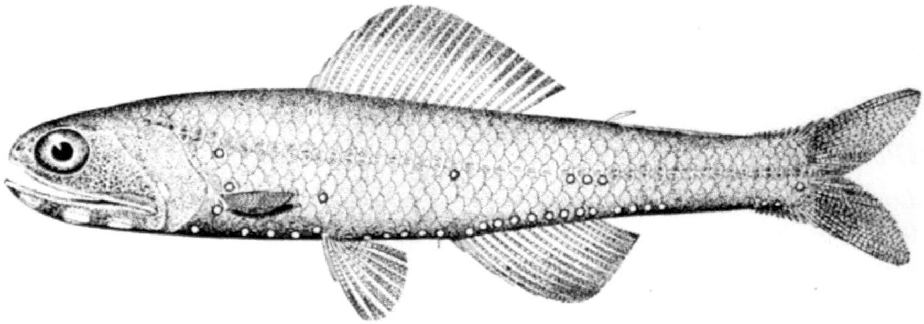

Source: Public domain

Common name(s): Lancet fish.

Global distribution: Atlantic Ocean.

Ecology: This pelagic and oceanodromous species dwells at a maximum depth of 1000 m and it is known to make vertical migration at night.

Biology: It feeds chiefly on crustaceans and attains a maximum length of 14.3 cm.

Luminous organs and bioluminescence: It has many small photophores on its sides and flanks, and they are arranged in a unique fashion that is characteristic to this species. Mature males have a gland consisting of eight or nine luminous scale-like segments on the dorsal surface of the caudal peduncle, but they lack the luminescent patches on the cheek and above the eye which are the characteristic features of some members of the genus (Anon., https://en.wikipedia.org/wiki/Notoscopelus_kroyeri).

10. Scopelopsis multipunctatus

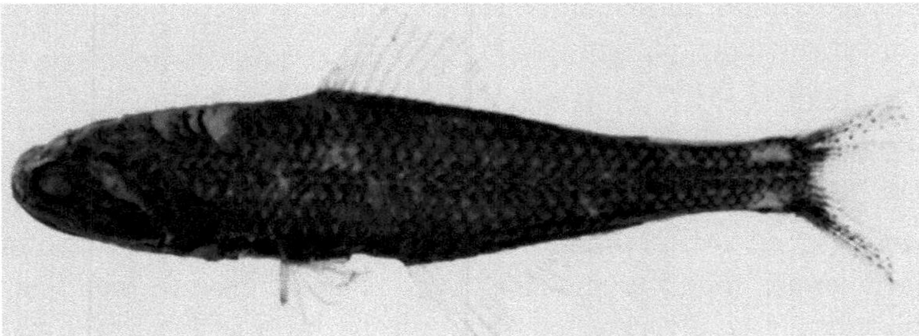

Image credit: Museums Victoria, https://fishesofaustralia.net.au/home/species/3971. https://creativecommons.org/licenses/by/3.0/au/.

Common name(s): Multispotted lanternfish.

Global distribution: Southern Hemisphere, Atlantic, Indian, and Pacific Oceans.

Ecology: This bathypelagic and oceanodromous species has a depth range of 3–2000 m. It can also be found in both warm and cold waters of the ocean.

Biology: It has a long and slender body with round eyes. A forked homocercal caudal fin is seen. It is known to feed on copepods; amphipods and euphausiids; larvae of mollusks, ostracods, and polychaetes; and siphonophores; and salps. It grows to a maximum length of 9.3 cm.

Luminous organs and bioluminescence: The primary photophores of this species cover the head in a definite pattern, and some of these photophores are considerably larger and more distinct than others. Many regular rows of secondary light organs are also seen on its body (Moser and Ahlstrom, 1972).

c) Subfamily: Lampanyctinae

1. Bolinichthys longipes

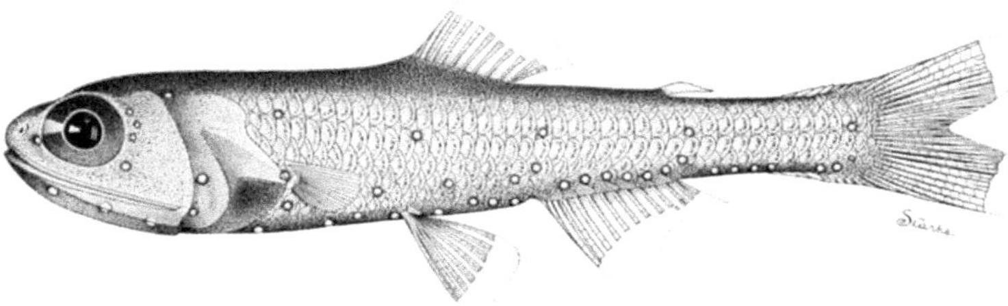

Bolinichthys sp.

Source: Public domain

Common name(s): Popeye lampfish.

Global distribution: Indian and Pacific Oceans.

Ecology:This mesopelagic- bathypelagic and oceanodromous species has a depth range of 50–1021 m. It is found in 525–725 m during day time and in 10–150 m at night time.

Biology: It grows to a maximum length of only 5 cm.

Luminous organs and bioluminescence: In this species, its subventral luminous organs are found touching or very near the lateral line. The second ventral organ is highly elevated about on the level of the lower end of the pectoral fin base. There are luminous scales at the base of its dorsal- and anal-fin rays (Cruz-Agüero, 2016).

2. Bolinichthys supralateralis *(= Lepidophanes supralateralis)*

Common name(s): Stubby lanternfish.

Global distribution: Eastern and Western Atlantic; Western Indian Ocean; and Pacific Ocean.

Ecology: This mesopelagic – bathypelagic but non-migratory species has a depth range of 40–850 m. It is also known to occur between 375 and 750 m during the daytime and between 40 and 650 m at night.

Biology: It grows to a maximum length of 11.7 cm.

Luminous organs and bioluminescence: It has 8–10 anal luminous organs and 3–6 small elongate patches of luminous tissue along the base of its anal fin. Its supracaudal luminous gland is short, with 2–3 scales and infracaudal gland is with 3–5 scales (Anon., https://www.fishbase.se/summary/7407).

3. Ceratoscopelus maderensis

Common name(s): Madeira lantern fish.

Global distribution: Eastern, Western and Northwest Atlantic Ocean.

Ecology: This bathypelagic- mesopelagic and oceanodromous species has a depth range of 51–1480 m. It has also been recorded at depths of 650–700 m during the day time and between 51 and 250 m at night.

Biology: It feeds on copepods and adult or larval forms of planktonic crustaceans. It grows to a maximum length of 8.1 cm.

Luminous organs and bioluminescence: The functional parts of the photophores of this species are the photogenic cells which are involved in the light production process, and a set of dioptric annexes viz. accessory scales, lens-scale, reflector and pigmented layer. The photocytes of the photophores are elongated, thin and flattened and are located inside a cup-shaped structure, made of the reflector and the pigmented layer. The lens scale which is covered by accessory scales has a protection function as well as a dioptric role. The innervation of the lens-scale of this species and the other structures which are having a neurotransmitter function are said to be implicated in the regulation of intensity and direction of light emission besides playing an important role in the counterillumination camouflage (Cavallaro et al., 2019).

4. Ceratoscopelus warmingii

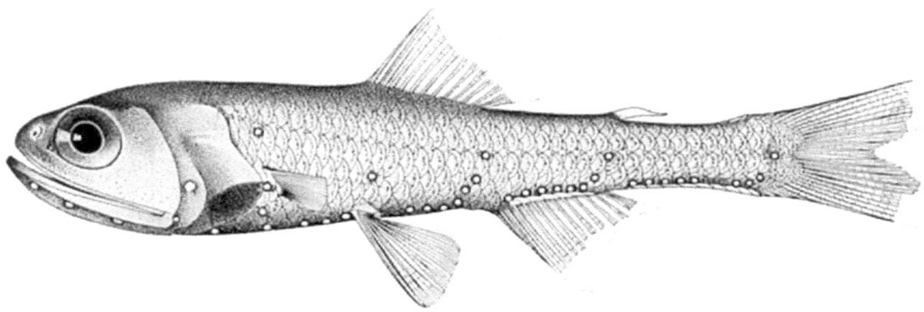

Source: Public domain.

Common name(s): Warming's lantern fish.

Global distribution: It is found distributed circumglobally in both hemispheres; Atlantic, Pacific, and Indian Oceans.

Ecology: This bathypelagic and oceanodromous species has a depth range of 391–2056 m. It is often found between 700 and 1,500 m during the day time and between 20 and 200 m at night.

Biology: It grows to a maximum length of 8 cm.

Luminous organs and bioluminescence: In this species, the photophores which are patches and scale-like specks of bioluminescent materials are found as spots on the sides and belly of the fish.- (Anon., https://www.amnh.org/explore/news-blogs/research-posts/bioluminescent-fish-flashing-patterns-might-facilitate-mating).

5. Lampadena luminosa

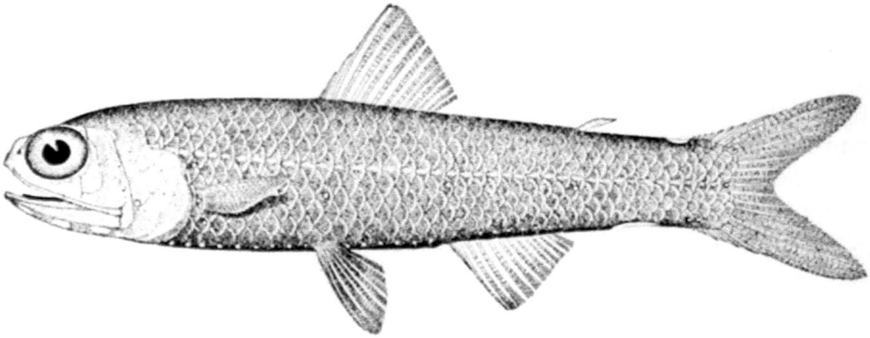

Lampadena sp.

Source: Public domain.

Common name(s): Luminous lanternfish.

Global distribution: Atlantic, Indian, and Pacific Oceans.

Ecology: This bathypelagic- mesopelagic and oceanodromous species has a depth range of 50–1021 m. It is often found between 425–850 m during the day time and in the upper 100 m depth at night.

Biology: It grows to a maximum length of 20 cm.

Luminous organs and bioluminescence: In this species, caudal luminous glands which are translucent, enlarged luminous tissues are seen (Paxton, 1972).

6. Lampadena urophaos

Common name(s): Sunbeam lampfish.

Global distribution: Eastern and Western Pacific.

Ecology: It is epipelagic to mesopelagic at the depths of 50–600 m during night and at 500–1000 m during the day.

Biology: The body of this species is robust. Head is bluntly rounded with slightly pointed opercular margin. The eyes are moderate in size. It grows to a maximum length of 20 cm.

Luminous organs and bioluminescence: The species of *Lampadena* are characterized by the presence of elevated dorsonasal and postorbital photophores. Caudal luminous glands are also seen (Paxton, 1963).

7. Lampanyctus ritteri

Common name(s): Broadfin lampfish.

Global distribution: British Columbia, Canada to Baja California and Mexico in the Northeast Pacific Ocean.

Ecology: This bathypelagic species has a depth range of 20–1095 m.

Biology: Its body coloration is almost black to metallic blue dorsally, lighter on sides, and black in the mouth and gill cavity. Fins are dusky and are marked with fine wavy lines. It feeds chiefly on chaetognaths (Sagitta), young fishes, copepods, and euphausiids. It grows to a maximum length of 12 cm.

Luminous organs and bioluminescence: It has 15–16 anal organs (photophores) and 35–38 lateral line luminous organs (https://www.fishbase.in/summary/5371). Hubbs and Wisner (1964) reported that the species of this genus has only 4 ventral luminous organs, and among them one organ is elevated in some species.

8. Lampanyctus vadulus

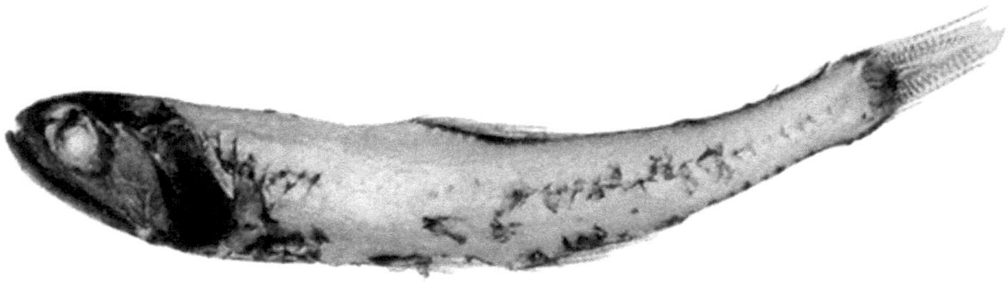

Image credit: Keiichi Matsuura.

Common name(s): Not designated.

Global distribution: Eastern Atlantic Ocean, Eastern and Western Pacific Ocean and Indian Ocean.

Ecology: This mesopelagic s- bathypelagic and oceanodromous species has a depth range of 0–370 m. It is known to occur at depths of 50–300 m at night.

Biology: It grows to a maximum length of 9 cm.

Luminous organs and bioluminescence: In this species, the supralateral luminous organs are found reaching anterior anal fin.

9. Lepidophanes guentheri

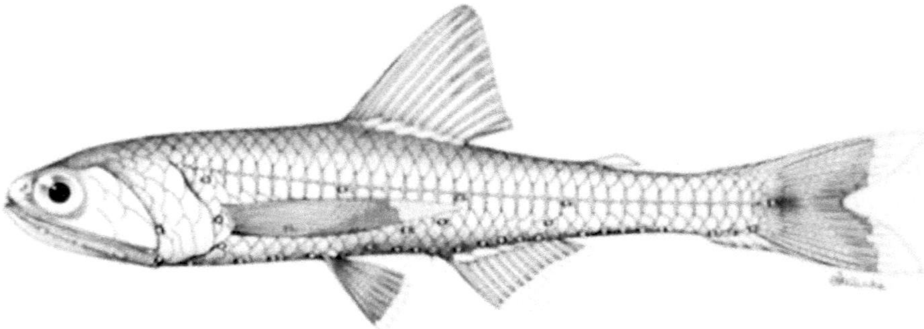

Image credit: World Register of Marine Species (WoRMS)

Common name(s): Günther's lanternfish.

Global distribution: Atlantic Ocean.

Ecology: This pelagic-oceanodromous has a depth range of 40–750 m. It performs diurnal vertical migrations and is found between 425 and 750 m during the day time and between 40- and 25 m at night.

Biology: It grows to a maximum length of 7.8 cm.

Luminous organs and bioluminescence: In this species, minute photophores have been reported to occur in its body in addition to the primary photophores. While the paired light-producing, primary photophores) are located along the ventral sides and at the head, the secondary organs occur at every scale pocket. The primary photophores help to mask the fish by counterillumination and are mainly used in intraspecific communication associated with shoaling, courtship, etc. (Anon., https://www.alamy.com/stock-photo-lepidophanes-guentheri-a-species-of-the-large-group-of-lanternfish-79235897.html).

10. Parvilux ingens

Common name(s): Giant lampfish.

Global distribution: Eastern Pacific.

Ecology: This mesopelagic -oceanic species has a depth range of 100–500 m.

Biology: The head of this species measures 3.5 times in standard length and the eye is small. Mouth is large and gill rakers are slender. The body coloration is very dark or black. It grows to a maximum length of 20 cm.

Luminous organs and bioluminescence: The photophores of this species are unusually small and are ovate to reniform in appearance. The third thoracic luminous organ is slightly elevated. The fifth and sixth ventral luminous organs are slightly elevated. The supralateral organ series is only slightly angulated. There are two precaudal organs. The four precaudal organs are either continuous with or separated from the posterior anal organ series. Both the supracaudal and infracaudal luminous glands comprise four scales each and are short. There are no secondary photophores, no luminous gland on the adipose fin, and no photophores on the cheek (Hubbs and Wisner, 1964).

11. Stenobrachius leucopsarus

Common name(s): Northern lampfish.

Global distribution: North Pacific.

Ecology: This bathypelagic species has a depth range of 31–3400 m.

Biology: This species has the following diagnostic features: small and slender adipose fin; and slender and short pectoral fins. The body coloration is gray to dark greenish blue dorsally and lighter ventrally. Operculum and fins are black. It grows to a maximum length of 13 cm. and feeds mainly on fishes, copepods and euphausiids.

Luminous organs and bioluminescence: It has numerous small round photophores that line the ventral surface of its head and body. (Anon., https://en.wikipedia.org/wiki/Northern_lampfish). Anctil and Case (1977) reported that its caudal luminous organs emit intense, short duration pulses of light

12. Taaningichthys bathyphilus

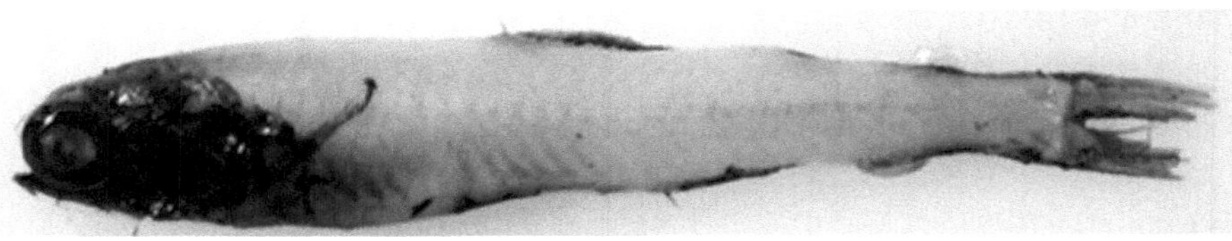

Image credit: Bill Mac Eachern.

Common name(s): Deepwater lanternfish.

Global distribution: It has circumglobal distribution between 43°N and 68°S in all oceans.

Ecology: This bathypelagic and non-migratory species has a depth range of 400–1550 m.

Biology: It grows to a total length of 8 cm.

Luminous organs and bioluminescence: The most striking feature of this species is the development of its caudal luminous organs which are associated with the enlarged and translucent luminous tissue (Paxton, 1972). McEachran (2021) reported that this species has 3–5 ventral luminous organs and 2–5 anal organs.

13. Taaningichthys minimus

Common name(s): Waistcoat lanternfish.

Global distribution: Atlantic Ocean, Indian Ocean; and Western, Central and Northeast Pacific Ocean.

Ecology: This bathypelagic- oceanodromous species has a depth range of 90–800 m. It is migratory and is found between 1 and 800 m during the day time and between 90 and 600 m at night.

Biology: It grows to a maximum length of 6.5 cm.

Luminous organs and bioluminescence: McEachran (2021) reported that this species has the characteristic 8–10 ventral luminous organs and 9–13 anal organs.

14. Triphoturus mexicanus

Common name(s): Mexican lampfish.

Global distribution: Eastern Pacific.

Ecology: This epipelagic and mesopelagic species dwells from a depth of 25 to 2,896 m. It is known to migrate vertically toward the surface at night to feed mainly on zooplankton.

Biology: It has a long and slender body that tapers toward the end. It has a mid-sized eyes and a large mouth that extends past the eyes. The body coloration is silvery black and is dark around its head and caudal fin base. Anal, caudal, and dorsal fins are transparent. It grows to a total length of 7 cm only.

Luminous organs and bioluminescence: In this species, the second and third ventral organs are highly elevated and displaced forward to before verticals from first and fourth ventral organs, respectively. Fourth subpectoral organ is well before vertical from second subpectoral organ. Suprapectoral organ is below but nearly touching the lateral line. The rest of upper photophores are slightly above the lateral line. Fourth thoracic and first and second supralateral organs form a straight line that passes well above second and third ventral organs. First, supralateral organ is well behind the vertical from the fourth ventral organ (Cruz-Agüero, 2016). Anon., https://mexican-fish.com/mexican-lampfish/) reported that it has a limited number of prominent photophores which are randomly positioned on its body.

15. Triphoturus nigrescens

Common name(s): Highseas lampfish.

Global distribution: Indian and Southeast Atlantic and Pacific Oceans.

Ecology: This mesopelagic – bathypelagic and oceanodromous species has a depth range of 100–1000 m. It is also often found in the upper 24 m at night.

Biology: It grows to a maximum length of 8.1 cm.

Luminous organs and bioluminescence: In this species, only the second ventral organ is elevated and is displaced forward the first ventral organ. The first subpectoral organ is almost directly below the second subpectoral organ. Only upper photophores are found touching the lateral line basally. The infracaudal luminous gland is short, culminating far behind the end of the anal base (Cruz-Agüero, 2016).

d) Subfamily: Myctophinae

1. Benthosema graciale

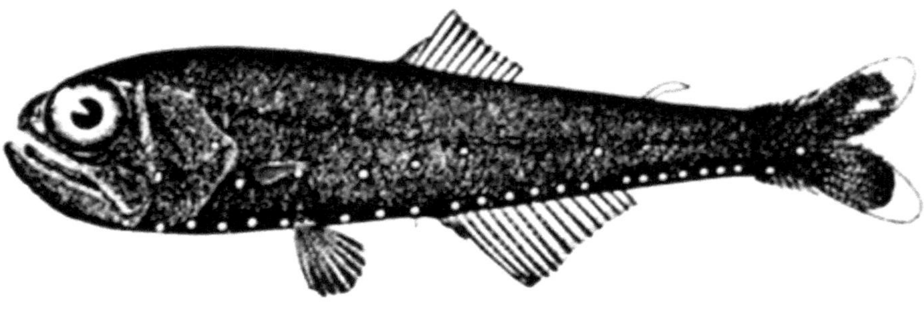

Source: Public domain.

Common name(s): Glacier lantern fish.

Global distribution: It is endemic to the Mediterranean and Atlantic.

Ecology: This high-oceanic and mesopelagic species is inhabiting at depths of 12–800 m and is known for its vertical migration. It has been reported to be common during the day at 375–800 m with its maximum at about 500 m.

Biology: It feeds on small invertebrates, including crustaceans such as copepods, krill and amphipods and attains a maximum length of 10.3 cm.

Luminous organs and bioluminescence: While males of this species possess single black-bordered supracaudal gland, females with two infracaudal luminous patches. Occasionally specimens may have both types of luminous structures (http://species-identification.org/species.php?species_group=fnam&id=2049).

2. Benthosema pterotum

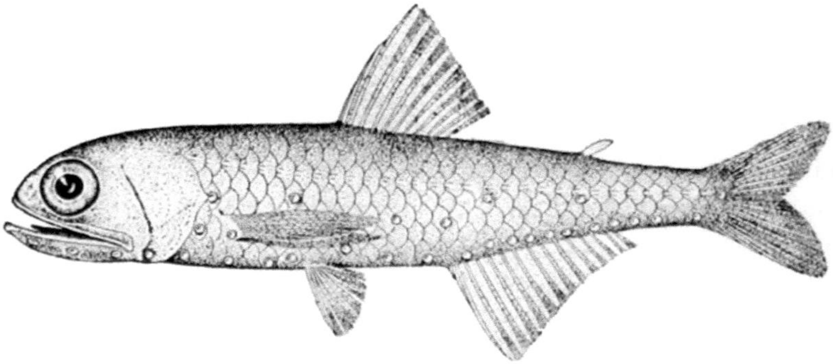

Source: Public domain.

Common name(s): Skinnycheek lanternfish or opaline lanternfish.

Global distribution: Indo-West Pacific and Southeast Atlantic.

Ecology: This benthopelagic species inhabits at depths of 10–300 m. It is however, found at 130–300 m during the day and at 10–200 m in night.

Biology: It has a slender compressed body which is covered with deciduous, cycloid or ctenoid scales. Head is bluntly rounded with large elliptical to round eyes. It feeds mainly on copepods and various crustacean larvae and attains a maximum length of 7 cm only.

Luminous organs and bioluminescence: In this species, photophores are seen in front of the eye, on the lower jaw under the eye, and on the opercular organ in juveniles. In adults, photophores have been reported under the pelvic fin, on the anal fin, and below the lateral line of the anal fin (Lee et al., 2020).

3. *Centrobranchus andreae*

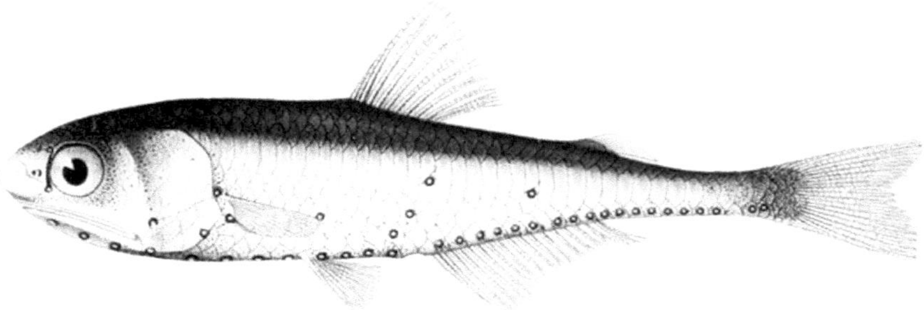

Source: Public domain.

Common name(s): Andre's lanternfish.

Global distribution: Indo-Pacific.

Ecology: It is a neustonic and mesopelagic species inhabiting at 0–165 m at night and at 640–650 m during the day.

Biology: It grows to a maximum length of 6.5 cm only.

Luminous organs and bioluminescence: In this species, the photophores lie in a straight oblique line with the supralateral organs, above or slightly behind the ventral photophores (Gago and Lavenberg, 1992).

4. *Dasyscopelus obtusirostris* (= *Myctophum obtusirostre*)

Image credit: Brandi Noble, NOAA Federal

Common name(s): Bluntsnout lanternfish.

Global distribution: Eastern and Western Atlantic and Western Pacific.

Ecology: It is a mesopelagic, bathypelagic and oceanodromous species inhabiting depths of 1–700 m. It is neustonic (inhabiting on or just below the surface of water) at 1–125 m at night and is found between 325–700 m during the day.

Biology: It grows to a maximum length of 9.0 cm.

Luminous organs and bioluminescence: This species which is with large, broadly-spaced photophores which are under neural control. It has been reported that this fish adjusts its ventral light emission in response to changes in downwelling light intensity with a range up to 15 000-fold (Davis et al., 2020). Its counterillumination helps in escaping from its predators swimming below.

5. Dasyscopelus spinosus *(= Myctophum spinosum)*

Common name(s): Spiny lantern fish.

Global distribution: Tropical and subtropical waters of Atlantic, Indian, and Pacific Oceans.

Ecology: This bathypelagic and oceanodromous species inhabits at depths of 0–700 m.

Biology: The body of this species is dark gray, and its sides and venter are paler or with metallic lustre and silvery, white, blue and violet reflections. Iris is bright silvery white. All fins are pale or gray and caudal base is gray black. It attains a maximum length of 9 cm.

6. Diogenichthys atlanticus

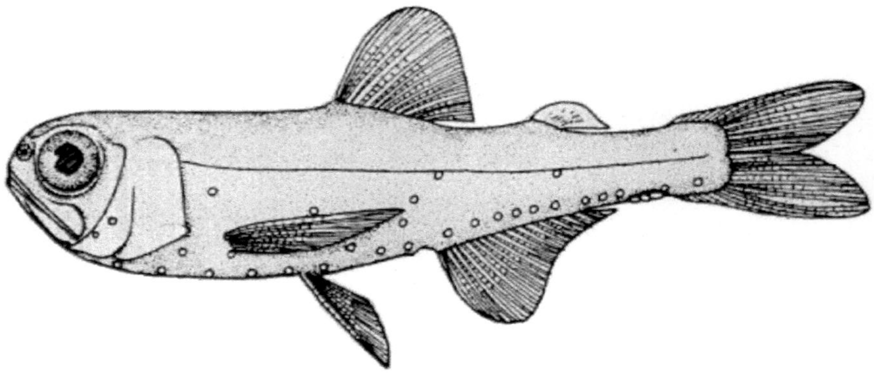

Image credit: Canadian Museum of Nature

Common name(s): Longfin lanternfish.

Global distribution: Atlantic, Pacific, and Western Indian oceans.

Ecology: This epipelagic to mesopelagic and bathypelagic and oceanodromous species inhabits at depths of 0–1050 m.

Biology: Its mouth is terminal and snout is not projecting. Jaws are short, and are extending less than 1/2 eye diameter behind orbit. Outer dentary teeth are flattened and hooked forward. It attains a maximum length of 2.9 cm only.

Luminous organs and bioluminescence: Its photophores are arranged as follows. Anterior anal photophores are divided into two groups viz. anterior anal anterior and anterior anal posterior. There are 1–2 subpectoral photophores located in a horizontal line. Its two ventral photophores are found elevated in position. There are two precaudal photophores. There are no photophores above lateral line (Anon., https://ncfishes.com/marine-fishes-of-north-carolina/diogenichthys-atlanticus/).

7. Electrona antarctica

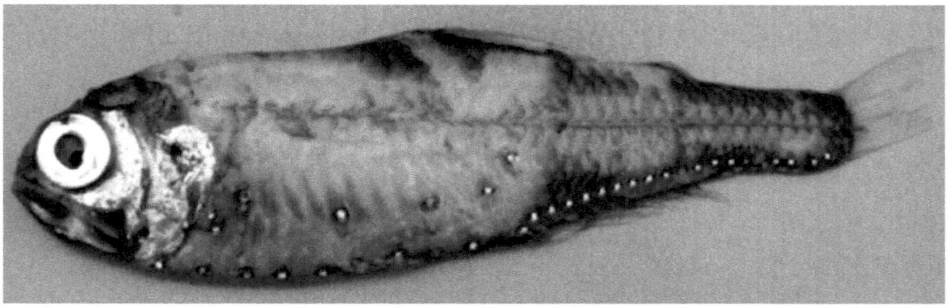

Source: Public domain.

Common name(s): Antarctic lanternfish.

Global distribution: It is the most common myctophid species occurring south of the Antarctic Polar Front.,

Ecology: This demersal oceanodromous species inhabits at depths of 1–1010 m.

Biology: Species in this genus may be identified by their larger size, round head, and horseshoe-shaped scale pockets along their lateral line. This fish grows to a maximum length of 12.5 cm and feeds more euphausiids and also prey upon polychaetes.

Luminous organs and bioluminescence: The substrate for its light reaction viz. coelenterazine is believed to be acquired through the food chain and this substrate is present in this species in two stabilized storage forms viz. enol-sulfate and dehydro- coelenterazine within various tissues of this species. Further, the presence of coelenterazine in the gonads of this species also suggests the potential maternal transfer of the bioluminescence capability to its offsprings (Duchatelet et al., 2019). Its wavelength of light maximum emission (λmax) has been reported to be 488 nm (Herring, 1983).

8. Electrona carlsbergi

Common name(s): Electron subantarctic lanternfish

Global distribution: It has circum-global distribution between the subtropical convergence and Antarctic Polar Front.

Ecology: This bathypelagic and oceanodromous species has a depth range of 1–1008 m.

Biology: This species grows to a maximum length of 11.2 cm and it feeds on copepods, hyperiids, euphausiids, ostracods, and gastropods.

Luminous organs and bioluminescence: In this species, a large number of photophores (light-producing organs) are present in ventrolateral rows of its body and head (Vacquié-Garcia et al., 2012).

9. Electrona risso

Common name(s): Electric lantern fish, chubby flashlight fish or Risso's lantern fish.

Global distribution: Eastern, Western and Northwest Atlantic; and Indian Ocean, Eastern and Western Pacific.

Ecology: This bathypelagic, epipelagic to mesopelagic and oceanodromous species has a depth range of 90–1485 m.

Biology: It is a deep-bodied black lanternfish with silvery sides and clear fins, Its main food items are copepods and it grows to a maximum length of 8.2 cm.

Luminous organs and bioluminescence: It has mainly 10–13 anal photophores. (Anon., https://fishesofaustralia.net.au/home/species/3100). Further, males are with supra-caudal or infra-caudal gland or both; and females have one supra-caudal and one or two infra-caudal glands or rarely with either supra- caudal or infra-caudal gland (Anon., http://species-identification.org/species.php?species_group=fnam&id=2069).

10. Gonichthys barnesi

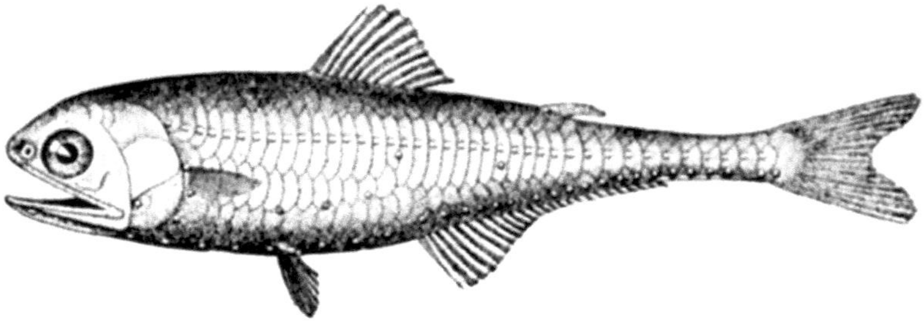

Gonichthys sp.

Source: Public domain.

Common name(s): Barne's lanternfish.

Global distribution: It is a subtropical convergence species found in all oceans.

Ecology: This bathypelagic and oceanodromous species has a depth range of 0–1000 m.

Biology: It grows to a maximum length of 5.0 cm and migrates closer to the surface for feeding at night.

Luminous organs and bioluminescence: This species is known for its small light organs which are found dotted along its underside. Its lights may attract mates and see its prey as well as to hide its own silhouette against the light. It is also capable of turning on and off as the surrounding light varies (Anon., https://teara.govt.nz/en/photograph/5276/slender-lanternfish).

11. Hygophum benoiti (= *Scopelus benoiti*)

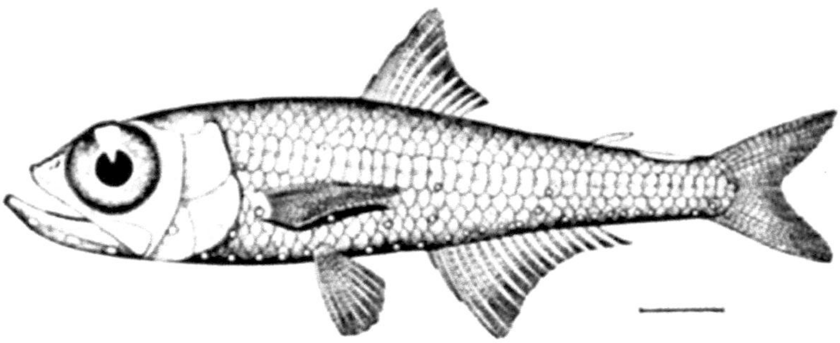

Source: Public domain.

Common name(s): Benoit's lanternfish.

Global distribution: Eastern and Western Atlantic.

Ecology: This bathypelagic and oceanodromous species has a depth range of 51–700 m.

Biology: Its mouth is oblique with a slightly protruding lower jaw. Operculum is almost evenly rounded. It grows to a maximum length of 5.5 cm.

Luminous organs and bioluminescence: It has cup-shaped ventral photophores each with a small group of long flattened light-producing cells called photocytes. Caudal organ of this species also possesses more abundant flattened photocytes. Further, the scales over the photocytes of this species are found modified to serve like a lens (Deshmukh, 2017). Anon. (http://species-identification.org/species.php?species_group=fnam&id=2071) reported that the males of this species have a single black-rimmed supra-caudal luminous gland while females with three infra-caudal luminous glands. Anctil (2018) reported that this species possesses large light organs behind its dorsal fin. According to him, these photophores, a variety of composite ocellar organs, may be termed "stern chasers" due to their projected light beam, which was found directed backward.

12. Hygophum hygomii

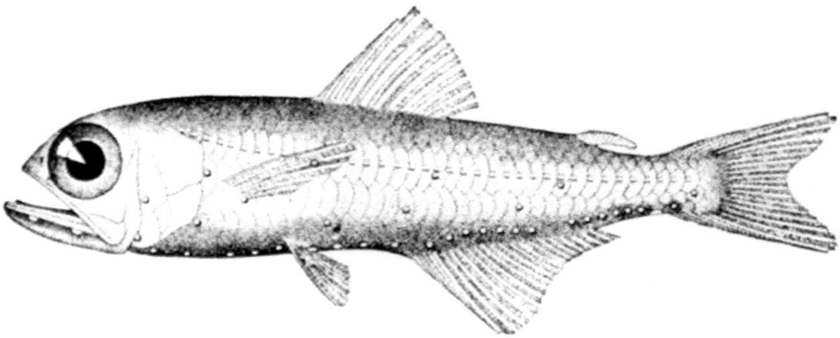

Source: Public domain.

Common name(s): Bermuda lantern fish.

Global distribution: Eastern and Western Atlantic Ocean and Indian and Pacific Oceans.

Ecology: This bathypelagic and oceanodromous has a depth range of 0–1485 m. It is nyctoepipelagic at surface and migrates down to about 300 m.

Biology: It grows to a maximum length of 6.8 cm.

Luminous organs and bioluminescence: In this species, males are with single supra-caudal luminous gland and females with 1–3 infra-caudal glands (Smith and Heemstra, 2012). Its wavelength of light maximum emission (λmax) has been reported to be 448 nm.

13. Krefftichthys anderssoni

Common name(s): Rhombic lanternfish.

Global distribution: It has circum-global distribution in the southern oceans.

Ecology: This bathypelagic and oceanodromous species inhabits depths up to 2700 m.

Biology: It grows to a maximum length of 7.1 cm and feeds mainly on copepods, euphausiids and amphipods.

Luminous organs and bioluminescence: In this species, its coelenterazine which is present in the form of enol-sulfate and dehydro- coelenterazine is believed to be obtained through its food chain and occurs in its tissues. Further, the coelenterazine of its gonads is largely responsible for the occurrence of bioluminescence in its offspring (Duchatelet et al., 2019).

14. Loweina interrupta

Common name(s): Not designated.

Global distribution: It is a bitemperate species in the Atlantic and in southern Indian and Pacific Oceans.

Ecology: This High-oceanic, mesopelagic, bathypelagic, and oceanodromous species has a depth range of 60–800 m.

Biology: It grows to a maximum length of 3.9 cm.

Luminous organs and bioluminescence: Males of this species have been reported to possess only a single large supracudal luminous gland which occupies half or more of supracaudal space (Nafpaktitus et al., 2018).

15. Metelectrona ahlstromi

Common name(s): Not designated.

Global distribution: Southeast Pacific.

Ecology: This bathypelagic species occurs up to a depth of 2000 m.

Biology: The body of this species is short, deep, and moderately compressed. The characteristic feature of this species is its shorter caudal peduncle.

Luminous organs and bioluminescence: This species has been reported to possess caudal luminous glands and small-sized 1 or 2 low posterolateral photophores. The development of precaudal group of luminous organs and moderately but distinctly elevated second precaudal luminous gland are very significant features of this genus.

16. Myctophum affine

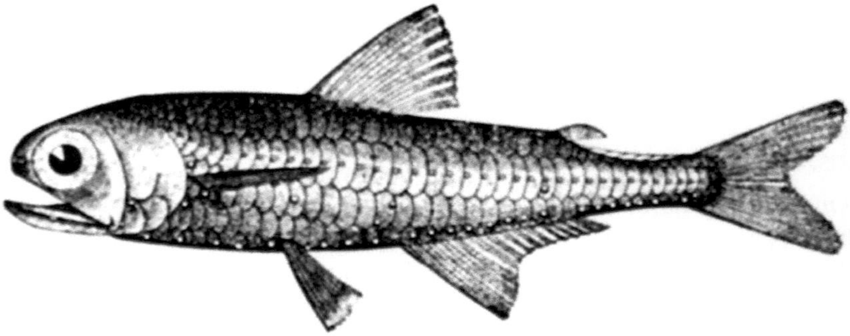

Source: Public domain.

Common name(s): Metallic lantern fish.

Global distribution: Eastern and Western Atlantic.

Ecology: This marine and brackish water species is bathypelagic and oceanodromous inhabiting at a depth range of 0–600 m.

Biology: In this silvery black colored species, the dorsal profile of the head is much arched. The body is moderately flattened sidewise and is tapering gently backward to the rather deep caudal peduncle. Its enormous eye is situated close to the tip of the longer snout, and it has a small mouth and deeply forked tail. It feeds on plankton and grows to a maximum length of 7.9 cm.

Luminous organs and bioluminescence: In this species, its luminescent organs which are pale green or blue are present on its snout and along its sides in the form of small dots (Anon., http://gma.org/fogm/Myctophum_affine.htm).

Its wavelength of light maximum emission (λmax) has been reported to be 470 nm (Herring, 1983). Clarke et al. (1962) reported that when the fishes of this species were attracted by a lamp and stimulated with short bursts of a.c. 80 V, they emitted a steady glow of light from all their photophores and this glow was found to last for a maximum of 4 sec during each stimulus.

17. Myctophum phengodes

Common name(s): Bright lanternfish.

Global distribution: Southeast and Southwest Atlantic.

Ecology: It is a mesopelagic and bathypelagic and oceanodromous species are also seen occasionally at the surface.

Biology: It grows to a maximum length of 9.3 cm.

Luminous organs and bioluminescence: Supernumerary photophores have been found in this species, displaced on the dorsal surface of the fish (Anon., https://www.proquest.com/docview/1860272652).

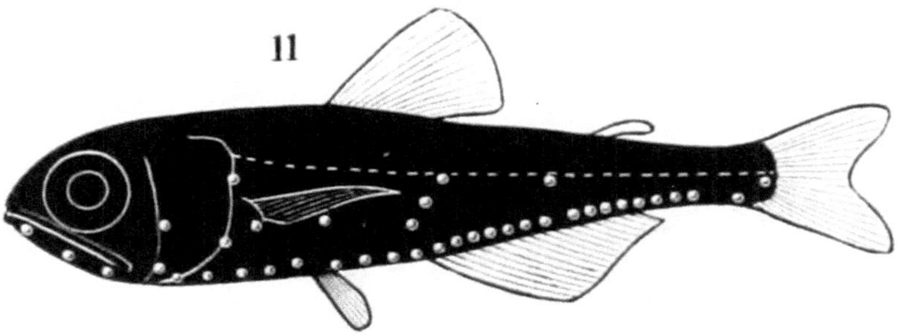

Source: Public domain.

18. Myctophum punctatum

Source: Public domain.

Common name(s): Spotted lanternfish.

Global distribution: Eastern and Western Atlantic.

Ecology: This bathypelagic and oceanodromous species has a depth range of 0–1000 m.

Biology: It feeds on copepods, euphausiids, crab zoeae and fish fry; and it grows to a maximum length of 11 cm.

Luminous organs and bioluminescence: In this species, males have 2–4 overlapping luminous scales supra- caudally and females are provided with 4–5 partially overlapping luminous patches infra-caudally. (Anon., http://species-identification.org/species.php?species_group=fnam&id=2095).

Deshmukh (2017) reported that it has cup-shaped ventral photophores each with a small group of long flattened light-producing cells called photocytes. Caudal organ of this species also possesses more abundant flattened photocytes. Further, the scales over the photocytes of this species are found modified to serve like a lens.

19. Protomyctophum bolini

Common name(s): Bolin's lanternfish.

Global distribution: It has circumpolar distribution between Antarctic Divergence and Subtropical Convergence zone. It is also found distributed in Southeast Atlantic and Western Indian Oceans.

Ecology: This bathypelagic and oceanodromous species has a depth range of 364–728 m.

Biology: The presence of 21–22 gill rakers on the first arch of this species is its characteristic feature. It feeds on copepods and larval stages of euphausiid krills and it grows to a maximum length of 6.7 cm.

Luminous organs and bioluminescence: In this species the opercular photophore is on the same level with lower margin of eye and it is distinctly above the line connecting suprapectoral (pectoral lateral) and supraventral (ventral lateral) photophores. The supralateral photophores are straight. The four anal photophores are located behind the last anal-fin ray. All the photophores of this species are without metallic shine; 21–22 gill rakers on the first arch (Prokofiev, 2004).

20. Protomyctophum tenisoni

Common name(s): Tenison's lanternfish.

Global distribution: It has circumglobal distribution between the subtropical convergence and the Antarctic Polar Front. It is also found in the Southern Ocean.

Ecology: This pelagic and oceanodromous species has been reported to occur from a depth of 96 m.

Biology: It grows to a maximum length of 5.4 cm and is believed to feed mainly on the mackerel icefish *Champsocephalus gunnari.*

Luminous organs and bioluminescence: This species is producing its own luciferin viz. coelenterazine for its intrinsic bioluminescence. Further, the occurrence of coelenterazine in the gonads of this species suggests the potential maternal transfer of the bioluminescence capability to its offspring (Duchatelet et al., 2019).

21. Symbolophorus californiensis

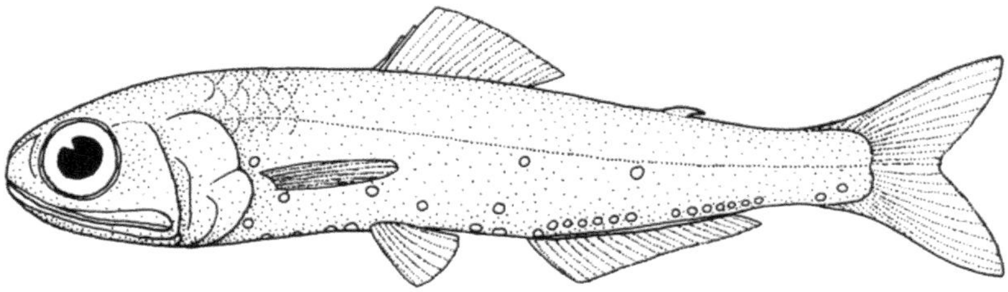

Symbolophorus sp.

Image credit: Tony Ayling. Wikipedia. https://creativecommons.org/licenses/by-sa/1.0/

Common name(s): Bigfin lanternfish.

Global distribution: North Pacific.

Ecology: This pelagic-oceanic species has a depth range of 557–1497 m.

Biology: It has the following diagnostic features: short based and slender adipose fin; and long pectoral fins which are reaching beyond pelvic insertion but are not reaching anus. The body coloration is black dorsally, silvery on sides and dusky on bases of fins. It grows to a maximum length of 11 cm.

Luminous organs and bioluminescence: In this species, its light-producing photophores are located under the lateral line and the ventral surface. (Anon., https://noaateacheratsea.blog/tag/bioluminescence/). The light from these photophores has been reported to be pale green (Anon., https://www.fishbase.se/summary/5351).

22. Tarletonbeania crenularis

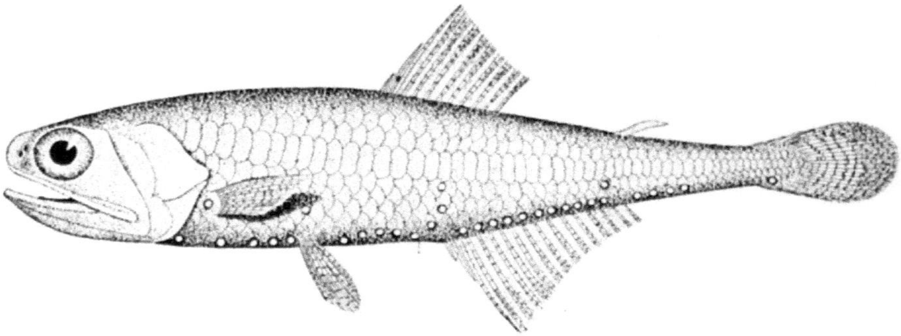

Source: Public domain

Common name(s): Blue lanternfish.

Global distribution: Eastern Pacific.

Ecology: This bathypelagic species has a depth range of 0–710 m.

Biology: Its adipose fin is small, short based, and slender. Pelvic fins are with 8 or 9 rays, including a rudimentary ray. The body coloration is bright metallic blue dorsally and bright silvery ventrally. It feeds on euphausiids and grows to a maximum length of 12.7 cm.

Luminous organs and bioluminescence: In this species, the light organs are regularly arranged in rows along the ventral surfaces of its head and trunk in both sexes. Males have a large luminous sternchaser on the dorsal surface of its caudal peduncle in addition to the primary photophores. The structure of the light organs from different regions of the body of this species is uniform. Its preferred direction for light emission from the ventral photophores is ventrally. When the accessory scales surrounding a photophore were removed while the animal was emitting light, the color of the light so emitted was found to change to green, yellow or white (Lawry Jr., 1973). Davis et al. (2020) reported that this species uses its eye-facing photophore as a reference standard that it adjusts to match downwelling light.

e) Subfamily: Notolychninae

1. Notolychnus valdiviae

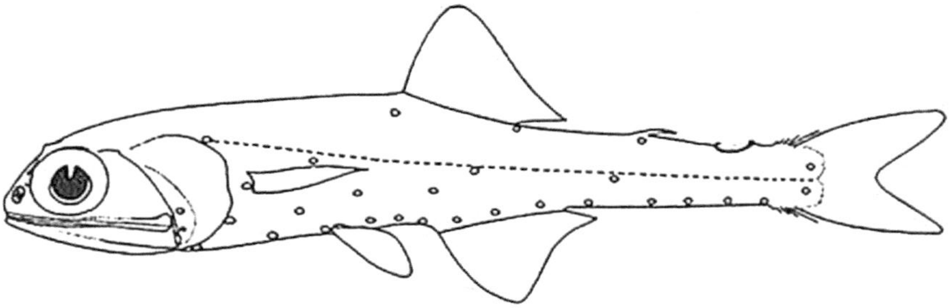

Image credit: South African Institute of Aquatic Biodiversity (SAIAB). https://creativecommons.org/licenses/by-nc/3.0/

Common name(s): Topside lampfish or topside lanternfish.

Global distribution: It has worldwide distribution in tropical, subtropical and temperate waters of all oceans.

Ecology: This bathypelagic and oceanodromous has a depth range of 25–700 m. It occurs between 375 and 700 m during the day time and between 25 and 350 m at night.

Biology: It grows to a maximum length of 6.3 cm and it feeds mainly on copepods, conchoecid ostracods and euphausiids.

Luminous organs and bioluminescence: It has 8 anal organs. A single deeply set translucent supracaudal luminous gland only is present in both sexes but infracaudal luminous gland is absent. Males have a longer supracaudal gland than females (https://www.fishbase.se/summary/4553). Cruz-Agüero (2016) reported that this monotypic genus is different from all other related genera by the high position of supraventral light organ, third supralateral organ and precaudal organ all of which are positioned close to the dorsal contour of the body. Further, its second precaudal organ is well above the lateral line and directly above the first precaudal organ.

3.1.2.8.2 Blackchins (Neoscopelidae)

1. *Neoscopelus macrolepidotus*

Image credit: Museum Victoria. https://fishesofaustralia.net.au/home/species/4023 South African Institute of Aquatic Biodiversity (SAIAB). https://creativecommons.org/licenses/by/3.0/au/

Common name(s): Large-scaled lantern fish.

Global distribution: Atlantic, Indian and Pacific Oceans.

Ecology: This bathypelagic species inhabits mainly continental shelves.

Biology: In this species, its eye is moderately large and upper jaw is expanded posteriorly, reaching to or slightly beyond the posterior margin of the eye. Teeth of outer jaw are small and villiform. The inner row of jaw teeth is enlarged and depressible. Its ventral surface is dark grey; head is greyish-silvery; and fins are pinkish-red. It grows to a total length of 25 cm.

2. *Neoscopelus microchir*

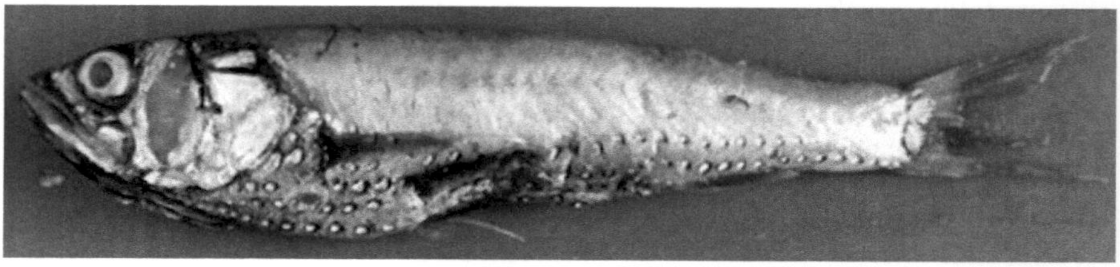

Image credit: Renju Ravi. https://creativecommons.org/licenses/by/3.0/deed.en

Common name(s): Shortfin neoscopelid.

Global distribution: Eastern Atlantic Ocean, Western Indian Ocean and Western Pacific Ocean.

Ecology: This bathypelagic species occurs in continental and island slopes at a depth range of 250–700 m.

Biology: The body of this species is fusiform and compressed. Dorsal profile of the head is straight with large eyes and large and terminal mouth. Teeth are small. The body coloration is red or blood-red on sides of head and body and paler below. The belly is silvery white with bluish reflections. It attains a maximum length of 30.5 cm.

3. Neoscopelus porosus

Image credit: Australian National Fish Collection, CSIRO. https://creativecommons.org/licenses/by/3.0/au/

Common name(s): Spangleside neoscopelid.

Global distribution: Indo-Pacific: western Australia and Japan; and South China Sea.

Ecology: This benthopelagic species inhabits continental slope areas at a depth range of 454–642 m.

Biology: Its eyes are moderately large and upper jaw is expanded posteriorly, reaching to or slightly beyond the posterior margin of the eye. Teeth of outer jaw are small and villiform; and inner row of jaw teeth are found enlarged and depressible. It grows to a maximum length of 18.3 cm.

Luminous organs and bioluminescence: Neoscopelid fishes

The luminescent species of this family have been reported to possess primary photophores on the ventral surface of the body and tongue in horizontal rows (Suntsov et al., 2008). While the larger photophores of the tongue serve as lure to attract small animals directly into the mouth of these fishes, the photophores of the ventral surface of the fish are believed to be used to help in ventral counter-illumination (Shimomura, 2012; Anon., https://en.wikipedia.org/wiki/Neoscopelus_macrolepidotus).

3.1.2.9 ORDER: HETEROMI (NOTACANTHIFORMES)

3.1.2.9.1 Eel-Shaped Fishes or Sea Lizards (Halosauridae)

1. Halosauropsis macrochir

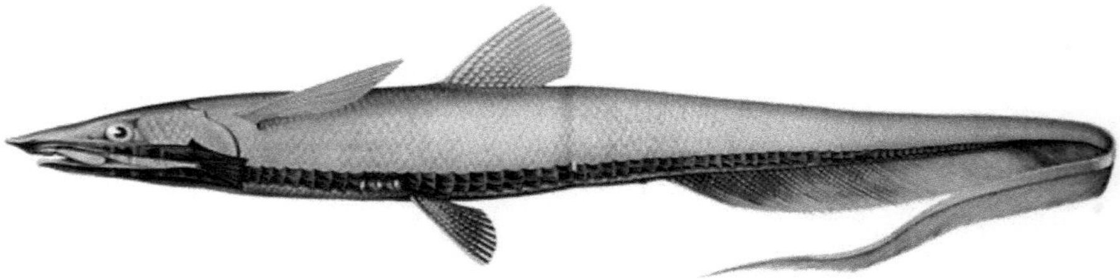

Image credit: Public domain.

Common name(s): Abyssal halosaur.

Global distribution: Eastern and Western Atlantic Ocean and Western Indian Ocean.

Ecology: This bathydemersal species has a depth range of 1100–3300 m.

Biology: The body of this species is moderately compressed. Head is broad, wedge-shaped and parallel-sided in its dorsal view. Snout is fairly short. Lateral line possesses a deeply pigmented sheath. Scales are absent on the top of the head. Pyloric caeca are long and unpigmented. The body coloration is greyish-black to bluish black. Top of head, gill isthmus and branchiostegal membranes are found to be black. It is an opportunistic feeder, on epibenthos and small fishes. It grows to a maximum length of 90 cm.

Luminous organs and bioluminescence: In this species, the luminescent organs vary in their number and location on the body from the head. They are found distributed in one or two rows laterally extending from the head to tail. These organs are only the specialized gland cells of the epidermis (Anon., https://www.scribd.com/document/93770802/pramodsingh).

3.1.2.10 ORDER: SACCOPHARYNGIFORMES

3.1.2.10.1 Deepsea Gulpers (Eurypharyngidae)

1. *Eurypharynx pelecanoides*

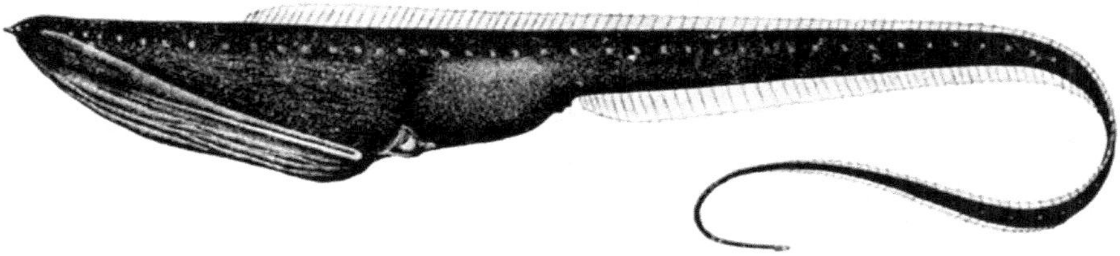

Source: Public domain.

Common name(s): Pelican eel, gulper eel, pelican gulper or umbrella-mouth gulper.

Global distribution: Temperate and tropical areas of all oceans.

Ecology: It is primarily a coastal species, but it is found to have a depth range of 500 to 3,000 m.

Biology: The characteristic feature of this species is its large mouth which is much larger than its body. Lower jaw is found hinged at the base of the head. The eyes are very small and the tail is very long and whip-like. Pelvic fins, swim bladders, and scales are absent in this species. The body coloration is black or olive. It feeds primarily on small crustaceans and grows to a maximum length of 1 m.

Luminous organs and bioluminescence: This genus has only one species which is luminescent. A complex luminous organ with numerous tentacles is seen at the end of its tail which glows pink normally and gives off bright-red flashes occasionally. The end of the tail bears a complex organ with numerous tentacles, which glows pink and gives off occasional bright-red flashes. The colors on its tail are displayed through its light-emitting photophores (Anon., https://en.wikipedia.org/wiki/Pelican_eel). Suntsov et al. (2008) reported that this species also possesses white line organs or nuchal troughs which run dorsally from just behind the head down the length of the body and tail of this species. These organs are capable of emitting brilliant luminescent flashes of short duration.

3.1.2.10.2 Gulper Eels or Swallowers (Saccopharyngidae)

1. *Saccopharynx* spp.

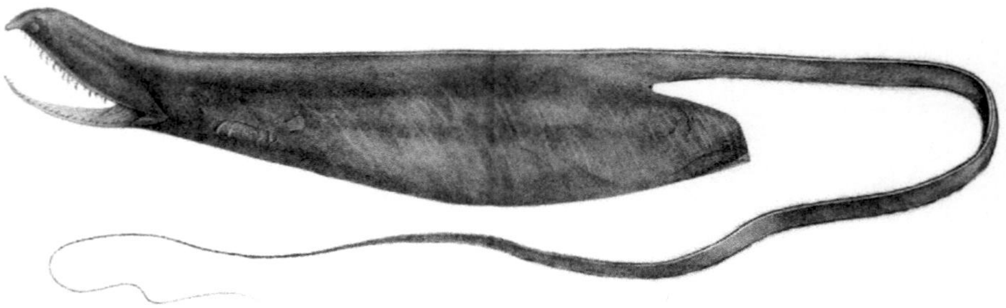

Saccopharynx ampullaceus

Image credit: Public domain.

Common name(s): Gulper eels or swallowers.

Global distribution: Atlantic and Pacific Oceans.

Ecology: It is a mesopelagic species occurring at depths of 1000–3000 m.

Biology: All the fishes of this genus have a long whip-like tail and their main body measures only 30 cm. They are also known for their large mouth and distensible abdomen. These species feed mainly on fishes. Among the various species of *Saccopharynx, Saccopharynx ampullaceus* has been reported to grow to a maximum length of 1.5 m.

Luminous species: The only genus of this family viz. *Saccopharynx* has 10 species which are all luminescent. While 7 species namely *Saccopharynx ampullaceus, Saccopharynx thalassa, Saccopharynx trilobatus, Saccopharynx harrisoni, Saccopharynx hjorti, Saccopharynx paucovertebratis,* and *Saccopharynx ramosus* are found in the Atlantic Ocean and the remaining three species namely *Saccopharynx lavenbergii, Saccopharnyx schmidti,* and *Saccopharnyx berteli* are from the Pacific Ocean.

Luminous organs and bioluminescence: In the species of *Saccopharynx,* there is a looped luminous organ which is found to extend along each side of the dorsal fin as far as its posterior extremity. This loop organ may begin either just behind the skull or just in front of the anterior extremity of the dorsal fin (Haneda, 1950). Anon. (https://en.wikipedia.org/wiki/Saccopharynx) reported that its tail which is tipped by a luminous, bulb-shaped organ acts as a lure for prey.

Suntsov et al. (2008) reported that in all nine species of the genus *Saccopharynx*, a terminal photogenic and non-bacterial caudal organ with numerous filaments serves as a bioluminescent lure for attracting prey. Further, these species also have white line luminous organs or nuchal troughs which are probably under neural control and run dorsally from just behind the head down the length of the body and tail.

3.1.2.11 ORDER: SCOMBRIFORMES

3.1.2.11.1 Snaketooth Fishes or Swallowerfishes (Chiasmodontidae)

***1. Pseudoscopelus* spp.**

Pseudoscopelus altipinnis

Image credit: Museums Victoria, https://fishesofaustralia.net.au/home/species/2579; https://creativecommons.org/licenses/by/3.0/au/

Common name(s): Snaketooth fishes or swallowerfishes.

Global distribution: It has a worldwide distribution in the Atlantic, Pacific, Indian and Southern Oceans, from equatorial to subpolar zones. But this genus is absent in the Mediterranean and Red Seas.

Ecology: It is a meso- and bathypelagic species.

Biology: These fishes are called swallowers because they swallow fishes larger than themselves. They possess large mouths which are capable of being opened very wide. The very elastic bodies and stomachs of these fishes can be greatly distended. These fishes are capable of swallowing other fishes larger than themselves. The species of this genus possess a highly developed acousticolateralis system (lateral line and associated pores and nerves) to help in their feeding. It is also believed that these fishes "listen with their face." (Anon., https://oceanexplorer.noaa.gov/explorations/19biolum/background/

midnight-zone/midnight-zone.html). These fishes have an average length of about 25 cm.

Luminous species: The luminous species of this genus include *Pseudoscopelus altipinnis, Pseudoscopelus astronesthidens, Pseudoscopelus australis, Pseudoscopelus cordilluminatus, Pseudoscopelus lavenbergi, Pseudoscopelus obtusifrons, Pseudoscopelus odontoglossum, Pseudoscopelus paxtoni, Pseudoscopelus pierbartus, Pseudoscopelus sagamianus, Pseudoscopelus scutatus, Pseudoscopelus scriptus,* and *Pseudoscopelus stellatus.*

Luminous organs and bioluminescence: All the luminescent species of *Pseudoscopelus* possess distinct photophores on their head and body and these light organs produce a clear, bright light (Melo et al., 2007).

Melo (2009) also reported on the five species groups of this genus based on the distribution of photophores and these groups are detailed below.

Pseudoscopelus altipinnis group: This group of *Pseudoscopelus* species has its transverse ventral-fin photophores in a single row of widely spaced organs. Medial cluster of organs are absent. The maxillary photophores which are widely spaced, are found in a single irregular row. The anteropreopercular photophores are present and there are six supraorbital pores.

Pseudoscopelus scriptus group: This group of *Pseudoscopelus* species has its transverse ventral-fin photophores in a single row where these photophores are closely spaced. The anteropreopercular photophores are absent and there are five supraorbital pores.

Pseudoscopelus australis group: This group of *Pseudoscopelus* species has its transverse ventral-fin photophores, which are short, interrupted and are not connected to anteropreopercular photophores.

Pseudoscopelus scutatus group: In this group of *Pseudoscopelus* species its transverse ventral-fin photophores, maxillary photophores, posteropreopercular photophores, pectoral-fin photophores, pectoral-fin axillary photophores, ventral-fin photophores and ventral-fin axillary photophore are all absent. While the supraventral-fin photophores are present the preventral-fin photophores and postventral-fin photophores are fused. The anteromaxillary photophores are restricted to space between pores three and four of mandibular canal. The supranasal pore of supraorbital canal is with three openings

Pseudoscopelus aphos group: This group of *Pseudoscopelus* species its photophores either as discrete organs or replaced by white lines. Some times these photophores may even be absent. The supranasal pore of supraorbital canal is with four openings.

3.1.2.12 ORDER: STOMIIFORMES

3.1.2.12.1 Bristlemouths, Lightfishes, or Anglemouths (Gonostomatidae)

***1. Cyclothone* spp.**

Cyclothone microdon

Image credit: Museums Victoria, https://fishesofaustralia.net.au/home/species/3396; https://creativecommons.org/licenses/by/3.0/au/

Common name(s): Bristlemouths.

Global distribution: Tropical to temperate latitudes of Atlantic, Indian, and Pacific Oceans. The species Cyclothone microdon may however be found in Arctic waters.

Ecology: These mesopelagic-open ocean species are found at depths exceeding 350 m.

Biology: The species of this genus are fairly elongate with large mouth large. They are feeding mainly on small crustaceans and small fishes. The genus, *Cyclothone* is remarkable because of its abundance in both density and weight. Even though, the members of *Cyclothone* spp. are small (about 6 cm), it is believed that this genus contains more number of individuals, and more weight, than any other genus of the fishes in the world, including the herrings of the genus Clupea (Anon., https://www.britannica.com/animal/Cyclothone).

Luminous species: The recognized bioluminescent species of this genus are: *Cyclothone acclinidens, Cyclothone alba,Cyclothone atraria, Cyclothone braueri, Cyclothone kobayashii, Cyclothone livida, Cyclothone microdon, Cyclothone obscura, Cyclothone pallida, Cyclothone parapallida, Cyclothone pseudopallida, Cyclothone pygmaea,* and *Cyclothone signata* (Anon., https://en.wikipedia.org/wiki/Cyclothone).

Luminous organs and bioluminescence: Cyclothone spp.

Suntsov et al. (2008) reported that the species of this family have large primary photophores on their head and ventral surface of the body separately or in groups. Further, some species of this family may also possess secondary photophores (Frank et al., 1984). The autogenic(intrinsic) luminescence of the species of this genus is due to coelenterate-type luciferin. The distribution of photophores in the different species of *Cyclothone* is given below.

In *Cyclothone braueri*, the photocytes of the photophores exhibit numerous secretory granules, which are found embedded in a well-developed endoplasmic reticulum. Further, the lens of photophores is composed of polyhedral cells; and the reflector is made of cells containing rich guanine crystals. The light emitted by the light organs of this species is of glandular nature (Davis et al., 2016). In *Cyclothone signata,* the luminescent spots are found distributed in the following manner. One on the head; 1 close below the eye; 2 on each gill cover; and 10 between the branchiostegal rays. In this species, there are also 2 longitudinal rows of photophores along each side of the body and among them, the lower row of 13 photophores is from throat to ventral fins, 4 from ventral fins to anal fin, and 13 from anal fin to caudal, fin and the upper row of 7 photophores are reaching as far back as the ventral fins (Anon., http://www.gma.org/fogm/Cyclothone_signata.htm). In *Cyclothone microdon*, the distribution of photophores is as follows: postorbital photophore 1; opercular photophores, 2; series on the branchiostegal membranes, 9–10; ventral series between pelvic-fin base and origin of anal fin, 5; ventral series posterior to anal-fin origin, 14–15; and lateral series, 8 (Anon., http://species-identification.org/species.php?species_group=fnam&id=1306).

2. Gonostoma atlanticum

Common name(s): Atlantic fangjaw.

Global distribution: It has worldwide distribution and is very common in the Eastern and Western Atlantic Ocean and Indo-Pacific and Eastern Pacific Ocean.

Ecology: This bathypelagic species has a depth range of 50–1352 m.

Biology: It is dark dorsally with silvery finishing on its translucent flanks. It attains a maximum length of 6.6 cm.

3. Gonostoma denudatum

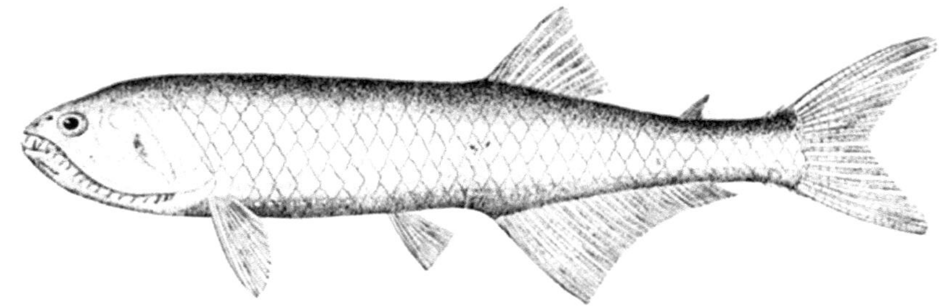

Source: Public domain

Common name(s): Bristlemouth.

Global distribution: Eastern and Western Central Atlantic Ocean.

Ecology: It is a bathypelagic species with a depth range of 100–700 m.

Biology: It is dark dorsally with silvery flanks. A distinctive gular pigment patch is seen. Dorsal adipose fin is present. It reaches a maximum length of 14 cm.

Luminous organs and bioluminescence: Haneda (1950) reported that the species of this genus possess several luminous dots in the underlying skin in addition to their regular series of well-developed, prominent ventral photophores. While in *Gonostoma atlanticum*, the photophores are found developing in fishes over 12–20 mm SL (https://www.fishbase.de/summary/5052), in *Gonostoma denudatum*, these photophores develop in fishes over 18–35 mm SL (https://www.fishbase.se/summary/1799).

4. Sigmops bathyphilus

Source: Public domain

Common name(s): Spark anglemouth.

Global distribution: Eastern, Northwest and Western Central Atlantic Ocean.

Ecology: This bathypelagic and non-migratory species has a depth range of 700–3000 m.

Biology: The body of this species is compressed; with black coloration. It feeds mainly on crustaceans and attains a maximum length of 15 cm.

*5. **Sigmops elongatus*** *(= Gonostoma elongatum)*

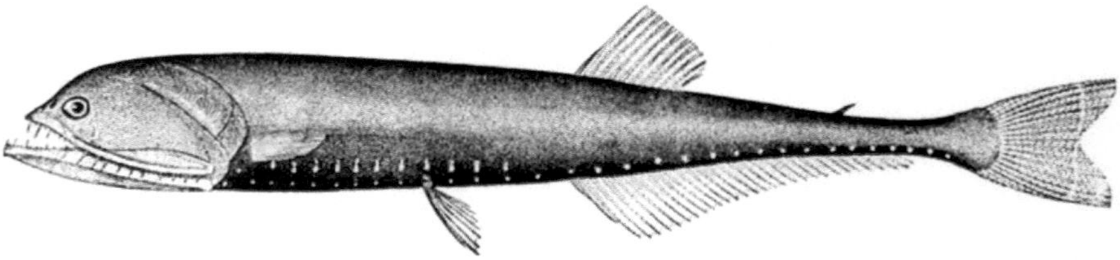

Source: Public domain

Common name(s): Elongated bristlemouth fish.

Global distribution: It is distributed worldwide.

Ecology: This bathypelagic species has a depth range of 25–4740 m.

Biology: Its body coloration is black with slightly silvery laterally. Fins are dusky and are with small black spots. It feeds mainly on crustaceans and small fishes and attains a maximum length of 27.5 cm.

6. Sigmops gracilis

Common name(s): Slender fangjaw.

Global distribution: Pacific Ocean.

Ecology: This bathypelagic species has a depth range of 0–4389 m.

Biology: The body of this species is moderately compressed and elongated. Head is moderate in size and eyes are round. Mouth is very large with hooked teeth. It attains a maximum length of 13.3 cm.

Luminous organs and bioluminescence: Sigmops spp.

Sigmops elongatus: In this species, the primary photophores have been reported to develop in fishes in the size range of 6–22 mm (Anon., https://www.fishbase.se/summary/Gonostoma-elongatum.html). Copeland (1991) reported that this species has numerous photophores which are located on the body as an upper serial row. Suborbital photophores and caudal photophores are also seen. Further, there is also a ventral serial row of photophores which produce a bluish light in this species.

Sigmops bathyphilus: It has very small photophores and they are often obscure. Lateral series (OA) photophores are irregularly arranged, most close to mid-line of flank (Anon., https://en.wikipedia.org/wiki/Sigmops_bathyphilus).

Sigmops gracilis: In this species, a row of 6 widely separated photophores is seen near the dorsal profile of the body. The lateral series (OA) of photophores are irregularly arranged in this species (Anon., https://www.fishbase.se/summary/22719).

7. Triplophos hemingi

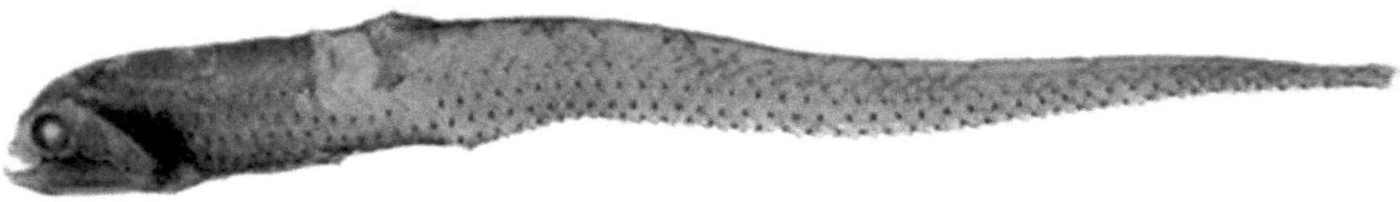

Image courtesy: Kwang-Tsao Shao. Taiwan Fish Database.

Common name(s): No common name.

Global distribution: Tropical waters of all oceans.

Ecology: This bathypelagic and oceanodromous species has a depth range of 200–2000 m.

Biology: It has a short trunk and long tail. The body coloration of this species is yellowish-brown. Black on the head and abdomen are black and fins are colorless. It attains a maximum length of 36 cm.

Luminous organs and bioluminescence: In this species, the primary photophores are seen on its isthmus. Further, a series of additional head photophores and a single photophore above the maxilla are also seen in this autogenic bioluminescent species.

3.1.2.12.2 Lightfishes (Phosichthyidae)

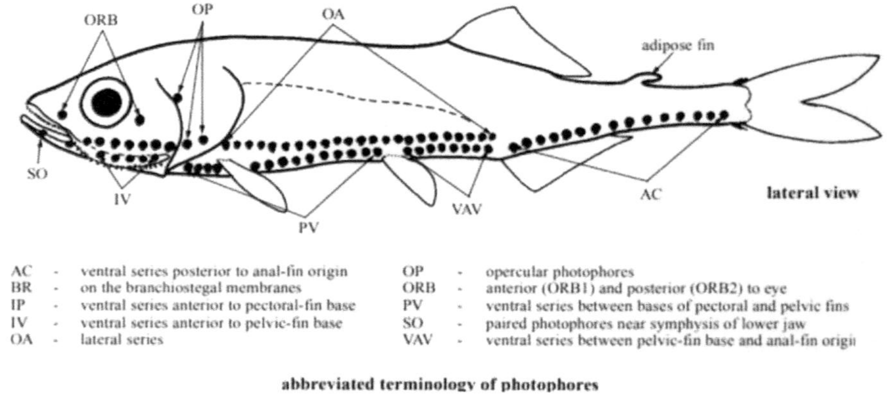

Image credit: FAO, https://www.fao.org/3/y4161e/y4161e21.pdf

1. Ichthyococcus ovatus

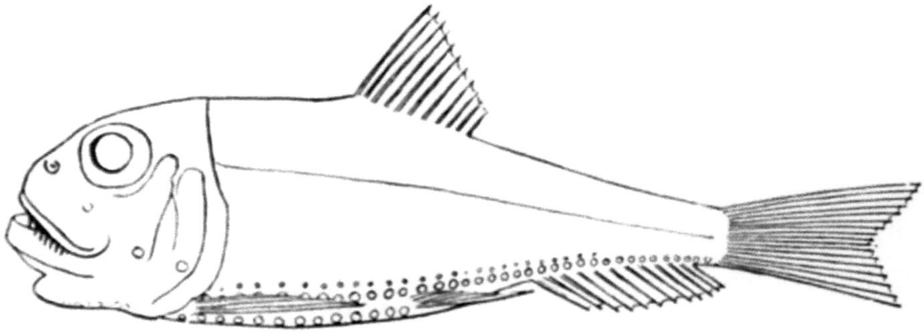

Source: Public domain

Common name(s): Lightfish.

Global distribution: Circumglobal except North Pacific.

Ecology: This bathypelagic species has a depth range 0–2500 m.

Biology: Body coloration of this species is brown-yellow and franks are silvery. Fin bases and edges of scale pockets are black. Both dorsal and ventral adipose fins are present . Other characteristic features are : a short, deep body; and microscopic teeth and tubular eye. It feeds mainly on euphausiid krills.

Luminous organs and bioluminescence: In the species of Phosichthyidae, large primary photophores are most seen on the head and ventral surface of the body separately or in groups. Further, some species of this family may also possess secondary photophores (Suntsov et al., 2008).

In *Ichthyococcus ovatus*, the development of photophores comes to an end at the size range of 15–17 mm SL and the photophores are present along its sides (Anon., https://www.fishbase.se/summary/1804).

2. Phosichthys argenteus

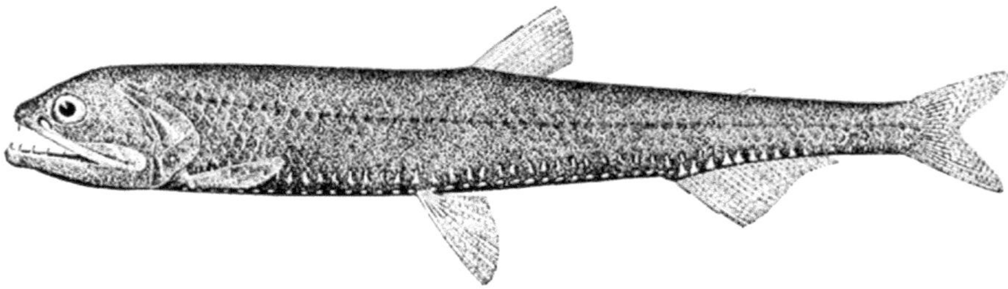

Image credit: Tony Ayling & Geoffrey Cox. https://creativecommons.org/licenses/by-sa/1.0/deed.en

Common name(s): Silver lightfish.

Global distribution: Eastern Atlantic Ocean, Indian Ocean and Pacific Ocean.

Ecology: This bathypelagic species has a depth range of 300–1050 m, and it is more common in the continental slope areas.

Biology: The body of this species is elongated and compressed. Dorsal fin is located well in front of its anal fin. Mouth has large canine teeth. The body coloration is dark brown above, lighter brown to pink on sides and black ventrally. Portions of the head are silvery. It feeds mainly on planktonic invertebrates, especially, the euphausiid krills. It grows to a maximum length of 30 cm.

Luminous organs and bioluminescence: This species has several rows of light organs which run along the ventral surface of its body. In addition to these photophores, two light organs are present near the eye (one in front and one behind) (McMillan et al., 2011; Anon., https://australian.museum/learn/animals/fishes/silver-lightfish-phosichthys-argenteus/).

3. Pollichthys mauli

Source: Public domain

Common name(s): Stareye lightfish.

Global distribution: Eastern, Western, and North Atlantic Ocean and the western Pacific Ocean.

Ecology: This bathypelagic and oceanodromous species has a depth range of 100–600 m. It is known to make daily migration with the "deep scattering layer."

Biology: It has a slender body with longer anal fin and tail set. The body coloration is dark dorsally with silvery flanks. It feeds mainly on small crustaceans and attains a maximum length of 6 cm only.

Luminous organs and bioluminescence: It has a well-developed series of light organs which are running almost the entire length of the body (Anon., https://www.sciencesource.com/archive/Stareye-Lightfish-SS2713156.html).

4. Polymetme corythaeola

Common name(s): Rendezvous fish.

Global distribution: Indo-West and Eastern Pacific Ocean and Atlantic Ocean.

Ecology: This benthopelagic species has a depth range of 165–800 m.

Biology: It has dark dorsum and silvery flanks. Black pigment is seen on its outer caudal rays. It grows to a maximum length of 26 cm.

5. Polymetme elongata

Common name(s): Not designated.

Global distribution: Western Pacific Ocean.

Ecology: This benthopelagic species occurs at a maximum depth of 350 m.

Biology: It has a more fusiform and elongated body with a smaller head. It attains a maximum length of 18.4 cm.

6. Polymetme illustris

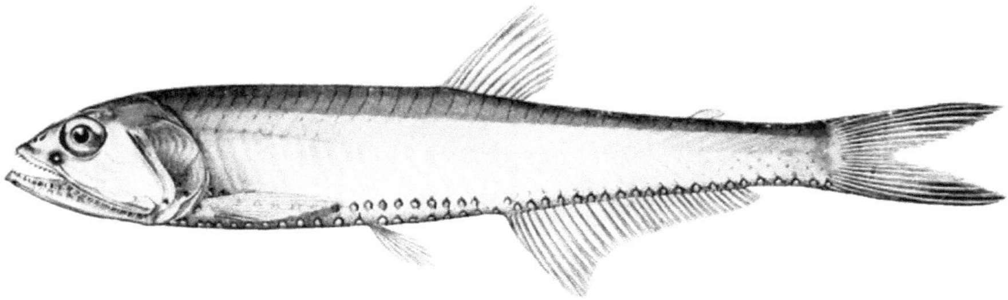

Image credit: Museums Victoria, https://fishesofaustralia.net.au/home/species/3934; https://creativecommons.org/licenses/by/3.0/au/

Common name(s): Brilliant lightfish.

Global distribution: Eastern Indian Ocean to Western Pacific Ocean.

Ecology: This bathydemersal species has a depth range of 230–822 m and it is common in the continental shelf and slope areas.

Biology: It is a light colored fish with a dark brown back which is distinctly separated from its silvery sides. Gill cover is blackish. It reaches a maximum length of 14.8 cm.

Luminous organs and bioluminescence: Polymetme spp.

In the species of *Polymetme,* the primary photophore consists of two components, viz. a bulb-shaped body which is lying within the muscle of the fish and a funnel-shaped structure which is in contact with the surface of the body. (Haneda, 1952). The location of these photophores is also somewhat different between the species of this genus. In *Polymetme corythaeola*, there are two rows of photophores on the body and are seen above the anal fin 15–17 (Anon., https://www.fishbase.de/summary/Polymetme-corythaeola.html). In *Polymetme elongata*, the photophores are present above anal fin 16–18. (Anon., https://www.fishbase.in/summary/23624). In *Polymetme illustris*, the photophores are located above anal fin 16–17 (Anon., https://www.fishbase.de/summary/Polymetme-illustris.html).

3.1.2.12.3 Hatchetfishes (Sternoptychidae)

1. Argyripnus brocki

Common name(s): Brock's bristle-mouth fish.

Global distribution: Hawaiian Islands in the Eastern Central Pacific Ocean.

Ecology: This benthopelagic species has a depth range of 180–380 m.

Biology: Its shorter snouts and longer upper jaws are the characteristic features of this species. A well developed black patch is seen on its caudal peduncle. It attains a maximum length of 8.9 cm.

2. Argyripnus ephippiatus

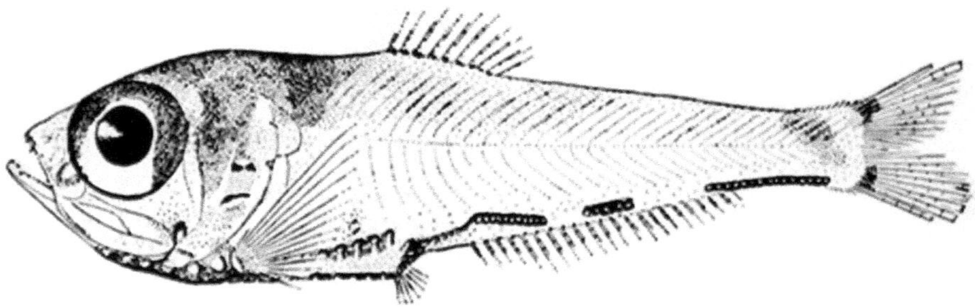

Image credit: Museums Victoria, https://fishesofaustralia.net.au/home/species/3936; https://creativecommons.org/licenses/by/3.0/au/

Common name(s): Gilbert & Cramer's bristle-mouth fish.

Global distribution: Pacific Ocean.

Ecology: This benthopelagic species has a depth range of 325–518 m and it is common in continental slope areas.

Biology: It has an extremely compressed body with a deep keel, large eyes, and an upturned mouth. A patch of black pigmentation is seen on the upper portion of its body just behind head. Further, a poorly developed black patch on its caudal peduncle. The body musculature is translucent. It attains a maximum length of 8.9 cm.

3. Argyripnus iridescens

Common name(s): Pearlyside lightfish or brilliant pearlside.

Global distribution: Southwest Pacific Ocean.

Ecology: This bathypelagic species occurs commonly in the continental slope areas.

Biology: It has a short snout and long upper jaw Unlike its related species, the black patch on the caudal peduncle is poorly developed. It attains a maximum length of 14 cm.

Luminous organs and bioluminescence: Argyripnus spp.

The fishes of the family Sternoptychidae possess groups of highly developed photophores along their ventral surface and lower sides of the head and body. These luminous organs are of a more complex structure, and each of such organ is consisting of a luminous body, reflector, lens, and color filter. In addition to these primary photophores, there may also be secondary luminous organs on the jaws and near the eyes (Haneda, 1950; Suntsov et al., 2008). The number of VAV (ventrally, from pelvic fin base to anal fin origin) +ACA (anterior photophores from anal fin origin to caudal fin base) and PAC (posterior AC) photophores showed marked difference between the above mentioned three species. In *Argyripnus brocki*, the number of VAV+ACA photophores and PAC photophores is 17 and 9–11, respectively. On the other hand, the number of these photophores is more number in the species. *Argyripnus ephippiatus* and *Argyripnus iridescens,* i.e., 18–28 and 12–18, respectively (Struhsakeri. 1973).

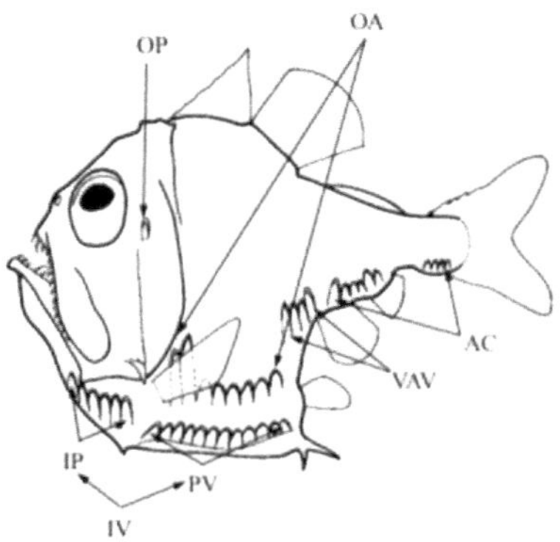

AC	-	ventral series posterior to anal-fin origin	OP	-	opercular photophores
BR	-	on the branchiostegal membranes	ORB1	-	anterior to eye
IP	-	ventral series anterior to pectoral-fin base	PV	-	ventral series between bases of pectoral and pelvic fins
IV	-	ventral series anterior to pelvic-fin base	SO	-	paired photophores near symphysis of lower jaw
OA	-	lateral series	VAV	-	ventral series between pelvic-fin base and anal-fin base

abbreviated terminology of photophores

Luminous organs in *Argyropelecus*

Image credit: FAO, https://www.fao.org/3/y4161e/y4161e21.pdf

4. Argyropelecus aculeatus

Common name(s): Lovely hatchetfish.

Global distribution: Eastern, Western and Northwest Atlantic Ocean, Western and Southeast Pacific ocean and Indian Ocean.

Ecology: This bathypelagic species has a depth range of 100–2056 m.

Biology: It has a deep body. It has been reported to be a selective dusk-time feeder mainly on conchoecid ostracods. Its other food items include copepods, decapod larvae, fish larvae, etc. It attains a maximum length of 8.3 cm.

Luminous organs and bioluminescence: Schultz (1961) reported that the number of organs in groups of photophores in the species of this genus as follows: subopercular 1; supra-abdominal 6; suprapectoral 2.abdominal 12; anal 6; branchiostegal 6; isthmus 6; preanal 4; preopercular 1; preorbital 1; postorbital 1; and subcaudal 4.

In *Argyropelecus aculeatus*, the adult photophores have been reported to develop at about 15 mm SL (Anon., https://www.fishbase.se/summary/Argyropelecus-aculeatus.html). Davis et al. (2020) reported that the dorsal and ventral sides of its eye-facing photophore are enclosed by two layers of connective tissue. Further, the anterior and lateral sides of the photophore are protected by a layer of melanin granules. Its ventral photophores are believed to help in counterillumination and associated defence from its predators. The wavelength of the light emission maximum of this species has been reported as 470 nm (Herring, 1983).

5. Argyropelecus affinis

Common name(s): Pacific hatchet fish.

Global distribution: Eastern, Western, and Northwest Atlantic Ocean, Eastern and Western Pacific Ocean, and Indian Ocean.

Ecology: This bathypelagic and oceanodromous species has a depth range of 1–3872 m.

Biology: It feeds mainly on euphausiids, salps, and chaetognaths, and it grows to a maximum length of 8.4 cm.

Luminous organs and bioluminescence: In this species, the adult complement of photophores develops at 14 mm SL (Anon., https://www.fishbase.in/summary/40). Widders et al. (1983) reported that the wavelengths of light emission maxima of this species were 482, 484, and 486 nm. This narrow bandwidth of the emission spectrum of this species was attributed to the presence of filters in the photophores. Davis et al. (2020) reported that the photophores of this species employ the guanine reflectors to match the angular distribution of the downwelling light.

6. Argyropelecus gigas

Common name(s): Hatchetfish.

Global distribution: Eastern and Western Atlantic Ocean, Eastern and Western Pacific Ocean, and Indian Ocean.

Ecology: This bathypelagic species has a depth range of 300–1000 m.

Biology: This species has upwardly-directed eyes. Its body margin is elevated behind the high dorsal blade. Tail base is deep and ventral keel scales extend well below the abdominal photophores forming flaplike process. It attains a maximum length of 12 cm.

Luminous organs and bioluminescence: In this species, the adult complement of photophores develops at 14 mm SL (Anon., https://www.fishbase.se/summary/Argyropelecus-gigas). Its supra-abdominal, preanal, anal, and subcaudal photophores are seen in an almost continuous straight line. A distinct gap is present between each subcaudal photophore (Anon., https://fishesofaustralia.net.au/home/species/3939).

7. Argyropelecus hemigymnus

Common name(s): Half-naked hatchetfish.

Global distribution: Eastern, Western and Northwest Atlantic Ocean and Eastern Pacific Ocean.

Ecology: This bathypelagic species has a depth range of 0–2400 m. Adults are known to make significant vertical migrations.

Biology: It is an opportunistic feeder on calanoid copepods and small fishes, and it attains a maximum length of 5.1 cm.

Luminous organs and bioluminescence: This species has been reported to emit flashes which are short, fatigable with short latency (Baguet and Marechal, 1976). Krönström et al. (2005) reported that nitric oxide plays an important role in the control of light emission from the photophores of this species.

8. Argyropelecus olfersii

Common name(s): No common name.

Global distribution: Eastern Atlantic Ocean and South Pacific Ocean.

Ecology: This bathypelagic and oceanodromous species has a depth range of 100–800 m.

Biology: It feeds mainly on crustaceans and small fishes and it grows to a maximum length of 9 cm.

Luminous organs and bioluminescence: In this species, the adult photophore complement develops at a size of 16–17 mm SL (Anon., https://www.fishbase.se/summary/Argyropelecus-olfersii.html). Herring (1983) reported its wavelengths of light emission maxima were 470–475 nm.

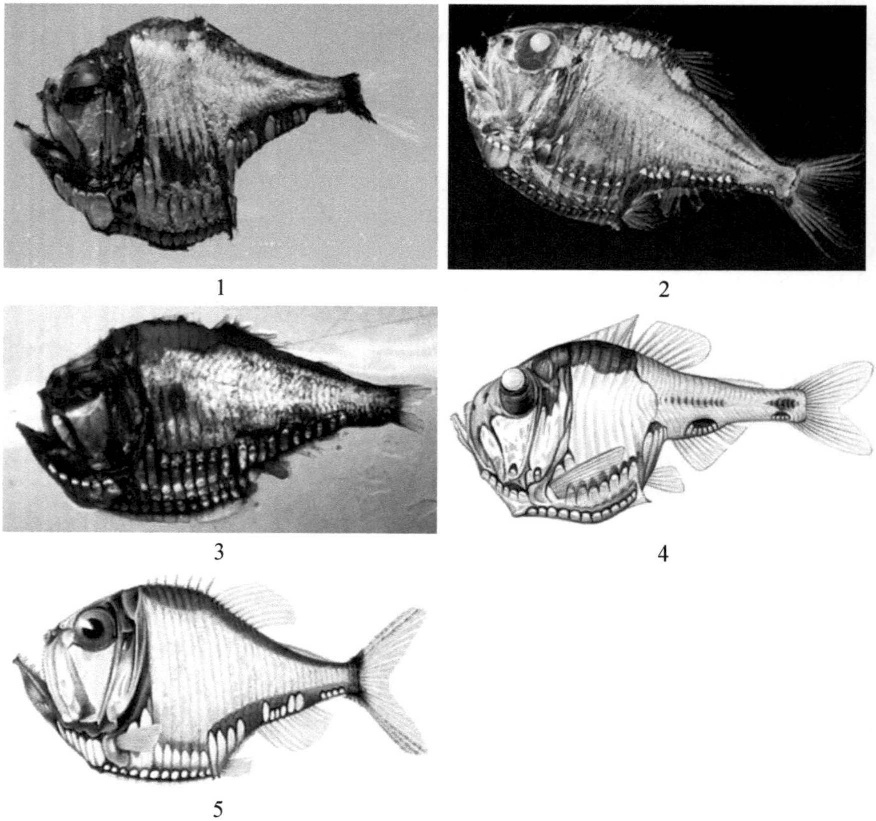

1. Argyropelecus aculeatus; 2. Argyropelecus affinis; 3. Argyropelecus gigas; 4. Argyropelecus hemigymnus; 5. Argyropelecus olfersii.

Source: Public domain.

9. Maurolicus spp.

Common name(s): Pearlsides.

Global distribution: It has circumglobal distribution and is found in all non-polar oceans.

Ecology: It is a mesopelagic species.

Biology: The body of this species is oblong and compressed and is covered with silvery pigment. Scales are normally absent, and when present, they are very thin and caducous. Head is compressed; Cleft of mouth is wide and very oblique and lower jaw is prominent. Both jaws are with minute teeth; Eyes are large and pectoral and ventral fins are well developed. Air bladder is absent.

Luminous organs and bioluminescence: The species of this genus possess a series of luminous spots along the lower side of their head and tail. The photophores located over and behind the anal opening are nearly as a continuous series. However, the luminous spots appear as simple impressions on the skin. In *Maurolicus muelleri*, there are 12 pairs of luminous spots along its belly before ventrals. Of these, 9 pairs are in an upper series between pectoral fins and ventral fins (Anon., 1896). In most of the luminescent species of this genus, significant variation in the number of photophores from the anal fin base to caudal fin base (AC). In the species, *Maurolicus australis, Maurolicus breviculus,* and *Maurolicus javanicus*, the AC photophores have been reported to be 22–25 (Anon., https://www.fishbase.se/summary/51609) 20–24 (Anon., https://www.fishbase.se/summary/51616)., and 24–27 (Anon., https://www.fishbase.in/summary/Maurolicus-javanicus.html) respectively; and in *Maurolicus mucronatus* and *Maurolicus weitzmani*, 22–26 (Anon., https://fishbase.mnhn.fr/summary/Maurolicus-mucronatus.html).

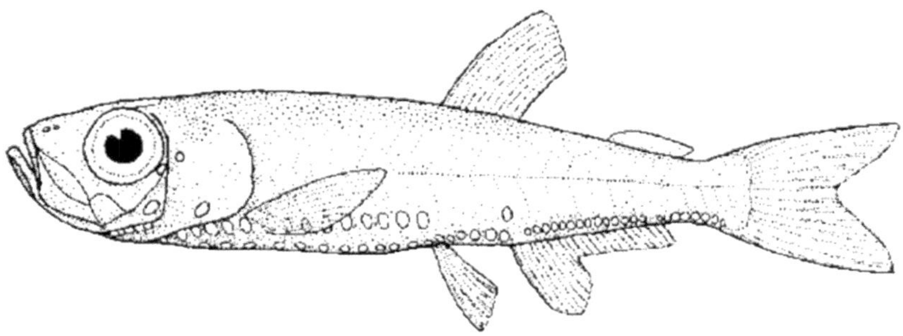

Maurolicus muelleri

Source: Public domain

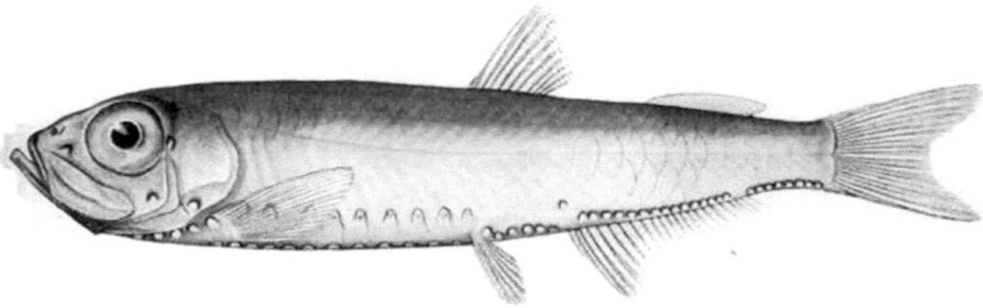

Maurolicus australis

Image credit: McCulloch, Wikipedia.

Maurolicus javanicus

Image credit: Museums Victoria, https://fishesofaustralia.net.au/home/species/3942; https://creativecommons.org/licenses/by/3.0/au/

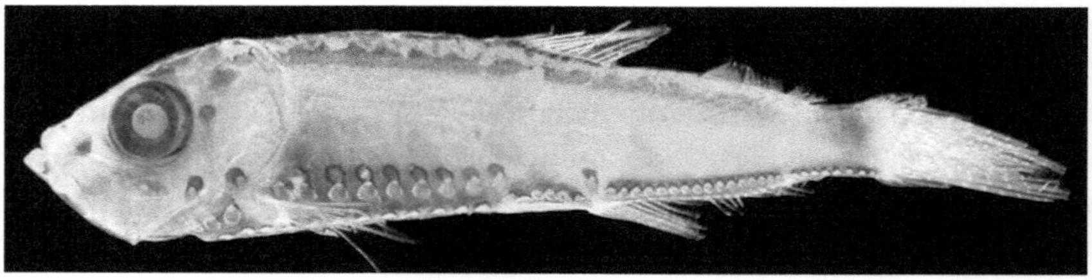

Maurolicus weitzmani

Image credit: EOL, https://eol.org/pages/49672998/media; https://creativecommons.org/licenses/by-nc-sa/3.0/

10. Polyipnus spp.

Common name(s): Marine hatchetfishes.

Global distribution: It has worldwide distribution.

Ecology: This benthopelagic species has a depth range of 50 to 1,500 m.

Biology: The body of this species is rounded and is extremely compressed laterally. Mouth which is located at the tip of the snout is directed straight downwards. The eyes are large and tube-shaped.

Luminous organs and bioluminescence: Polyipnus spp. Of the 32 species of this genus, 8 species have been reported to be luminescent as given in the following table. The number of organs in groups of photophores in the fishes of this genus are as follows: preopercular 1, subopercular 1, preorbital 1, postorbital 1, suprapectoral 3, branchiostegal 6, isthmus 6, abdominal 10, supra-abdominal 3, lateral 1, preanal 5, anal 4 to 17, supra-anal 3 (when present), and subcaudal 4 (Schultz, 1961). The number of anal photophores in the different luminescent species of this genus are given below.

Species	**Number of anal photophores**
Polyipnus polli	6 or 7
Polyipnus laternatus	10 to 1 3
Polyipnus triphanos	8 to 10.
Polyipnus fraseri	4
Polyipnus unispinus	10 to 17
Polyipnus tridentifer	13 to 17
Polyipnus spinosus	12 to 14
Polyipnus stereope	*10 to 13*

Source: Schultz, 1961; Anon., https://eol.org/pages/46563342/articles.

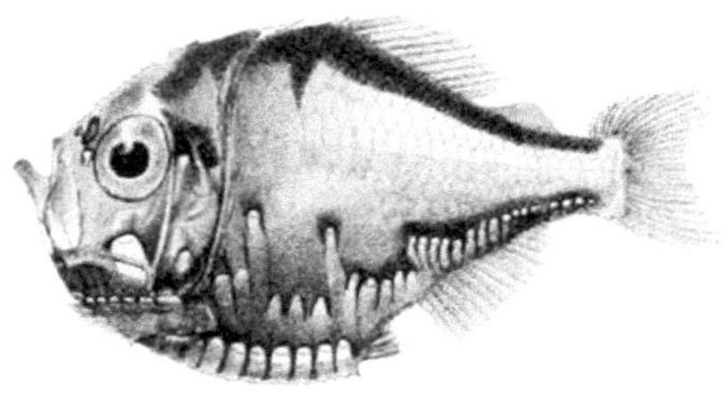

1

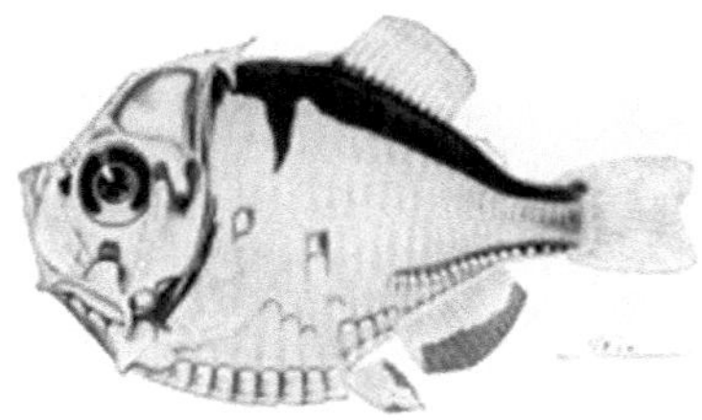

2

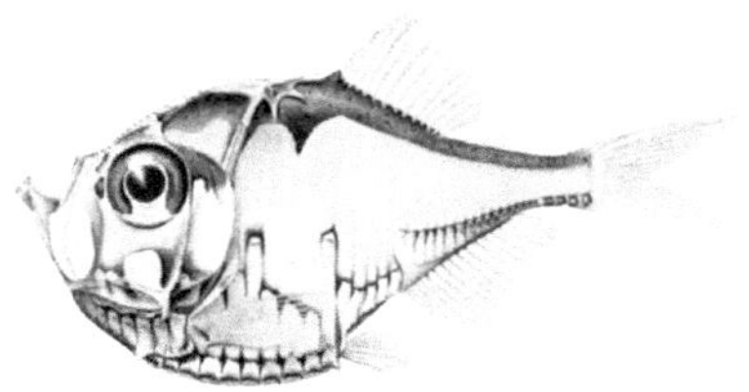

3

1. Polyipnus spinosus; 2. Polyipnus stereope; 3. Polyipnus tridentifer.

Source: Public domain.

11. Sternoptyx spp.

Common name(s): Marine hatchetfish.

Global distribution: Temperate and tropical waters of all the world's oceans.

Ecology: This species has been reported to have a depth range of 200–2000 m.

Biology: It has a silvery, high, and laterally compressed body with large, non-telescopic eyes. All the species of this genus are relatively small and even the largest species viz. *Sternoptyx pseudobscura* and *Sternoptyx pseudodiaphana* are not exceeding 6 cm in their total length.

Luminous species: Sternoptyx diaphana, Sternoptyx pseudobscura obscura, Sternoptyx pseudobscura, and *Sternoptyx pseudodiaphana.*

Luminous organs and bioluminescence: The number of organs in groups of photophores in the species of this genus are as follows: preopercular 1, subopercular 1, preorbital 1, postorbital 1, branchiostegals 3, isthmus 5, suprapectoral 3, abdominal 10, supra-abdominal none, preanal 3, supra-anal 1, anal 3, and subcaudal 4 (Schultz, 1961). Anon. (2017) reported that *Sternoptyx pseudobscura* showed a new form of light-reflection produced in guanine crystal platelets present in the skin. Herring (1983) reported that the wavelength of light emission maximum was 473 nm for the species, *Sternoptyx diaphana.*

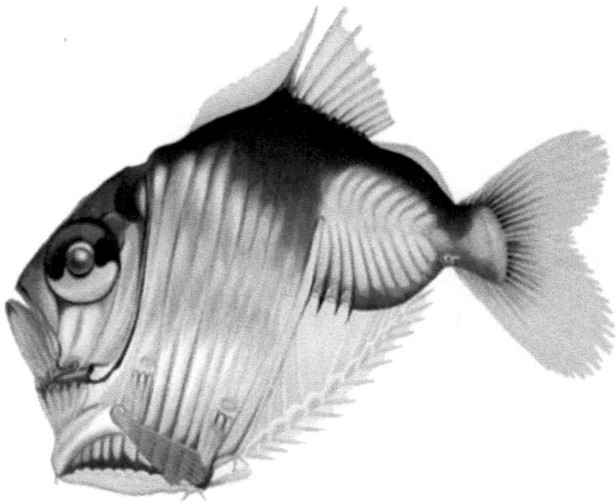

Sternoptyx diaphana

Image credit: Wikipedia.

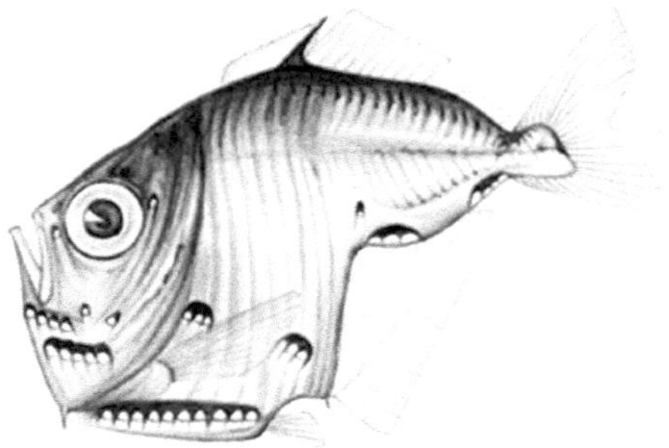

Sternoptyx obscura

Image credit: Kwang-Tsao Shao. Taiwan Fish Database.

12. Valenciennellus tripunctulatus

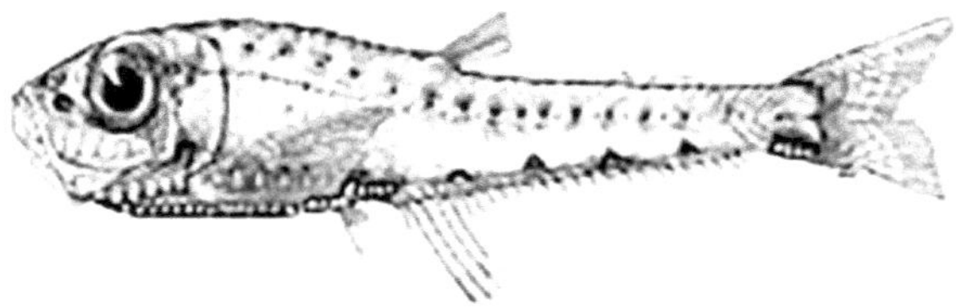

Distribution of luminous organ in *Valenciennellus tripunctulatus*

Image credit: NMNH Research & Collections. https://creativecommons.org/licenses/by/4.0/deed.en

Common name(s): Constellationfish.

Global distribution: It has a worldwide distribution in tropical to temperate waters.

Ecology: This bathypelagic; and non-migratory species has a depth range of 100–1000 m.

Biology: Body of this species is elongated, strongly compressed and scaleless. Head is strongly compressed. Gape of mouth is wide. Caudal is emarginate. The body coloration is silvery during the day and darker at night. It is a selective daytime feeder mainly on copepods (*Pleuromamma* spp.) and ostracods (*Conchoecia* spp.). It attains a maximum length of 4 cm.

Luminous organs: In this species, unique arrangements of the photophores are seen on the tail. There are five distinct groups containing 2–4 light organs in each (Anon., https://www.fishbase.se/summary/Valenciennellus-tripunctulatus.html).

3.1.2.12.4 Dragonfishes (Stomiidae)

In barbeled fishes of the family Stomiidae are known for their highlight organ diversity. In addition to the normal lateral and ventral series of luminous organs, these fishes possess luminescent barbels and minute secondary photophores which are scattered over the body. Furthermore, sexually dimorphic pre, sub and postorbital luminous organs and species-specific luminous patches are also present in some of the species of this family (Suntsov et al., 2008). Kenaley (2010) reported that the loosejaw dragonfishes of this family possess accessory cephalic photophores (AOs). The fishes of the genera, *Aristostomias, Malacosteus*, and *Pachystomias*, produce far-red, long-wave emissions (>650nm) from their AOs, which is believed to be a character unique among vertebrates. While *Aristostomias* and *Malacosteus* possess a single far-red AO, *Pachystomias* possesses anterior and posterior far-red AOs which are located in their ventral margins.

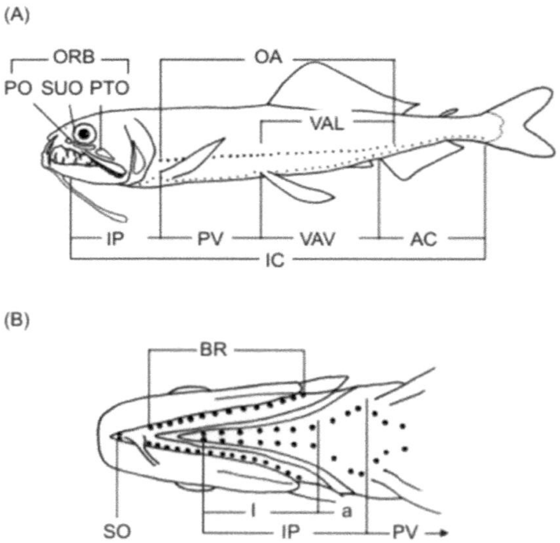

Photophores of stomiatoid fishes: IC, entire ventral row of photophores from the anterior end of the isthmus to posterior of the caudal peduncle (IC = IP + PV + VAV + AC); IP, ventral row of photophores from anterior of the isthmus to a ventral line at the pectoral fin origin; PV, ventral row of photophores between vertical lines at the origins of the pectoral and pelvic fins; VAV, ventral row of photophores between a vertical line at the origins of the pelvic and anal fins; AC, posterior part of the IC series, from posterior of the VAV series to the posterior part of the caudal peduncle; OA, all large photophores of the lateral series. BR, branchiostegal photophores; ORB, orbital photophores: SO, located near the anterior end or the symphysis of the lower jaw; PO, preorbital photophores; SUO, suborbital photophores; PTO, postorbital photophores.

Image courtesy: Kwang Tsao Shao. Taiwan Fish Database.

The stomiid fishes emit both blue and red light. Of the 57 species of the family, Stomiidae, all the valid 10 species of the 4 genera viz. *Chirostomias (1) Malacosteus (2), Pachystomias (1),* and *Aristostomias* (6) have been reported to produce and see red light (Anon., https://www.mindat.org/taxon-2347391.html) in addition to their normal blue light production. These genera have been reported to possess numerous small body photophores in addition to their large suborbital photophores that emit red light and postorbital ones that emit blue light. While the red light- emitting suborbital photophore of *Malacosteus* is similar in general organization to their other photophores, the red-emitting photophores of *Pachystomias* and *Aristostomias* have a unique organization in that the bulk of such photophores comprises a gland which lies outside the thick reflector (Herring and Cope, 2005; Shimomura, 2012).

1. Aristostomias spp.

Common name(s): Barbeled dragonfishes.

Global distribution: These fishes are native to the Pacific, Atlantic, and Indian Oceans.

Ecology: These species inhabit the mesopelagic zone.

Biology: Body of these fishes is moderately long and its depth is 4 or more times in standard length (SL). Head is large and its length is about 3–4 times in SL. Snout is longer than eye diameter and is with two nostrils on each side. Chin barbela are present. These fishes are the dominant predators.

Luminous species: All the six valid species of this genus viz. *Aristostomias grimaldii, Aristostomias lunifer, Aristostomias polydactylus, Aristostomias tittmanni, Aristostomias scintillans,* and *Aristostomias xenostoma* are luminescent.

Luminous organs and bioluminescence: Aristomias spp.

Arrangement of photophores in the fishes of the genus Aristostomias: The serial photophores present in the species of this genus are arranged as follows: 45–57; photophores in the ventral row from front of isthmus to tail; and 29–38 photophores in lateral series to above anterior end of anal fin. All these photophores are either evenly spaced with few gaps or in widely spaced groups of 1–4. Two rows of photophores are present on each side of the isthmus; and pale luminous spots or lines are present on head (Anon., http://species-identification.org/species.php?species_group=fnam&id=280&menuentry=groepen).

Variation in photophores of the different species of *Aristostomias*: Kenaley (2010) reported that far-red light producing accessory cephalic photophores (AOs) are present in *Aristostomias grimaldii*. The wavelengths of such light emission maxima were reported to be >650 nm. In *Aristostomias lunifer* a chin barbel with distinct serial phorophores is present (Anon., http://species-identification.org/species.php?species_group=fnam&id=1320). Shimomura (2012) reported that the wave lengths of luminescence emission maxima of *Aristostomias scintillans* were 479 nm and 703 nm for the blue and red light emitted from the postorbital light organ and suborbital light organ, respectively. O'Day and Fernandez (1974) reported that this species had two photo-pigments with wavelengths of absorbance maxima at 551 and 526 nm, respectively. The species *Aristostomias tittmanni* possesses a third pigment with the wavelengths of luminescence maxima in the range of 588–595 nm.

Aristostomias lunifer

Image credit: EOL, https://eol.org/pages/46563659 PD

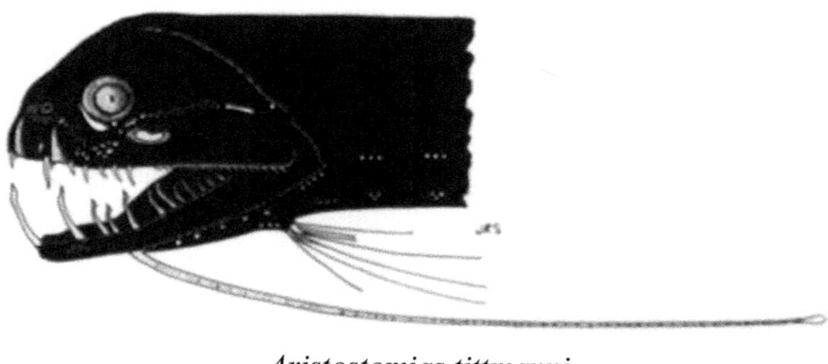

Aristostomias tittmanni

Image credit: EOL, https://eol.org/media/13449003

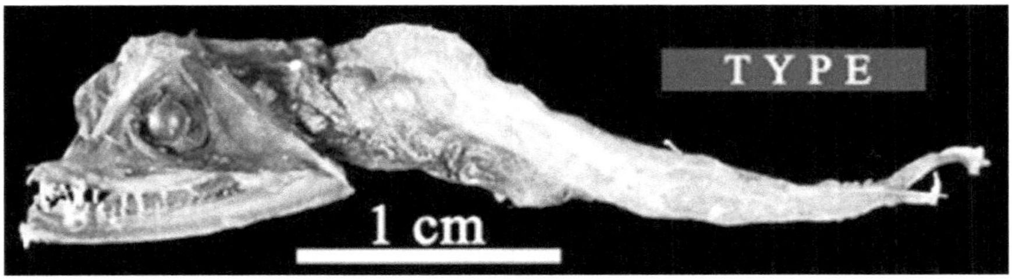

Aristostomias xenostoma

Image credit: gbif.org. https://www.gbif.org/occurrence/583567746. https://creativecommons.org/licenses/by/4.0/

2. ***Chirostomias pliopterus*** (= *Chirostomias lucidimanus*)

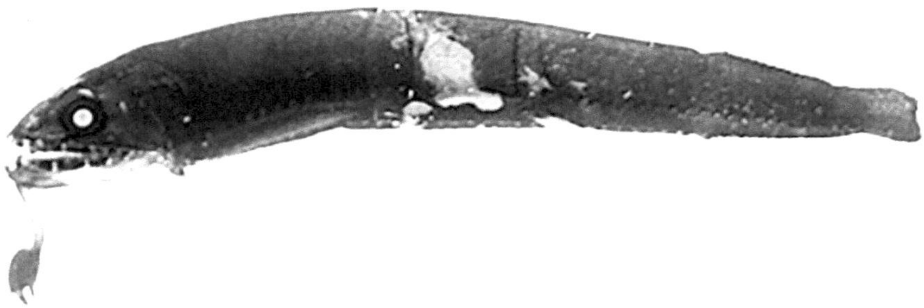

Credit: Peter Rask Møller. https://www.inaturalist.org/people/peterraskmoller

Common name(s): Barbeled dragonfish.

Global distribution: Eastern, Western, and NW Atlantic Ocean.

Ecology: This bathypelagic species inhabits up to a maximum depth of 500 m.

Biology: The body of this species is slender, compressed and is much deeper than wide. Head has a fairly long snout which is tapering almost to a point. The lower jaw is no longer than the upper jaw and is not curved strongly upward. Barbel is short. Both head and body are black with green, blue or bronze iridescence. There is no pale loop or line on its flanks. It attains a maximum length of 20.5 cm.

Luminous organs and bioluminescence: In this species, a maximum of 28 photophores are present in ventral row between pectoral and pelvic fin insertions (PV). Its blue light-emitting post-orbital organ is moderately large in males and small in females (http://species-identification.org/species.php?species_group=fnam&id=1327). Its chin barbel is short. Stem of barbel is black or dark brown. While bulb's basal portion is black, its terminal luminous portion and filaments are yellowish.

3. Malacosteus australis

Common name(s): Southern stoplight loosejaw.

Global distribution: It has got circumglobal distribution.

Ecology: It is a mesopelagic species inhabiting at depths more than 500 m. This species does not regularly migrate on a diel basis out of its living zone.

Biology: It has a long body with large eyes. Snout is blunt and short with a single round nostril. Smaller jaws and a fleshy orbit are seen. Barbed, curved, knife-shaped teeth are seen in the jaws. It grows to a maximum length of 253 cm. In both the species of *Malacosteus,* a barbel which is luminous at the end is present (Haneda, 1950).

4. Malacosteus niger *(= Malacosteus choristodactylus; Malacosteus indicus; Malacosteus danae)*

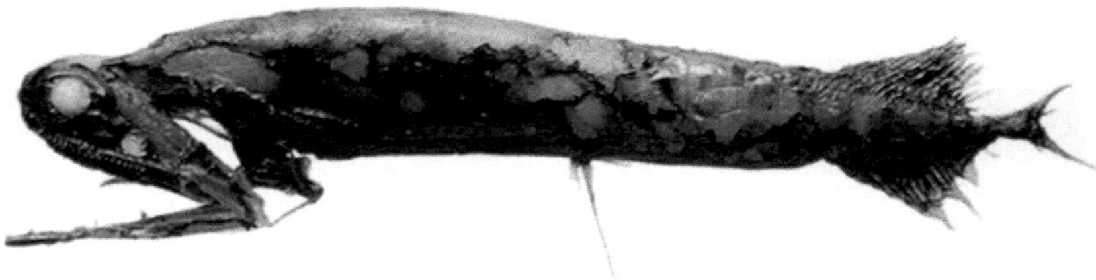

Image credit: Keiichi Matsuura. https://creativecommons.org/licenses/by-sa/3.0/deed.en

Common name(s): Black loosejaw.

Global distribution: Atlantic Ocean and Indo-West Pacific Ocean.

Ecology: This is a mesopelagic – bathypelagic species.

Biology: The body of the fish is elongate and compressed. Its jaws are considerably longer than the skull. Head and body are in black coloration.

Luminous organs and bioluminescence: Malacosteus spp.

These loosejaw dragonfishes have accessory cephalic photophores (AOs) which emit far-red light with long-wave length emissions (>650 nm) (Kenaley, 2010). Anon. (https://bie.ala.org.au/species/37110003) reported that in addition to its numerous blue light producing small body photophores, large red light emitting suborbital photophores are present in this species. In *Malacosteus australis*, the lateral photophores are few and its blue-light emitting postorbital photophore which is present in both sexes is small (Anon., https://bie.ala.org.au/species/37110003). Shimomura (2012) reported that the wavelengths of luminescence emission maxima of *Malacosteus niger* were at 469 nm and 702 nm for the blue and red light emitted from the postorbital light organ and suborbital light organ, respectively

*5. **Pachystomias microdon*** (= *Pachystomias atlanticus*)

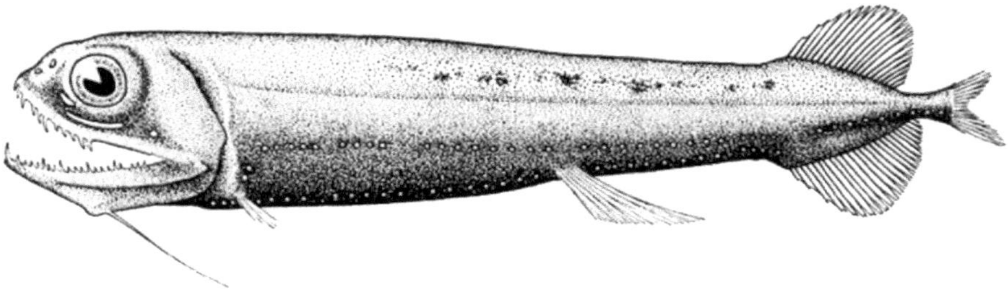

Image credit: Museums Victoria, https://fishesofaustralia.net.au/home/species/4434; https://creativecommons.org/licenses/by/3.0/au/

Common name(s): Smalltooth dragonfish.

Global distribution: Eastern and Western Atlantic Ocean and Indian and Pacific oceans.

Ecology: This bathypelagic; species has a depth range of 660–4000 m.

Biology: Body of this species is moderately elongated and slightly compressed. Snout is shorter than eye. Jaws are slender and the gape is slightly curved upward. Chin barbel is slender and is tapering without a bulb. The body coloration is velvety dark brown to black. It attains a maximum length of 22.1 cm.

Luminous organs and bioluminescence: This species has a large curved reddish photophore under its eye. As the species of the genera *Malacosteus* and *Aristostomias*, this small tooth dragonfish also produces red light with long-wavelengths. It is believed that these fishes shine this red light on prey that are unable to see long-wave spectra. Unlike all other deep-sea fishes, these fishes have retinal pigments tuned to their own red light-like having "night-vision" (https://fishesofaustralia.net.au/home/species/4434). Herring and Cope (2005) reported that this species has a red-emitting preorbital photophore additionally

6. *Astronesthes* spp.

Common name(s): Snaggletooth fishes.

Global distribution: It is found distributed in most temperate and tropical seas.

Ecology: This is a mesopelagic or bathypelagic species.

Biology: The body of these species is moderately sized and is with small eyes. Mouth is large with large, fanglike teeth. Chin barbel is present with its terminal bulb which is of variable length. Caudal fin is forked. The body coloration is black but silvery pigmentation may be present on its flank. Diets of astronesthids consist of other mesopelagic fishes and crustaceans.

1

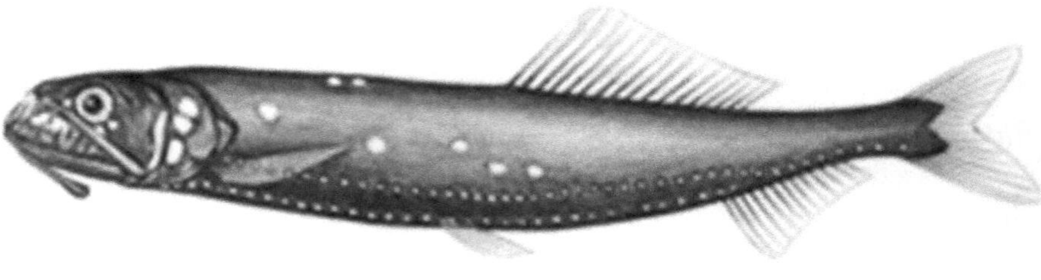

2

3

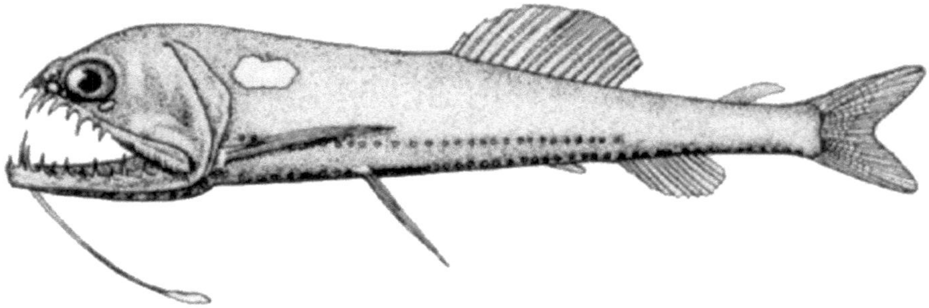

4

Credits: 1: Public domain; 2: Image credit: Shao, Kwang-Tsao, Taiwan Fish Database; 3: Image credit: Keiichi Matsuura. https://creativecommons.org/licenses/by-sa/3.0/; ; 4: Public domain.

1. *Astronesthes splendida*; 2. *Astronesthes formosana*; 3. *Astronesthes macropogon*; 4. *Astronesthes niger*

Luminous organs in Astronesthes *spp*: In the blue-light emitting species of the genus *Astronesthes,* suborbital and postorbital photophores are seen. Patches of luminous tissue are present on the gill cover or body in some species. Ventral body has two rows of photophores and numerous small photophores are found scattered on the head and body (Liao et al., 2006). The distribution of photophores in nine luminescent species of this genus is e presented here.

Photophores in the different species of Astronesthes *(Liao et al., 2006).*

Astronesthes trifibulata: Photophores: IP 10 or 11, PV 16 or 17, VAV 20–24, AC 11, IC 58–62, OA 37–41.

Astronesthes splendida-:Photophores: IP 10, PV 16 or 17, VAV 19–21, AC 11 or 12, IC 56–59, OA 36 or 37

Astronesthes lucifer: Photophores: IP 10, PV 19 or 20, VAV 21–24, AC 11 or 12, IC 61–65, OA 38–40

Astronesthes indica: Photophores: IP 5, PV 6, VAV 7–9, AC 8 or 9, IC 26–28, OA 12

Astronesthes chrysophekadion: Photophores: IP 10; PV 17–19; VAV 21–23; AC 5+2+4 (rarely 5) = 11 or 12; IC 59–63; OA 37–39.

Astronesthes indopacifica: IP 9 or 10, PV 13 or 14, VAV 17–21, AC 9–11, IC 50–54, OA 31–34.

Astronesthes formosana: IP 8–10, PV 12–15, VAV 17–21, AC 10–12, IC 49–54, OA 31–34.

Astronesthes macropogon: IP 9 or 10, PV 13 or 14, VAV 17–19, AC 10 or 11, IC 51–53, OA 32–34.

Astronesthes lamellosus: IP 9, PV 12 or 13, VAV 17 or 18, AC 9–11, IC 48–50, OA 29–31.

Explanation for the abbreviations- IP: ventral row of photophores from anterior of the isthmus to a ventral line at the pectoral fin origin; PV: ventral row of photophores between vertical lines at the origins of the pectoral and pelvic fins; VAV: ventral row of photophores between a vertical line at the origins of the pelvic and anal fins; AC: posterior part of the IC series, from posterior of the VAV series to the posterior part of the caudal peduncle; IC: entire ventral row of photophores from the anterior end of the isthmus to posterior of the caudal peduncle; VAL: from above pelvic insertion to end of row, usually before anal origin, but occasionally slightly behind it; OA: all large photophores of the lateral series.

7. Borostomias abyssorum

Common name(s): Not designated.

Global distribution: Northeast Atlantic Ocean.

Ecology: It is a pelagic-oceanic species.

Biology: It grows to a maximum length of 7.5 cm

8. Borostomias antarcticus

Common name(s): Snaggletooth.

Global distribution: Western Mediterranean and Atlantic Ocean.

Ecology: This bathydemersal and oceanodromous species has a depth range of 300–2630 m.

Biology: This species which grows to a maximum length of more than 35 cm feeds mainly on midwater fishes and crustaceans. Head and body coloration is black.

9. Borostomias elucens

Common name(s): Not designated.

Global distribution: Eastern and Western Atlantic ocean and Indian and Pacific oceans.

Ecology: This bathypelagic species has a depth range of 500–2500 m.

Biology: The body of this species is moderately slender and elongated. Its serial photophores are not broken in small groups. The body coloration is black. It grows to a maximum length of 35 cm and it feeds on midwater fishes and crustaceans.

10. Borostomias mononema

Common name(s): Sickle snaggletooth.

Global distribution: Eastern and Western Atlantic Ocean, Western and NW Pacific Ocean, and Indian Ocean.

Ecology: This bathypelagic species has been reported to live in waters up to a depth of 658 m.

Biology: The body of this species is moderately elongated and slender. It has enlarged fanglike teeth in its jaws. It has a moderate chin barbel with a black bulb at its proximal part and distal part is narrow and pale. It grows to a maximum length of 31 cm and feeds on midwater fishes and crustaceans.

11. Borostomias pacificus

Common name(s): Not designated.

Global distribution: It is known only from the Suruga Bay (Japan) in the Northwest Pacific Ocean.

Ecology: This bathypelagic species has a depth range of 350–1400 m.

Biology: It has its characteristic chin barbel which is long and is about two times as head length. Stem of the barbel is sparsely pigmented and is ending in a black swollen part which is with a prominent photophore on each side. This swollen part is continuous with a large ovate white bulb with a long filament.

12. Borostomias panamensis

Common name(s): Panama snaggletooth.

Global distribution: Eastern Pacific Ocean.

Ecology: This bathypelagic and non-migratory has a depth range of 500–1800 m.

Biology: It grows to a maximum length of 30 cm and feeds mainly on crustaceans.

Borostomias elucens

Image credit: Keiichi Matsuura

Borostomias mononema

Image credit: Keiichi Matsuura.

Luminous organs and bioluminescence: Borostomias spp.

The species of this genus have many rows of light organs (photophores) along the lower sides of their body. Further, these fishes possess a bioluminescent lure that dangles at the end of a short rod (a barbel) attached to the chin. The distribution of photophores and bioluminescence in the different species of this genus are given below.

Borostomias mononema: It has a row of photophores (AC) which begin above the anal fin and end on the tail. There is a maximum of 49 serial photophores in lateral row (OA) of which about 25 photophores are located behind pelvic fin insertion (VAL). Further, there is a maximum of 72 photophores in ventral series to tail (IC). (http://species-identification.org/species.php?species_group=fnam&id=1296).

Borostomias elucens: The number of photophores distributed in the different parts of the body of this species has been reported as: IP 11; PV 23; VAV 15; AC 11; IC 61; and OA 38. While the AC photophores are found greatly arched behind the anal fin base, the serial photophores are not found broken into small groups (Liao et al., 2006; https://www.fishbase.se/summary/5130). Herring (1983) reported its wavelengths of emission maxima as 470 and 475 nm.

Borostomias antarcticus: The row of photophores which begin above the anal fin and end on the tail (AC) of this species are found curved upward behind the anal fin base. There is a maximum of 46 serial photophores in lateral row (OA) of which a maximum of 22 photophores is seen behind the pelvic fin insertion (VAL). Furter, there is a maximum of 68 photophores in ventral series to tail (IC) (http://species-identification.org/species.php?species_group=fnam&id=1294).

Borostomias abyssorum: In this species, the postorbital luminous organ is large. (https://www.fishbase.se/summary/57903).

Borostomias panamensis: Widder et al. (1983) reported on its emission maximum wavelength of this species as 484 nm.

13. Chauliodus spp.

Common name(s): Viperfish.

Global distribution: It is found distributed in tropical and temperate waters of the world oceans.

Ecology: This mesopelagic and bathypelagic species has a depth range of 1000–5000 m.

Biology: This species is characterized by the presence of long, needle-like teeth and hinged lower jaws. These fishes vary in their coloration from green, silver, to black. They prey on small pelagic organisms such as shrimp, small fish, and squid and grow to a maximum length of 60 cm.

Chauliodus macouni

Source: Public domain

Chauliodus sloani

Source: Public domain

Luminous organs and bioluminescence: Chauliodus spp.

The photophores of viperfish may be of three types viz. microscopic spheres without a pigment layer; large spheres with a pigment coat, reflector ad lens;

and large bell-shaped organs resembling large spheres. While the first type of photophores is found scattered over the dorsal side of the fish, the second and third type light organs are normally grouped together in rows along the dorsal surface of the fish. Photophores may also be seen along the ventral and lateral surface of the fish in some species (Haffner, 1952; Anon., https://en.wikipedia.org/wiki/Viperfish). The distribution of photophores observed in the different species of *Chauliodus* is given below.

Chauliodus macouni: In this species, the photophores are found distributed on the dorsal fin, on the side and underneath its body (Anon., https://en.wikipedia.org/wiki/Pacific_viperfish).

Chauliodus sloani: It has light organs in its greatly elongated first dorsal-fin ray. Further, in its ventral part of the body, it has two rows of photophores one on each side (Anon., https://bie.ala.org.au/species/urn:lsid:biodiversity.org.au:afd.taxon:a48dd410-bde2–42cd-9f38–2f5272e38625). Davis et al. (2020) reported that the photophores of this species use guanine reflectors to match the angular distribution of downwelling light. Haneda(1950) observed that this fish has numerous luminous dots in the underlying skin apart from their regular series of luminous organs. Crescitelli (2013) reported on the presence of photophores in the mouth of the fishes of this species.

Chauliodus danae: A lateral photophore series is present below the fifth row of scale pockets (Llano, 1965).

Chauliodus minimus: In this species, dorsal fin-fin origin is over 5th–8th OV photophore. Further, its barbel is enlarged at its end (FAO, 2020).

Chauliodus schmidti: Total photophores of the ventral row (IC, from the anterior end of the isthmus to the posterior of the caudal peduncle) has been reported be 58–63 in this species.

14. Echiostoma barbetum

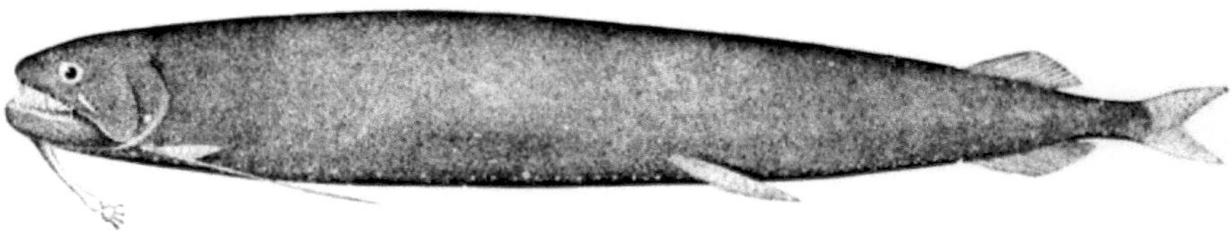

Source: Public domain

Common name(s): Threadfin dragonfish.

Global distribution: It is widespread through the tropical to temperate waters in all oceans.

Ecology: It inhabits mid to deep waters up to 2,000 m.

Biology: The body of this species is moderately long and its depth is less than 10 times in standard length. Head is with short, blunt, rounded snout. The lower jaw is not longer than the upper jaw and is not curved strongly upward. Its chin barbel is shorter than head. Head and body coloration is black. There are no pale loops or lines on its flanks. It grows to a size of 37 cm in length.

Luminous species: It is a monotypic genus (i.e., only one species in this genus).

Luminous organs and bioluminescence: In this species, the photophores in ventral row between the pectoral fin and pelvic fin insertions (PV) are ranging from 25 to 28. Its post-orbital luminous organ is large and is more or less triangular in shape. There is no pre-orbital or sub-orbital luminous organ. Similarly, there are no large patches of luminous material on its head behind the eye (Anon., http://species-identification.org/species.php?species_group=fnam&id=1328). The color of light produced by this species is yellow (Anon., https://www.scribd.com/document/93770802/pramodsingh). However, Anon (https://fishesofaustralia.net.au/home/species/1780) reported that these photophores emit blue and pinkish light. Anon. (https://www.sciencesource.com/archive/Threadfin-Dragonfish--Echiostoma-barbatum-SS22345173.html) reported the photophores present in the face of this species emit red light. Yellow lens have also been identified in the photophores of *Echiostoma* spp. producing red bioluminescence (Anon., https://en.wikipedia.org/wiki/Malacosteus_niger).

In addition to body photophores, a short bioluminescent barbel or tentacle is present beneath the lower jaw and a large triangular luminous organ is located behind its eye (Anon., https://fishesofaustralia.net.au/home/species/1780). Its bioluminescent lure that dangles at the end of its chin barbel helps to attract its prey. The massive teeth grip the prey items which are drawn to the lure. The red light of this species allows it to see red prey items, like shrimps. Though red is one of the first colors which drops out of the light spectrum at depths, in deep waters, this red turns to black and is invisible to all, except those fishes that can both produce and see red light. (Anon., https://www.sciencesource.com/archive/Threadfin-Dragonfish–Echiostoma-barbatum-SS22345173.html).

15. Eustomias spp.

Common name(s): Barbeled dragonfishes.

Global distribution: It is found distributed in the tropical and subtropical waters of all oceans and is very common in the Atlantic Ocean and adjacent seas; and the Pacific Ocean.

Ecology: This mesopelagic species has been reported to undertake diel vertical migrations between depths more than 300–400 m during the day time and in waters of less than 200–250 m at night.

Biology: The body of this species is strongly elongated. Dorsal and ventral outlines are almost straight at the abdomen, and are gradually becoming shallow at the tail. The body and head are fairly compressed laterally. Head is small and the snout is moderately pointed. The eye is moderately large and circular. Mouth is large and is weakly oblique. Chin barbels are present Body and head coloration is uniformly black. It is a predator preying on other small fishes. Barbel has a slender stem which ends with one to three small terminal bulbs which are with or without terminal filaments or projections.

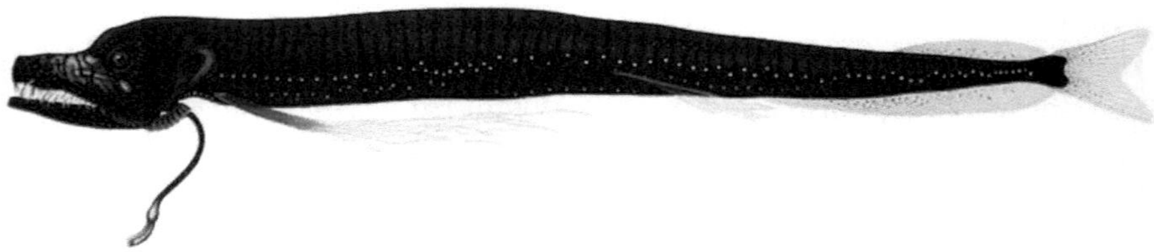

Eustomias braueri

Source: Public domain

Eustomias obscurus

Source: Public domain

Luminous species: It is interesting that all the 117 recognized species of this genus are luminescent (https://en.wikipedia.org/wiki/Eustomias).

Luminous organs and bioluminescence: The species of this genus are characterized by the presence of a serial photophores in both rows. There is a maximum of 77 Photophores in ventral series (IC); photophores in lateral series (OC) are varying between 64 and 71; and ventral row of photophores between a vertical line at the origin of the pelvic fin and anal fin is between 13 and 19 (https://eol.org/pages/46563535/articles). In addition to these photophores, there is a row of black spots associated with photophores along its barbel stem (Gibbs et al., 1983).

16. Grammatostomias spp.

Common name(s): Barbeled dragonfishes.

Global distribution: Endemic to Atlantic Ocean.

Ecology: These fishes live in extremely deep waters below 1600 m.

Biology: Body of these fishes is moderately long and its depth is less than 10 times in standard length. Head is with a tapering snout of moderate length. The lower jaw is not longer than the upper jaw and is not curved strongly upward. Its chin barbel is slender. It attains a maximum length of 20 cm.

Grammatostomias circularis

Image credit: Texas A&M University at Galveston. https://creativecommons.org/licenses/by/3.0/

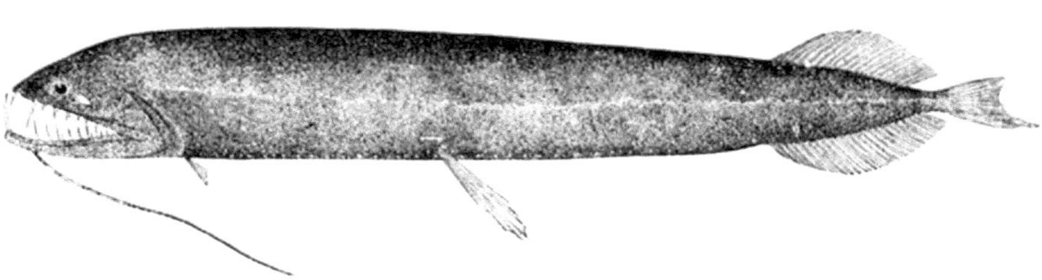

Grammatostomias dentatus

Source: Public domain.

Luminous organs and bioluminescence: Species of *Grammatostomias*

The number of photophores in ventral row between pectoral and pelvic fin insertions (PV) of these fishes is varying between 15 and 18. A pale loop or line present on the flanks. The post-orbital luminous organ is an elongate triangle and is much smaller than, to slightly longer than the eye diameter. There are no large patches of luminous material on head behind eye, but small spots or lines are sometimes present (http://species-identification.org/species.php?species_group=fnam&id=291&menuentry=groepen).

Grammatostomias circularis (= Lamprotoxu circularis): Haneda (1950) reported that this species has a closed loop of luminous tissue on each side of the body between the head and pelvic region in addition to a luminous antenna. Herring (1983) observed its blue light with a wavelength of its emission maximum of 470 nm.

Grammatostomias dentatus: Sides of this species show the presence of long-line luminous tissue from its operculum to beyond pelvic- fin base.

Grammatostomias flagellibarba: It possesses bioluminescent or light-emitting photophores along the sides of its body and in its barbel attached to the chin (http://www.seasky.org/deep-sea/dragonfish.html). Further, this species has an elongated luminous loop which extends posteriorly to the pelvic-fin base.

17. Idiacanthus spp.

Common name(s): Barbeled dragonfishes.

Global distribution: Idiacanthus atlanticus is found distributed in subtropical to temperate habitats of the southern hemisphere; *Idiacanthus fasciola* is more widely distributed, in both North and South Atlantic Ocean as well as the Indo-Pacific; and *Idiacanthus antrostomus* is common in the Eastern Pacific Ocean.

Ecology: All these species inhabit from the mesopelagic zone to the bathy-pelagic zone.

Biology: In these eel-like fishes, the body is highly elongated with a very long dorsal fin.

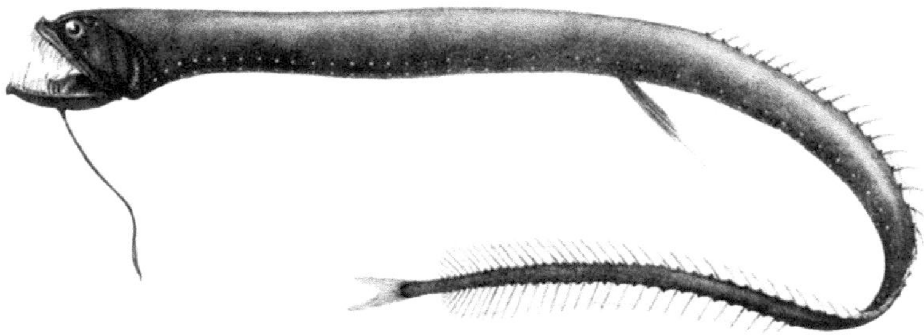

Idiacanthus atlanticus.

Source: Public domain

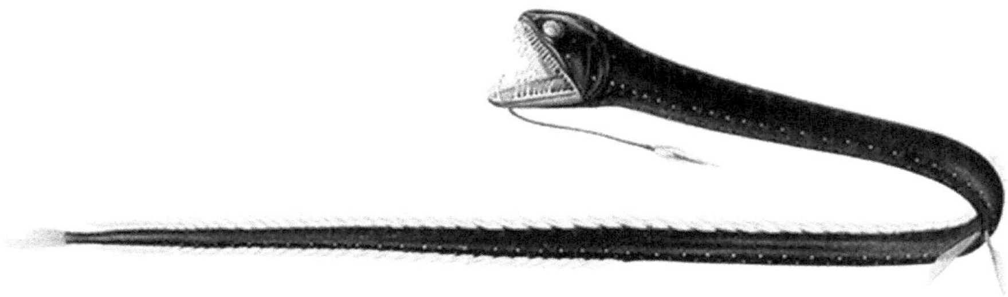

Idiacanthus Fasciola.

Image credit: Museums Victoria, https://fishesofaustralia.net.au/home/species/1811; https://creativecommons.org/licenses/by/3.0/au/

Luminous organs and bioluminescence: Idiacanthus spp.

The luminescent species of *Idiacanthus* possess a luminous spherical organ that hangs off their lower jaw as a thin barbel. A single strip of chevron-shaped spots of luminescent tissue is seen all along the dorsal and ventral sides of the body. Similar tissue is also seen radiating along their fins (Anon., https://en.wikipedia.org/wiki/Idiacanthus). The luminous organs of the different species of Idiacanthus are given below.

Idiacanthus atlanticus: The long slender black female fishes of this species have a long chin barbel with a bioluminescent tip. Further, two rows of photophores are also seen along each lower side (Anon., https://fishesofaustralia.net.au/home/species/1810).

Idiacanthus antrostomus: In this species, several rows of photophores are seen in its body. In addition to these photophores, a separate light organ is seen at the end of its long chin barbel (Anon., https://oceana.org/marine-life/ocean-fishes/pacific-blackdragon). Widder et al. (1983) reported that its wavelengths of light emission maxima as 471–472 nm.

Idiacanthus fasciola.: As in *Idiacanthus antrostomus*, the slender black females of this species have a long chin barbel with a bioluminescent tip. Further, two rows of photophores are seen on both sides that extend along the ventral part of the body. It has 14–16 photophores (VAO) ventrally from the pelvic fin base to the anal fin. Luminous material is also seen on the flanks of the dorsal fin and fin ray (Gibbs, 1964). A post-orbital luminous organ has also been reported in this species (Anon., https://animaldiversity.org/accounts/Idiacanthus_fasciola/).

18. Melanostomias spp.

Common name(s): Black dragonfish.

Global distribution: Atlantic Ocean.

Ecology: These species are found mainly in the mesopelagic and deep sea habitats.

Biology: These are scaleless fish with very sharp and fang-like teeth. The body is moderately long with a short head which has a blunt and rounded snout. The lower jaw is no longer than the upper jaw and is not curved strongly upward. Length of chin barbel is more or less equal to head length.

Melanostomias bartonbeani

Image credit: Andrey Dolgov (Polar Branch of the Russian Federal Institute of Fisheries and Oceanography (PINRO), Murmansk, Russia)

Melanostomias melanops

Source: Public domain

Luminous organs and bioluminescence: Species of *Melanostomias.*

In these fishes, tiny photophores are seen along their body and larger photophores are present below the eyes and in their chin barbels. In the different species of this genus, 23–30 photophores are seen in the ventral row between the pectoral and pelvic fin insertions (PV). The post-orbital luminous organ is large and is more or less triangular in shape (Anon., http://species-identification.org/species.php?species_group=fnam&id=293&menuentry=groepen). These dragonfish use their glowing barbel to lure fish and other prey. The dragonfish are also capable of emitting red light in addition to the production of their normal blue-green light, Red light helps the dragonfish to locate their prey in the dark. The luminous organs in the different species of *Melanostomias* are given below.

Melanostomias bartonbeani: In this species, the light organs are arrayed along its belly and it also has luminous bodies embedded in the gelatinous dermis and between fin rays.

Melanostomias biseriatus: In this species, there are 27–29 photophores in ventral row between the pectoral and pelvic fin insertions (PV) (Anon., http://species-identification.org/species.php?species_group=fnam&id=1344).

Melanostomias globulifer: In this species, there is a large, triangular luminous body which is located close above maxillary below and behind the eye. One photophore is also seen at the middle of front eye edge, at the middle of lower eye ridge, and at the middle of the opercle. In addition to these photophores, there are 25 upper lateral series photophores between the pectoral fin and ventral fin. Before the ventral fin, there are 36 lower or ventral series of photophores.

Melanostomias macrophotus: In this species, there are 26–29 photophores in ventral row between the pectoral and pelvic fin insertions (PV) (Anon., http://species-identification.org/species.php?species_group=fnam&id=1345).

Melanostomias margaritifer: In this species, the number of photophores in the different regions of the body are: IP, 8+2; PV, 27–28; VAV, 13 or 14; OV, 27; VAL,13; AC,10 (Bigelow et al., 2018).

Melanostomias melanopogon: In this species, the number of photophores in the different regions of the body are IP, 8+2; PV, 28; VAV,15; OV, 26; VAL,13; AC,9 (Bigelow et al., 2018).

Melanostomias melanops: In this species, the number of photophores in the different regions of the body are IP, 8+3; PV, 28; VAV,13 or 14; OV, 27 or 28; VAL,11 or 12; AC,9 or 10 (Bigelow et al., 2018).

Melanostomias niger; In this species, the number of photophores in the different regions of the body are D 15–17, A 17–20, and P 5. Photophores in ventral series: 59–64 (IP 8–11, PV 26–28, VAV 14–1`6, AC 9–11); and Photophores in lateral series: 40–42 (Ov 26–27, VAL 13–15) Smith and Heemstra (2012).

[VAV – ventrally, from pelvic fin base to anal fin origin; AC – from anal fin origin to caudal fin base; OV – laterally, from pectoral fin origin to pelvic fin origin; VAL – laterally, from pelvic fin origin to anal fin origin; PV-: ventral row of photophores between vertical lines at the origins of the pectoral and pelvic fins; IP – along isthmus to pectoral fin base commencing at mandibular symphysis or posteriorly halfway along isthmus length, respectively.]

Melanostomias paucilaternatus: In this species, its chin barbel has its terminal bulb with two ovoid luminous bodies. Stem of barbel is black and the barbel tip is flattened and pale in coloration. Further, two rows of distinctly separated photophores are seen along the body. In this species the number of photophores in the different regions of the body are D 13–16, A 16–18, and P 5. Photophores in ventral series, 57–59; Photophores in lateral series, 36–38 (Smith and Heemstra, 2012).

Melanostomias pauciradius: In this species, the luminescent barbel is present on its lower jaw. The barbel 's terminal bulb has ovoid luminous bodies. Further, two rows of distinctly separated photophores are seen along its body.

Melanostomias pollicifer: In this species, the barbel is present on its lower jaw and the terminal bulb of the barbel has a slender luminous organ. Two rows of distinctly separated photophores are also seen along its body.

Melanostomias stewarti: In this species, large, triangular luminous body is seen below and behind the eye. A distinct photophore is seen at middle of hind preopercle flange. There are 24 upper lateral series of photophores between the pectoral and ventral; and 36 lower or ventral series of photophores.

Melanostomias tentaculatus: In this species, there are two rows of distinctly separated photophores along the body.

Melanostomia valdiviae: In this species, there are 24–28 photophores in ventral row between pectoral and pelvic fin insertions (PV) (Anon., http://species-identification.org/species.php?species_group=fnam&id=1348).

Melanostomia vierecki: In this species, a slender triangular luminous body close above maxillary. A small photophore is seen at the middle of the lower eye ridge and on the middle of the opercle. There are 25 lateral series photophores between the pectoral and ventral fins, 9 photophores between the ventral fins and anal, and 3 photophores above the front of anal. In the lower or ventral series, there are 8 photophores on the isthmus, then 28 photophores to ventrals, 9 photophores between the ventrals and anal and 13 photophores from the front of anal fin to caudal fin.

19. Opostomias micripnus

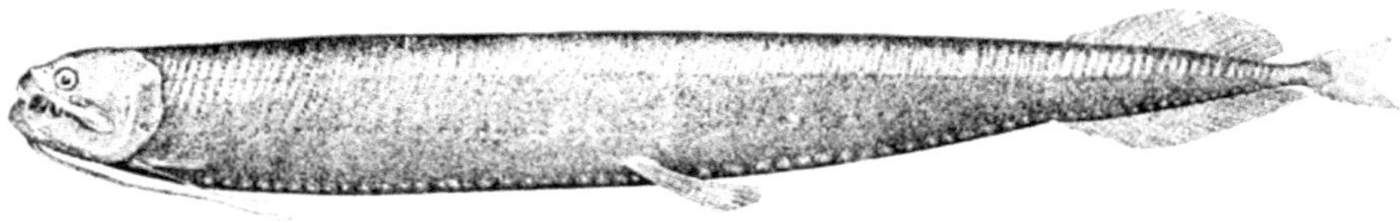

Source: Public domain

Common name(s): Obese dragonfish.

Global distribution: Atlantic, Indian and Pacific Oceans.

Ecology: This bathypelagic fish occurs up to a depth of 5000 m,

Biology: It is a scaleless, black bodied species with large mouth with long fang-like teeth, the anterior-most in the lower jaw piercing the upper jaw. It has a chin barbel, photophores (light-producing organs) in rows along the body and a photophore behind and below both eyes. The dorsal fin is similar to the anal fin and located directly above it on the dorsal margin. It grows to a maximum length of 55 cm and this species is the largest of the family Stomiidae.

Luminous species: Both the species of this genus viz. *Opostomias micripnus* and *Opostomias mitsuii* are luminescent.

20. Opostomias mitsuii

Common name(s): Pitgum lanternfish.

Global distribution: Northeast and Northwest Pacific Ocean.

Ecology: This mesopelagic–bathypelagic species has a depth range of 60–1366 m.

Biology: It grows to a maximum length of 36 cm.

Luminous organs and bioluminescence: Opostomias spp.: In the species of this genus, the light producing organs which are arranged in transverse bands on the body are in the form of sub-organs and are of large size. (Anon., https://www.scribd.com/document/93770802/pramodsingh). Further, a photophore is also seen behind and below both eyes. The distribution of photophores in the species of this genus as follows. IP, 8; PV,27; VAV,17; AC, 16; OV,27; and VA, 17.

21. Photonectes braueri

Image credit: Museums Victoria, https://fishesofaustralia.net.au/home/species/4436; https://creativecommons.org/licenses/by/3.0/au/

Common name(s): Brauer's dragonfish.

Global distribution: Eastern and Western Atlantic Ocean, Western Pacific Ocean and Western Indian Ocean.

Ecology: This mesopelagic- bathypelagic species has a maximum depth of 1000 m.

Biology: The species of Photonectes is separated from the other stomiid fishes by the following characters: scaleless body and the protruded, curved lower jaw which is bearing a single row of teeth. The body of *Photonectes braueri* is elongated and slender. Snout is short, blunt and non-protrusible. Its chin barbel stem is black with an unpigmented bulb. It grows to a maximum length of 28.2 cm.

Luminous organs and bioluminescence: In this species there are 22–24 photophores in ventral row between pectoral and pelvic fin insertions (PV) (Anon., http://species-identification.org/species.php?species_group=fnam&id=1350).

22. Photonectes gracilis

Image credit: Museums Victoria, https://fishesofaustralia.net.au/home/genus/1154; https://creativecommons.org/licenses/by/3.0/au/

Common name(s): Graceful dragonfish.

Global distribution: Eastern Indian Ocean and Western Central Pacific Ocean; and Western Central Atlantic Ocean.

Ecology: This meso- and bathypelagic species occurs up to a depth of 850 m.

Biology: It grows to a maximum length of 20.8 cm.

Luminous organs and bioluminescence: In this species, its chin barbel which is not exceeding eye in length has a thin unpigmented terminal filament splitting on tip into three short filaments. The barbel lacks the bulb. A solid longitudinal band of blue luminous tissue is present on its flanks between the gill opening and caudal-fin base.

23. Photonectes mirabilis

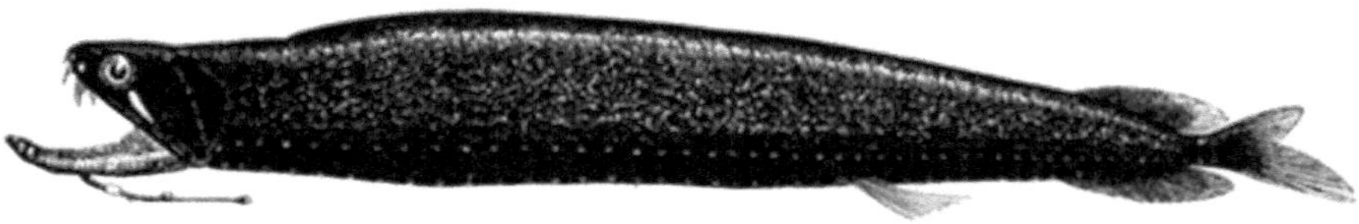

Image credit: Museums Victoria, https://fishesofaustralia.net.au/home/genus/1154; https://creativecommons.org/licenses/by/3.0/au/

Common name(s): Blueband dragonfish.

Global distribution: Eastern and Western Atlantic Ocean and Western Central Pacific Ocean.

Ecology: This meso- and bathypelagic species is oceanic.

Biology: The body of this species is slender and is slightly compressed. Its chin barbel has a posterior branch between the bulb and base. It grows to a maximum length of 16.4 cm.

Luminous organs and bioluminescence: In this species, the number of photophores in the body is as follows: VAV- 11–12; VAL- 10–12; IV 33–34. Blue luminous tissue is seen on the body (Klepadlo, 2011).

24. Photonectes parvimanus

Common name(s): Fleshyfin dragonfish.

Global distribution: Eastern and Western Atlantic Ocean and Indo-west-central Pacific Ocean.

Ecology: This bathypelagic species occurs up to a depth of 1463 m.

Biology: Its chin barbel has a round to ovoid terminal bulb bearing a distal appendage that is curved or straight. This appendage may be longer than bulb and is tapering to a slender point distally, with a few filaments along the appendage. Coloration of head and body is black. One or more small white spots are present on the upper side ('shoulder') behind the head. It grows to a maximum length of 25 cm.

Luminous organs and bioluminescence: In this species, there are more than 30 photophores in ventral row between the pectoral and pelvic fin insertions (PV) (Anon., http://species-identification.org/species.php?species_group=fnam&id=1352).

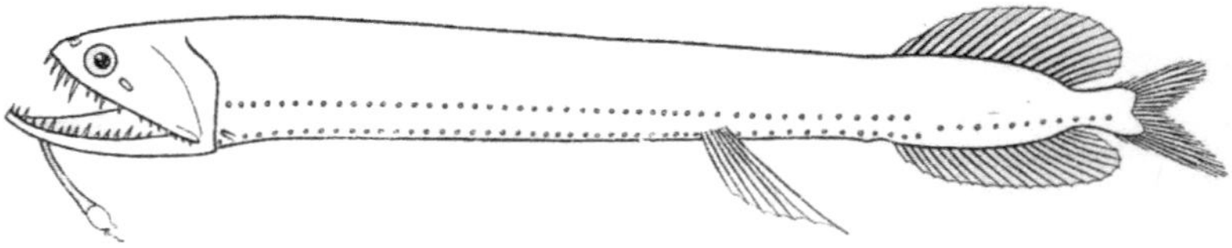

Distribution of photophores in *Photonectes parvimanus*.

Image credit: Museums Victoria, https://fishesofaustralia.net.au/home/genus/1154; https://creativecommons.org/licenses/by/3.0/au/

25. Photonectes achirus

Common name(s): Not assigned.

Global distribution: Northwest Atlantic Ocean and Western Pacific Ocean.

Ecology: Meso- and bathypelagic fish has a depth range of 75–1400 m.

Biology: The body of this species is elongated. Gill filament is very long, feathery and is extending slightly beyond the gill cover. Pectoral fins are absent. Dorsal and anal fins are not covered with black fleshy skin. Pelvic fins are inserted closer to the caudal fin than to the snout tip. The caudal fin is forked and its ventral lobe is longer than dorsal lobe. Coloration of the body is black. It grows to a maximum length of 9 cm.

Luminous organs and bioluminescence: In this species, a blue luminous tissue is present in a band which is extending from pectoral region to pelvic fins This tissue is with short transverse streaks between OV photophores, and in patches on sides of isthmus, under lower jaw and above end of maxillary. Its chin barbel is shorter than head length and is ending in a terminal appendage with a tiny bulb at the tip (Flynn and Klepadlo, 2012).

26. Photonectes paxtoni

Common name(s): Paxton's dragonfish.

Global distribution: It is known only from type locality off Madang, Papua New Guinea.

Ecology: It is shallow water species occurring at depths of 130–135 m.

Biology: It has no pectoral fins. Body of this species is elongated. Opercle is lobate and is slightly concave dorsally. Gill filaments on the first branchial arch are reduced and their length is less than arch depth. Tips of gill filaments are unpigmented. Color of the body is assumed to be black in life. It has a maximum length of 2.4 cm.

Luminous organs and bioluminescence: Its chin barbel is short, with a very enlarged ovoid terminal bulb. Terminal appendages and blue luminous tissue are absent.

27. Photonectes waitti

Common name(s): Not assigned.

Global distribution: Tropical western and central Pacific Ocean and Indian Ocean.

Ecology: It occurs at depths of 0–1200 m.

Biology: The body of this species is elongated and its depth is about eight times into length. This species differs from other species of Photonectes in the following characters: dorsal-fin rays, 11–14; anal-fin rays, 12–16; and length of gill filaments on first branchial arch is reduced and is less than depth of gill arch. Pectoral fins are absent. Dorsal and anal fins not covered with black fleshy skin. Pelvic fins are found inserted closer to the caudal fin than to snout. Chin barbel is shorter than head and its two bulbs are lavender to pink under white light. It has a maximum length of 7.1 cm.

Luminous organs and bioluminescence: A blue luminous tissue in a band is extending from the pectoral region to pelvic fins. Short transverse streaks are seen between OV photophores, and are in patches on sides of isthmus, under lower jaw and above end of maxillary (Anon., https://www.fishbase.se/summary/58267). In freshly caught specimens, these lateral bands have been reported to harbor minute secondary photophores and luminous patches. Flynn and Klepadlo (2012), however, reported on the absence of blue luminous tissue in this species. The number of photophores and their distribution in *Photonectes achirus, Photonectes paxtoni,* and *Photonectes waitti* are given below.

Photophores	Photonectes achirus	Photonectes paxtoni	Photonectes waitti
IP	8–11	11	\`11–13
PV	22–28	21	22–26
VAV	11–13	16	12–15
AC	10–11	13	10–13
OV	19–22	20	21–26
VAL	11–12	14	12–15

VAV – ventrally, from pelvic fin base to anal fin origin; AC – from anal fin origin to caudal fin base; OV – laterally, from pectoral fin origin to pelvic fin origin; VAL – laterally, from pelvic fin origin to anal fin origin; PV – ventral row of photophores between vertical lines at the origins of the pectoral and pelvic fins; IP – along isthmus to pectoral fin base commencing at mandibular symphysis or posteriorly halfway along isthmus length, respectively.

Source: Flynn and Klepadlo (2012).

28. Photostomias spp.

Common name(s): Loosejaw dragonfish.

Global distribution: These species are found distributed in tropical to temperate oceans almost worldwide.

Ecology: These species are found in mesopelagic depths.

Biology: The fishes of this genus possess the following characteristics: lacking skin between the mandibular rami; no hyoid barbel; and absence of pelvin fin after transformation. Further there is considerable sexual dimorphism in the size of postorbital photophores (PO) and accessory orbit.

Photostomias atrox

Source: Public domain

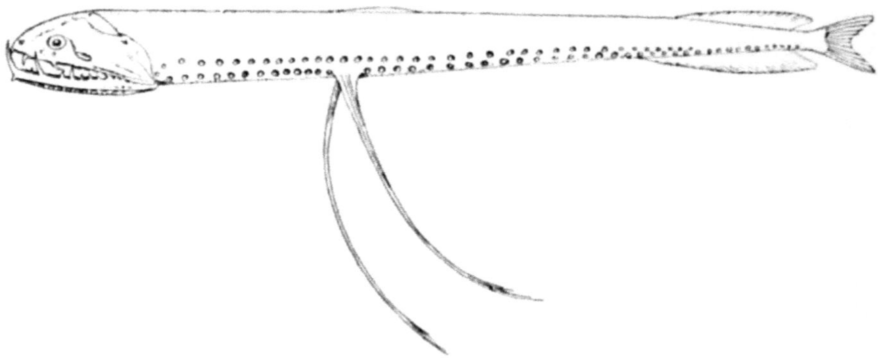

Photostomias guernei

Source: Public domain

Luminous organs and bioluminescence: Photostomias spp.

In the species of this genus, the serial photophores are distinctly seen. From front of the isthmus to tail (IC), there is a maximum of 60 photophores in the ventral row. In the lateral series, there is a maximum of 39 photophores above the anterior end of anal fin (OA). In both these rows, the photophores are evenly spaced. Further, a single row of photophores is present on each side of the isthmus. Pre-orbital luminous organ is present and sub-orbital organ is absent. Post-orbital organ is more or less triangular and is longer than wide (Anon., http://species-identification.org/species.php?species_group=fnam&id=282&menuentry=groepen). Kenaley (2009) reported that the species of *Photostomias* exhibit significant sexual dimorphism in the size of their pre-orbital luminous organ (PO). While the accessory orbital (AO) photophore is smaller in males and it is reaching a greater maximum size in females. The arrangement of photophores in the different species of this genus is given below.

Photostomias atrox: In this species, three cephalic photophores viz. AO, SO, and PO are seen near orbit. While AO is larger in individuals greater than 6 cm, SO is minute, round and is situated on the midventral margin of fleshy orbit in an elongate pocket with a lateral and larger dorsomedial opening. PO is an elongated triangle photophore which is pointing posteriorly (Kenaley, 2009).

Photostomias goodyeari: In both the males and females of this species, the PO is small to moderate (Kenaley, 2009).

Photostomias guernei: Anon. (https://www.pinterest.com/pin/33988172158016758) reported that this species has a "flashlight," which is a built-in bioluminescent system used to see in the dark. Anctil (2018) observed that there are about 1500 tiny light organs in this species. Anon. (https://www.cell.com/current-biology/pdf/S0960–9822(16)00046–4.pdf) observed a continuously emitting blue light in this species with a wavelength of emission maximum value of 530 nm.

Photostomias liemi: In this species, there are three cephalic photophores viz. AO, SO, and PO near its orbit. While the AO is moderately large in specimens greater than 60 mm, the SO is minute, round and is situated on midventral margin of the fleshy orbit in an elongate pocket with a lateral and larger dorsomedial opening. PO is an elongate triangle that is pointing posteriorly (Kenaley, 2009).

Photostomias lucingens: In this species, the three cephalic photophores viz. AO, SO, and PO are located near orbit. AO is sexually dimorphic and is distinct in males. On the other hand, this AO is smaller and subcutaneous in females. Further, the male AO is minute in specimens greater than 60 mm.

SO is minute, round and is situated on the midventral margin of the fleshy orbit in an elongate pocket with a lateral and larger dorsomedial opening. PO is sexually dimorphic and is enormous in males and is moderately large in females as an elongate, triangle-shaped organ pointing posteriorly (Kenaley, 2009).

Photostomias tantillux: In this species, the three cephalic photophores viz. AO, SO, and PO are located near orbit. AO is sexually dimorphic and is distinct in males and is smaller and subcutaneous in females (Kenaley, 2009).

29. Rhadinesthes decimus

Image credit: Museums Victoria, https://fishesofaustralia.net.au/home/species/4817; https://creativecommons.org/licenses/by/3.0/au/

Common name(s): Slender snaggletooth.

Global distribution: Tropical Indian Ocean, the southern Atlantic Ocean, the eastern and western south Pacific Ocean, and the central north Pacific Ocean.

Ecology: This mesopelagic and oceanic fish is living at depths around 500 m during the day and it migrates closer to the surface during the night.

Biology: It is a long and slender bodied fish reaching a maximum length of 41 cm.

Luminous organs and bioluminescence: It has a line of luminous material on its body. This material begins behind the post-orbital organ and is continues along the side of the body for most of its length Its chin barbel has also luminous material at its stem (Anon., https://animaldiversity.org/accounts/Rhadinesthes_decimus/).

30. Stomias spp.

Common name(s): Barbeled dragonfish.

Global distribution: All Oceans.

Ecology: These bathypelagic–mesopelagic species show diel vertical migration.

Biology: The genus *Stomias* includes elongate stomiatoid fishes in which the body length is 10–20 times their depth. There are five to six rows of closely spaced hexagons on their sides. Each hexagon is apparently occupied by a scale. Dorsal and anal fins are set far back toward their caudal fin. The chin barbel of these species is shorter to slightly longer than the head. Sexual dimorphism is quite evident in these species as males are generally smaller with larger eyes and larger postorbital photophores than their counterparts.

Stomias affinis

Image credit: Museums Victoria, https://fishesofaustralia.net.au/home/species/1809; https://creativecommons.org/licenses/by/3.0/au/

Stomias boa

Source: Public domain

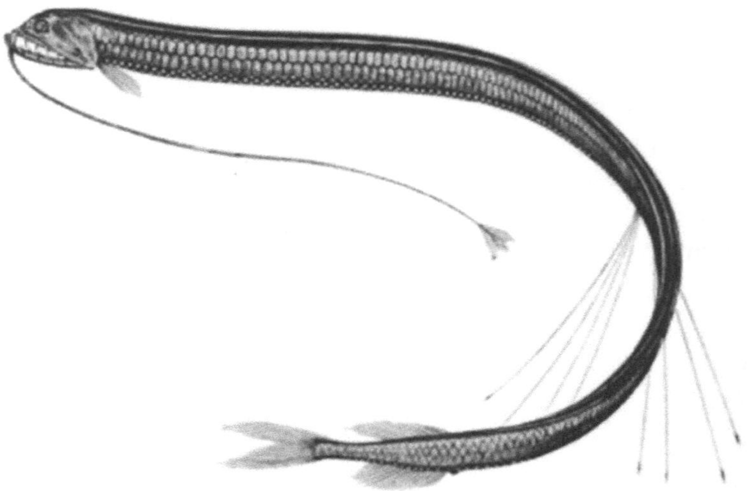

Stomias longibarbatus
Image credit: FAO (Applied for permission)

Luminous organs and bioluminescence: Stomias spp.: In these species, the luminous organs are logged in gelatinous corium of the epidermis (Anon., https://www.yourarticlelibrary.com/fish/anatomy-and-physiology/luminous-organs-or-photophore-of-the-fishes-with-diagram/88411).

Gibbs, Jr (1969) reported that in these species more than 100 photophores (IC) are seen in the ventral series. The arrangement of photophores in the different luminescent species of Stomias is given below.

Stomias affinis: Total photophores in ventral series (IC) 90 or fewer (73–90); lateral series (OA) 64 or fewer; PV 51 or fewer; OV 50 or fewer (Gibbs, Jr., 1969); VAO photophores, 5–8; VAL photophores,4–8.

Stomias atriventer: VAV photophores 10–14 (rarely 9); VAL 10–15; OV photophores 39–43; OA photophores 52–55; IC photophores 77–82 (Gibbs, Jr., 1969). Widder et al. (1983) reported that the wave length of its emission maximum of blue light was in the range of 476–483 nm. It also a possesses a bioluminescent chin barbel.

Stomias boa: OV photophores 43–50; OA photophores 57–64; IC photophores 79–90, usually 83 or more (Gibbs, Jr., 1969); VAO photophores, 10–14; VAL photophores,10–15.

Stomias brevibarbatus: 7 or more photophores in any hexagonal area in the 3 rows above lateral series of photophores; (Gibbs, Jr., 1969). Anon. (https://www.scribd.com/document/93770802/pramodsingh) reported that it produced blue light.

Stomias danae: 6 or fewer photophores in any hexagonal area in the 3 rows above lateral series of photophores; (Gibbs, Jr., 1969).

Stomias gracilis: Total photophores in ventral series (IC) more than 90 (94—102); in lateral series (OA) 65 or more; PV 53 or more; OV 52 or more (Gibbs, Jr., 1969).

Stomias lampropeltis: It has less than 100 IC photophores.

Stomias longibarbatus: It has more than 170 IC photophores.

Stomias nebulosus: Small uninterrupted serial photophores are seen along its body (Tatsuta et al., 2014).

31. Tactostoma macropus

Common name(s): Longfin dragonfish.

Global distribution: Pacific Ocean and Northwest and Eastern Pacific Ocean.

Ecology: It is a bathypelagic species inhabiting at depths of 30–2000 m.

Biology: Coloration of the entire body of this species is Jet black all over except for the dark gray on the posterior part of its lower jaw. However, when the thin epidermis of this species is rubbed off, the dermis shows lead graycoloration. It attains a maximum length of 34 cm.

Luminous organs and bioluminescence: This is the only known luminescent species of this genus. In addition to the presence of photophores under the lateral line and the ventral surface, it has a large photo organ which appears as a large green circle under its eye. (Anon., https://noaateacheratsea.blog/tag/photophores/).

32. Thysanactis dentex

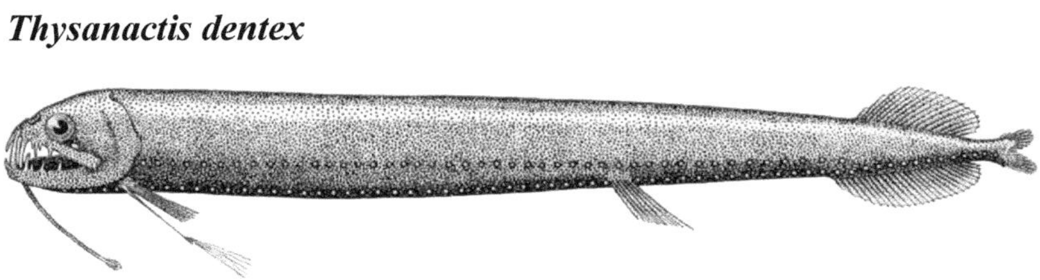

Image credit: Museums Victoria, https://fishesofaustralia.net.au/home/species/1805; https://creativecommons.org/licenses/by/3.0/au/

Common name(s): Broomfin dragonfish.

Global distribution: Pacific Ocean and Eastern and Western Atlantic Ocean.

Ecology: This mesopelagic and bathypelagic species has a depth range of 100–1000 m.

Biology: The body of this species is slender. Its chin barbel is as long as head, and is pigmented with white spots. Tip of the barbel is white. A small spindle-shaped bulb which is a minute and rounded body is seen at the terminal end of the barbel. It attains a maximum length of 18 cm.

Luminous organs and bioluminescence: In this species, the entire chin barbel and the first pectoral fin ray and its filaments are luminescent due to the presence of aggregations of photocytes which are of the same type as those found in the body and eyeball photophores (Jørgensen and Munk, 1979). Pringgenies, et al. (2001) reported that the lateral serial photophores of this fish are surrounded by a thin pigment mantle.

3.1.2.13 ORDER: EUPERCARIA INCERTAE SEDIS (UNCERTAINTY)

3.1.2.13.1 Drums or Croakers (Sciaenidae)

1. Collichthys lucidus

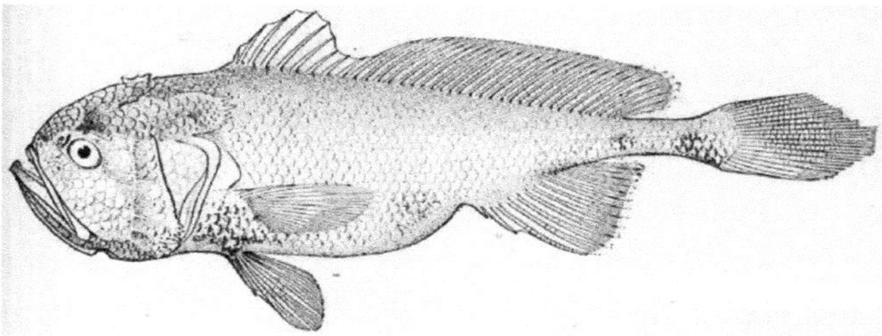

Source: Public domain

Common name(s): Spiny head croaker or big head croaker.

Global distribution: Northwestern Pacific Ocean.

Ecology: This demersal and oceanodromous species is found up to a maximum depth of 90 m. It is also reported from the sandy-muddy bottoms of shore waters and estuaries.

Biology: This species has been reported to exhibit apparent sex dimorphism on the growth rate and the females grow much faster than males. It attains a maximum length of 19.8 cm.

Luminous organs and bioluminescence: In this species, the luminescent organs or photophores are found distributed in its supermaxilla organ, buccal organ, opercular organs, isthmus organs, pectoral organs, antero-median organs, lateral organs, ventral organs and precaudal organs (Haneda, 1961).

2. Larimichthys crocea *(= Collichthyes croceus; Pseudosciaena crocea)*

Image credit: FAO, https://www.fao.org/fishery/en/species/3163

Common name(s): Large yellow croaker or croceine croaker.

Global distribution: Northwest Pacific Ocean

Ecology: This benthopelagic and oceanodromous has a depth range of 0–120 m. It is also found on the muddy-sandy bottoms of coastal waters and brackish waters including estuaries.

Biology: This gold colored fish gors to a maximum length of 80 cm and it feeds mainly on crustaceans and fish.

Luminous organs and bioluminescence: This self luminescent species has numerous small, yellow photophores which are found distributed on the ventral scales of the lower part of the body. The major function of its luminescence is counter-illumination which is helpful to avoid its predators (Suntsov et al., 2008; Anon., http://www.fao.org/fishery/species/3163/en).

3. Sonorolux fluminis

Common name(s): Estuary croaker.

Global distribution: Western Central Pacific Ocean.

Ecology: This benthopelagic and non-migratory species is found in brackish water areas, including estuaries.

Biology: It is a small species with a deep, rounded body. The eye is moderately large. Snout is steeply rounded, but it is not inflated or projecting. Mouth is large and oblique and its upper jaw is extending backward slightly beyond hind margin of eye. Lower jaw is upturned. Chin is prominent without barbel. It grows to a maximum length of 8 cm only.

Luminous species: It is a monotypic genus.

Luminous organs and bioluminescence: In this species, much of its ventral scales is covered with patches of yellow, luminous tissue (Anon., https://www.fishbase.in/summary/Sonorolux-fluminis).

3.2 MARINE FISHES WITH BACTERIAL BIOLUMINESCENCE

3.2.1 ORDER: ACANTHURIFORMES

3.2.1.1 PONYFISHES (LEIOGNATHIDAE)

Indirect light emission type is confined to the fishes of Leiognathidae which are Shallowater fishes differing from other luminous fish in possessing unusually large luminous areas. In fact, these fishes utilize half the muscles of their complicated body structure for this purpose (Haneda, 1950). The light organ system (LOS) of the luminescent leiognathid fishes comprises both internal and external components. Its unique internal light organ system (LOS) comprises a number of distinct anatomical structures in addition to the circumesophageal light organ (ring-like organ around the esophagus/pharynx), which houses a high concentration of luminescent bacteria, *Photobacterium leiognathi* (Vibrionaceae) (Chakrabarty et al., 2011). Each of these organs is made of tubules where the luminous bacteria are solidly packed inside the elongated thin-walled saccules. Each light organ tubule of the light organ is believed to be filled tightly with 1 to 20 saccules. These saccules have been reported to hold the luminous bacteria at a density of about 1 x 10^{11} cells per ml which is a density roughly 15 times greater than the estimated total volume of the light organ (Dunlap, 1984). Many leiognathid species possess strong sexual dimorphism related to their light organ system in which males have a larger light organ and associated features to intensify their light during sexual display (Sparks and Chakraborty, 2015). Males of some species of leiognathids have also been reported to possess transparent patches on their flanks, pectoral, buccal, or gular regions which are mainly used to concentrate light on their bodies (Chakrabarty et al., 2010). Chakraborty et al. (2011) also reported that a caudal luminous organ window oriented into the gas bladder and caudoventral clearing of the silvery gas bladder lining present in all leiognathid taxa and in both sexes aid in ventral counter-illumination.

1. Aurigequula fasciata *(= Leiognathus fasciatus)*

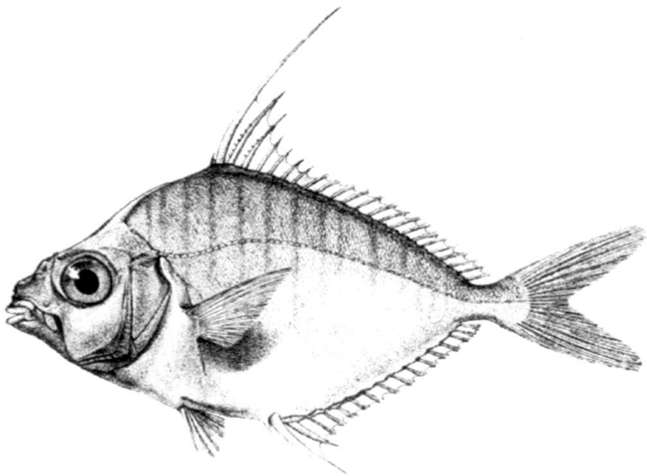

Source: Public domain

Common name(s): Striped ponyfish.

Global distribution: Indo-West Pacific Ocean.

Ecology: This marine and brackish water-demersal species has a depth range of 20–50 m; This coastal and euryhaline species may also enter semi-enclosed sea areas and estuaries and forms schools.

Biology: It is a deep and silvery bodied fish with characteristic vertical dark bars on its upper sides. There is a long filament on dorsal spine. Head is naked without scales but with a nuchal spine. Its protracted mouth is pointing downward. It grows to a maximum length of 21 cm and it feeds mainly on polychaetes and small crustaceans.

Luminous species: It is a luminescent monotypic genus.

Luminous organs and bioluminescence: Dunlap et al. (2007) reported that this species harbors the luminescent bacteria Photobacterium leiognathid in its light organ. The flashing of this species is from the continuous ventral luminescence and emission of its light may be intense or mild (Chakraborty et al., 2011).

2. ***Deveximentum insidiator*** *(= Secutor insidiator)*

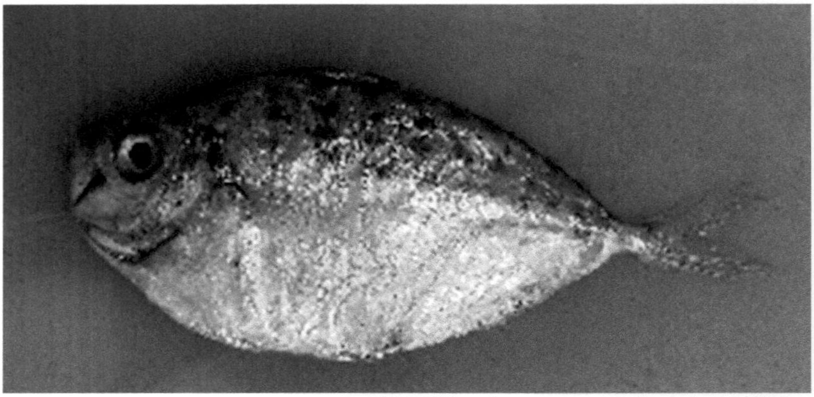

Bla1234l. https://creativecommons.org/licenses/by-sa/4.0/deed.en

Common name(s): Pugnose ponyfish or barred ponyfish.

Global distribution: Indian Ocean and western Pacific Ocean.

Ecology: This demersal and amphidromous species has a depth range of 10–150 m. This can occasionally be found in brackish waters; and are often seen in close to bottom areas.

Biology: It has a bare head with nuchal spines and its body is a distinctive, reflective silver. It has a protracted mouth pointing upward and the tip of the maxilla reaches below the level of the lower margin of the eye. It feeds on zooplankton, including fish larvae and crustaceans and it attains a maximum length of 11.3 cm.

Luminous organs and bioluminescence: Its light organ contains the luminous bacterial species, Photobacterium leiognathid (Tanet et al., 2020). It has a fairly larger luminous gland and its luminosity is greater than that of other members of the family Leiognathidae (Haneda, 1950). The males of this species have a significantly larger luminescent organ than females.

3. ***Equulites*** **spp.**

Common name(s): Ponyfishes or slipmouths.

Global distribution: These species are native to the Indian Ocean and the western Pacific Ocean.

Ecology: These fishes are restricted to shallow, coastal waters of less than 160 m and estuaries.

Biology: Body of these fishes is laterally compressed with highly protractible mouths, which protract either dorsorostrally, rostrally, or ventrorostrally. They are small and silvery fish and are often less than 20 cm in length.

Credits: 1: Bedo. https://creativecommons.org/licenses/by-sa/4.0/; 2: Image courtesy Laith Jawad; 3: Izuzuki. https://creativecommons.org/licenses/by-sa/3.0/; 4. Museums Victoria, https://fishesofaustralia.net.au. https://creativecommons.org/licenses/by/3.0/au/

1. *Equulites elongatus*; 2. *Equulites klunzingeri*; 3. *Equulites rivulatus*; 4. *Equulites moretoniensis*

Luminous organs and bioluminescence: Equulites spp.: Sparks and Chakrabarty (2015; Tanet et al., 2020) reported that the species of Equulites, are sexually dimorphic both internally (e.g., the volume of the light organ is enlarged in males) and externally (i.e., males have expansive translucent external patches, whereas in females, these are absent) with regard to features of the light organs (Chakrabarty et al., 2011). Further, the dorsolateral lobes of the light organ containing the luminous bacteria, *Photobacterium leiognathi* are also heavily pigmented, and they extend posteriorly well into the gas bladder. Sasaki et al. (2003) reported that the light emitted from the lateroventral portion of the body of these species seems to be bright rectangular-shaped luminescence patches which are turned on and off periodically. Its luminescence could be observed in the natural environment during nighttime diving. The variations in the structure of light organs and associated light emission of the different species of the genus *Equulites* are given below.

In the males of *Equulites antongil, Equulites moretoniensis,* and *Equulites stercorarius* a translucent lateral flank stripe is seen. On the other hand in the species, *Equulites absconditus, Equulites elongatus, Equulites laterofenestra,*

Equulites leuciscus, Equulites klunzingeri, and *Equulites rivulatus,* an expansive and species-specific translucent triangular-, polygonal-, or crescent-shaped flank patch is seen (Chakrabarty et al., 2011).

The light organ volume (mm3) of *Equulites rivulatus, Equulites laterofenestra* and *Equulites moretoniensis* was found to be 86,21 and 4 respectively (Chakrabarty et al., 2011).

The males and females of *Equulites rivulatus* harbored different strains of the luminescent bacterium *Photobacterium leiognathi* in June, but in no other month (Chakrabarty et al., 2011).

Nocturnal flashing behavior has been reported in male *Equulites elongatus* and the light was emitted from "the lateroventral portion of its body, as bright rectangular-shaped luminescence patches turned on and off periodically." Females, on the other hand, were found to be nonluminous and were followed by the males, which is believed to be a sort of lekking behavior. Luminescence this species was "bright greenish" with periodic on–off sequences that lasted from one to a few seconds. Males of this species were also reported to produce synchronous on–off flashing patterns (Chakrabarty et al., 2011).

*4. **Eubleekeria splendens*** (= *Leiognathus splendens*)

Source: Public domain

Common name(s): Blacktip ponyfish, black-tipped ponyfish, or splendid ponyfish.

Global distribution: Indo-West Pacific Ocean.

Ecology: This demersal;-amphidromous species has a depth range of 10–100 m. It is highly euryhaline and is seen in marine, brackish water freshwater areas.

Biology: The body of this species is compressed and fairly deep. Head is naked with a nuchal spine. Snout is (shorter than eye diameter. Mouth is pointing slightly downward when it is protracted. Pelvic fins are not reaching to anal fin origin. The body is silvery, and scales of lateral line, bases of pectoral fins, margins of dorsal and anal fins are bright yellow. It feeds mainly on fish, crustaceans, and bivalves and attains a maximum length of 17 cm.

Luminous organs and bioluminescence: The luminous organ of this species has been reported to contain the luminous *Photobacterium leiognathi* ATCC 25521 (Guerrero-Ferreira and Nishiguchi, 2007). Dunlap et al. (2007) reported that both this species and its related species *Eubleekeria jonesi* harbor the luminous bacteria *Photobacterium leiognathi* in their light organs.

5. *Gazza achlamys*

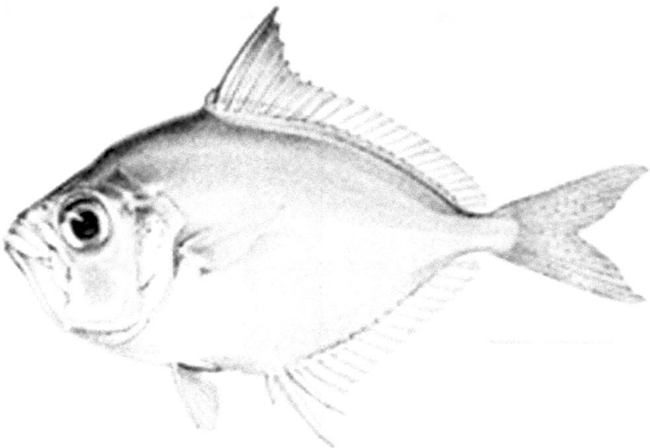

Source: Public domain

Common name(s): Smalltoothed ponyfish.

Global distribution: Indo-West Pacific Ocean.

Ecology: This reef-associated species is found over sandy or muddy bottoms of coastal inshore waters up to a depth of about 20 m. It is also found in brackish water areas and it forms schools.

Biology: The characteristic feature of this species is the membrane of its spinous dorsal fin which is black at its distal portion. It feeds mainly on small fishes, crustaceans and polychaetes and attains a maximum length of 17 cm.

6. Gazza minuta

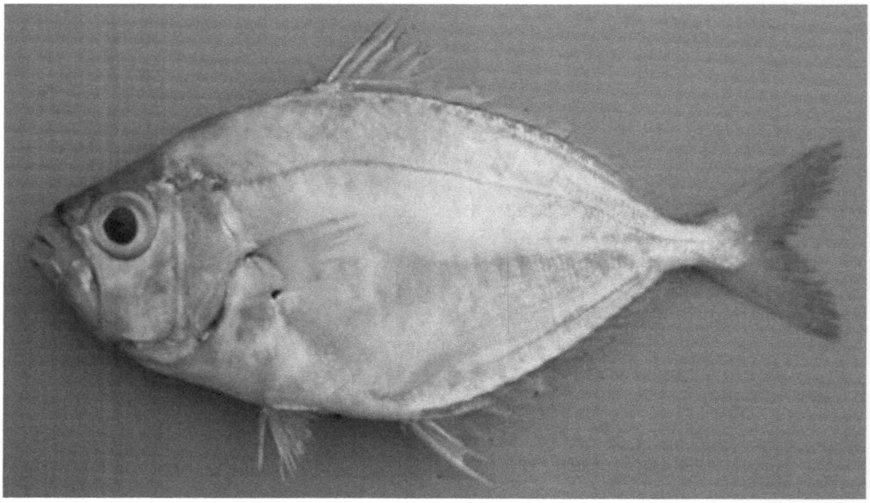

Common name(s): Toothpony.

Global distribution: Indo-Pacific Ocean.

Ecology: This marine, brackish and demersal species has a depth range of 10–110 m and it is often found in shallow inshore coastal waters with silty bottoms.

Biology: The body of this species is oval, moderately deep and laterally compressed. Dorsal and ventral profiles are convex. Mouth is pointing forward when it is protracted Both jaws are with distinct caniniform teeth. Head and breast are naked and scaleless. Head and body are silvery. Dorsally it is greyish or less bright with yellow irregular marks which are extending to below lateral line. Spinous dorsal-fin membrane is black-edged. Soft dorsal fin and anal fin are edged with grey. Anterior part of the anal fin is yellow. Caudal fin is yellowish and its hind margin is dusky. Pectoral and pelvic fins are colorless. Bottom of pectoral-fin base is with black dots. It feeds mainly on small fishes, shrimps, other crustaceans, and polychaetes and it attains a maximum length of 21 cm.

Luminous organs and bioluminescence: Species of *Gazza.*

The circumesophageal light organ (LO) of males of these species is enlarged compared with that of females and in both *Gazza minuta* and *Gazza achlamys*, the luminous organ contains *Photobacterium leiognathi* (Tanet et al., 2020).

Further, these species possess translucent external patches on the head and flank which are responsible for their species-specific luminescence displays (Chakraborty et al., 2011).

Gazza minuta: This species is known for its luminescence diversity as it has three new modes of luminescence viz. discrete projected luminescence (DPL), ventral body flash, and buccal luminescence which are all due to the internal location of its light organ (McFall-Ngai &Dunlap, 1983). Chakrabarty et al. (2011) reported that this species emits continuous, ventral, bluish white luminescence. Further, the bacteria in the luminous organ of this species emit a continuous glow up to several days even after the death of the host fish(Chakrabarty et al., 2011).

Gazza achlamys: In this species, the ventrolateral lobes of its light organs are hypertrophied (Chakrabarty et al., 2011).

*7. **Karalla daura** (= Leiognathus daura)*

Image credit: D.G.R. Wiadnya, https://creativecommons.org/licenses/by/3.0/deed.en

Common name(s): Goldstripe ponyfish.

Global distribution: Indo-West Pacific Ocean.

Ecology: Theis shallow water and demersal species has a depth range of 1–40 m. It seems to be common in predominantly over muddy bottoms and it forms schools.

Biology: It feeds mainly on polychaetes, bivalves, small crustaceans and sponges; and grows to a maximum length of 14 cm.

8. *Karalla dussumieri* (= *Leiognathus dussumieri*)

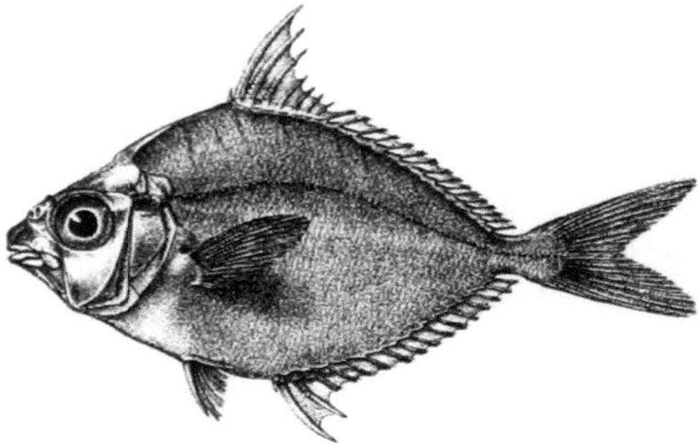

Image credit: Public domain.

Common name(s): Dussumier's ponyfish.

Global distribution: Indo-West Pacific Ocean.

Ecology: This marine and brackish – demersal species has a depth range of 10–40 m. It is often found in coral sand bottoms of coastal waters. It also enters estuaries and forms schools.

Biology: It feeds chiefly on small crustaceans, polychaetes, bivalves, gastropods and nematodes; and it grows to a maximum length of 14 cm.

Luminous organs and bioluminescence: Karalla.

In the species of *Karalla*, the luminous gland consists of a small luminous body and thick ventral cavity walls, which are moderately translucent. Owing to these, the brightness of the luminosity of this species is often found diminishing (Haneda, 1950).

9. *Leiognathus berbis*

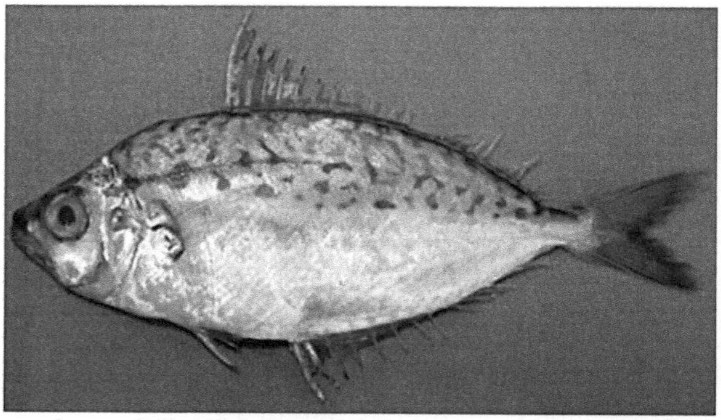

Image credit: Hamid Badar Osmany. https://creativecommons.org/licenses/by/3.0/deed.en

Common name(s): Berber ponyfish.

Global distribution: Indian and Pacific Oceans.

Ecology: This marine; brackish and demersal species has a depth range of 3–40 m. It is often found inhabiting inshore coastal waters.

Biology: It has a compressed body and elongated body, which is more than the depth of the body. Dorsal and ventral sides are equally convex, and mouth is tapering and projecting downward when it is protracted. The dorsal side of the fish is greenish with light gray with dark irregular vermiculations. Belly is gray and the base of anal and caudal fins are light yellowish in color. It feeds on small crustaceans and bivalves; and it grows to a maximum length of 11.0 cm.

10. Leiognathus equula *(=Leiognathus equulus; Leiognathus argenteus)*

Image credit: Hamid Badar Osmany. https://creativecommons.org/licenses/by/3.0/deed.en

Leiognathus berbis Leiognathus equula

Common name(s): Common ponyfish.

Global distribution: Indo-West Pacific Ocean.

Ecology: This marine, freshwater and brackish water species is demersal and amphidromous having a depth range of 10–110 m. It is often found in estuaries, muddy inshore areas and mangrove areas.

Biology: The body of this species is very deep and compressed with a strongly humped back. Mouth is pointing downward when it is protracted. Gill rakers are short and fleshy. Coloration of adults is back greyish with many parallel close-set faint bars. Belly is silvery and a dark brown saddle is seen on caudal peduncle. It feeds mainly on polychaetes, small crustaceans and small fishes and it attains a maximum total length of 28 cm.

Luminous organs and bioluminescence: Leiognathus spp.: The luminescent species of *Leiognathus* possess small luminous glands and each gland consists of small luminous body and thick ventral cavity walls. These luminous glands which are moderately luminous harbor the luminous bacterium *Photobacterium leiognathi* (Haneda, 1950; Barton, 2016).

*11. **Nuchequula nuchalis*** *(= Leiognathus nuchalis)*

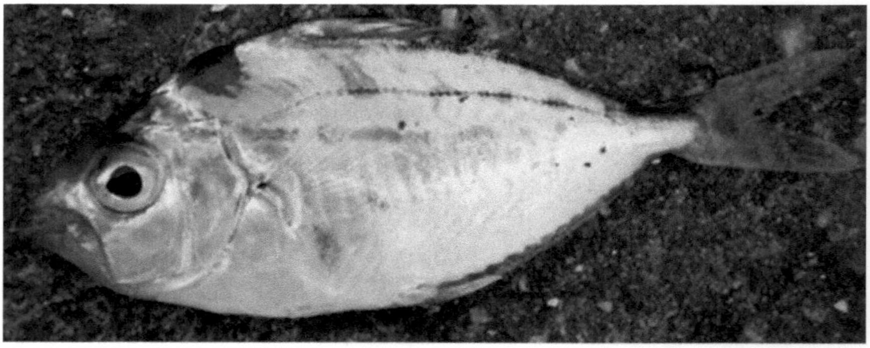

Image credit: Show_ryu. https://creativecommons.org/licenses/by-sa/3.0/deed.en

Common name(s): Spotnape ponyfish.

Global distribution: Northwest Pacific Ocean.

Ecology: This marine and brackish-pelagic species inhabits neritic, shallow coastal waters and estuaries. It is known to aggregate in schools.

Biology: This deep-bodied fish has a dark blotch on its nape with highly reflective silvery areas next to it. It attains a maximum length of 25 cm.

Luminous organs and bioluminescence: This species has a circumoesophageal light organ which ring shaped and is packed with the symbiotic luminous bacteria *Photobacterium leiognathi* inside. The light discharged by the luminous bacteria is emitted in the inside of a swimbladder through a transparent part of the luminous organ. Its light emission is believed to be most common in midnight (Azuma et al., 2005; Barton, 2016).

12. Photolateralis antongil

Common name(s): Not assigned.

Global distribution: Western Indian Ocean.

Ecology: It is a benthopelagic species.

Biology: It grows to a maximum length of only 7.5 cm.

13. Photolateralis moretoniensis

Image credit: Museums Victoria, https://fishesofaustralia.net.au/home/species/2217; https://creativecommons.org/licenses/by/3.0/au/

Common name(s): Moreton Bay ponyfish, Zig zag ponyfish.

Global distribution: Western Pacific Ocean.

Ecology: It is a demersal species with a depth range of 10–105 m. It is known to form schooling which occurs over the soft bottom of the continental shelf.

Biology: It grows to a maximum length of 13 cm.

14. Photolateralis polyfenestrus

Common name(s): Not assigned.

Global distribution: Western Indian Ocean: Oman.

Ecology: It is a reef-associated coastal water species.

Biology: This new species has a shallower body, and its pigmentation pattern which is above the lateral midline comprised of larger rounded blotches. It is characterized by a lower jaw which is deep and convex in lateral view. It grows to a total length of 6.7 cm.

15. Photolateralis stercorarius *(= Leiognathus stercorarius)*

Image credit: D.G.R. Wiadnya. https://creativecommons.org/licenses/by/3.0/deed.en

Common name(s): Oblong slipmouth.

Global distribution: Indo-West Pacific Ocean.

Ecology: This reef-associated species inhabits at a maximum depth of about 20 m. It is commonly found in the inner reef flats and silty coastal waters.

Biology: It is an elongated, shallow-bodied ponyfish growing to a maximum length of 10.2 cm.

Luminous organs and bioluminescence: Photolateralis spp.

All leiognathids possess a circumesophageal light organ that contains symbiotic luminescent bacteria (*Photobacterium* sp.), whose luminescence help these the fish in photic communication and in avoiding predators. The species of the genus *Photolateralis* is unique among leiognathid genera in having a species-specific translucent midlateral flank stripe which may be of either multiple independent translucent windows as in the species *Photolateralis stercorarius, Photolateralis moretoniensis,* and *Photolateralis polyfenestrus)* or a continuous translucent lateral band *(Photolateralis antongil)*. The species *Photolateralis polyfenestrus*, is different from its congeners by the presence of a short, composite midlateral stripe comprised of three small, rounded translucent windows peppered with chromatophores. In the species, *Photolateralis moretoniensis,* and

Photolateralis stercorarius, the translucent mid-lateral flank stripe, is a composite "stripe" comprised of numerous independent translucent windows; and in *Photolateralis antongil*, this translucent stripe is continuous (Sparks and Chakraborty, 2019).

16. Photopectoralis aureus

Image credit: Australian National Fish Collection, CSIRO. https://fishesofaustralia.net.au/home/species/531. https://creativecommons.org/licenses/by/3.0/au/

Common name(s): Golden ponyfish.

Global distribution: Western Pacific Ocean.

Ecology: This marine and brackish-demersal species inhabits coastal waters at a depth range of 70–140 m.

Biology: In this species, the body is rather slender and mouth protrudes forward. The snout is short and the ventral profile is more convex than the dorsal. Chest and nuchal region are asquamate and the remainder of the body is scaled except the head. The upper half of the fish is silvery with irregular grey-brown blotch and marbling. It is a medium-sized ponyfish growing to a maximum length of 10 cm only.

17. Photopectoralis bindus

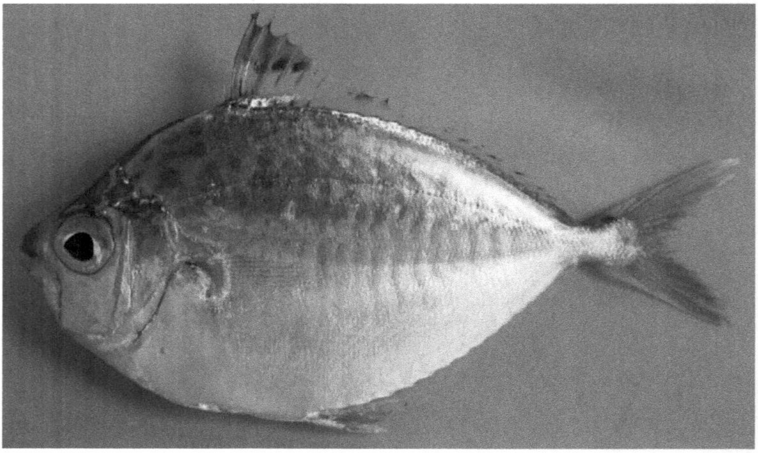

Image credit: D.G.R. Wiadnya. https://creativecommons.org/licenses/by/3.0/deed.en

Common name(s): Orangefin ponyfish.

Global distribution: Indo-West Pacific Ocean.

Ecology: This marine, brackish water species is demersal and amphidromous. It has a depth range of 2–160 m and is commonly found over muddy sand bottoms of coastal inshore waters. It is often entering estuaries and is forming schools.

Biology: It is a deep bodied and relatively laterally compressed species; 11.0 cm. Its ventral profile is much more convex than the dorsal profile. It has a short snout and its mouth extends forward when protracted about 15% of the body length. Lower jaw profile is straight. Chest and nuchal region are asquamate; and the remainder of the body except the head is scaled.

18. Photopectoralis panayensis

Common name(s): Panay ponyfish.

Global distribution: Western Central Pacific Ocean.

Ecology: It is a demersal species.

Biology: It is fairly a deep-bodied ponyfish with its mouth protruding forward. Snout tip is dark. A densely pigmented membrane is seen between the jaws and the cheeks. When the mouth is closed, a black line is visible between the lower front margin of the eye and the lower articulation. A large blotch is present on the nape, and irregular dark blotches are on the upper sides. It grows to a maximum length of 8 cm.

Luminous organs and bioluminescence: Photopectoralis spp.: In the species *Photopectoralis aureus*, the dorsolateral and ventral expansion have been reported to serve as the light organ (Chakrabarty et al., 2011). Dunlap et al. (2007) reported that the species *Photopectoralis bindus* and *Photopectoralis panayensis* contain the luminous bacterium, *Photobacterium leiognathi* in their light organs. However, Kaeding et al. (2007) observed the presence of luminous bacterium, *Photobacterium mandapamensis* as the light organ symbiont in the species of *Photopectoralis.*

19. Secutor megalolepis

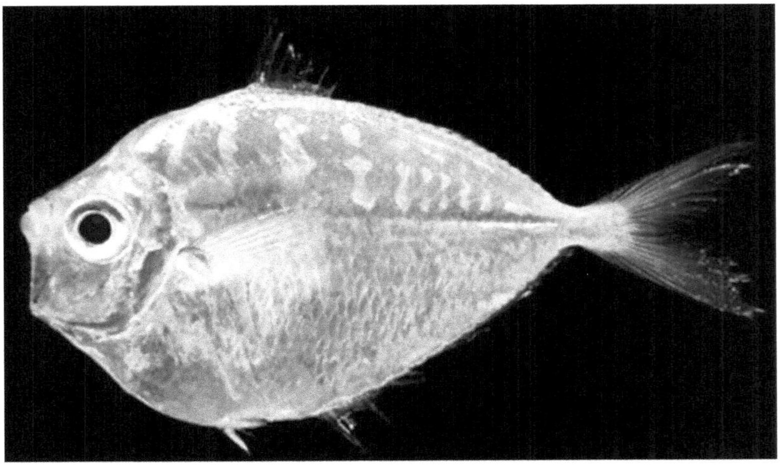

Image credit: S. Kimura. Atlas of living Australia. https://creativecommons.org/licenses/by/3.0/au/

Common name(s): Bigscale Ponyfish.

Global distribution: Western Pacific Ocean.

Ecology: This marine and brackish – demersal species occurs over the soft bottom of the continental shelf.

Biology: In this species, the mouth is angled upwards and its lower jaw anterior profile is almost vertical (when the mouth is closed). A black line is seen from the anteroventral margin of the orbit to the chin. It grows to a maximum length of 7.5 cm only.

20. Secutor ruconius

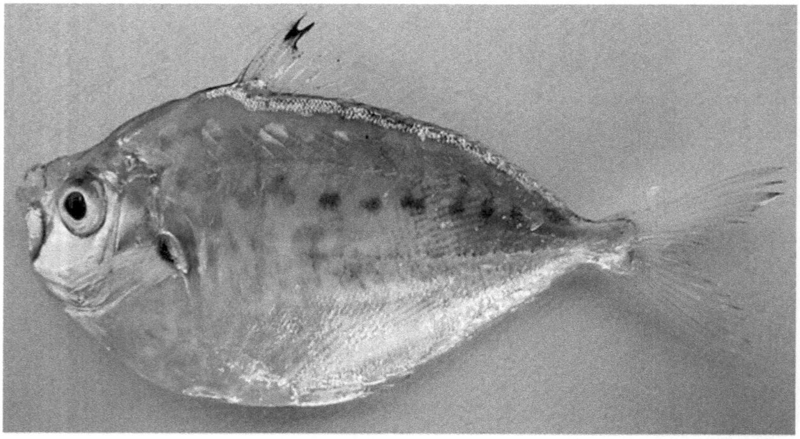

Source: Public domain

Common name(s): Deep pugnose ponyfish.

Global distribution: Indo-West Pacific Ocean.

Ecology: This marine, freshwater and brackish water species is demersal and amphidromous. It has a depth range of 3–60 m and is commonly found over sand or mud bottoms in coastal waters. It often enters estuaries and rivers and is forming schools.

Biology: It is a strongly laterally compressed and deep-bodied species of ponyfish. Ventral profile is more convex than the dorsal. Mouth is upwardly turned and lower jaw is straight and vertical. Mouth can extend about 20% of the SL. Snout is very short and eyes are small. There are no scales on the chest, head or nuchal region while the remainder of the body is scaled. It feeds mainly on small crustaceans and it grows to a maximum length of 8 cm only.

Luminous organs and bioluminescence: Secutor spp.

These species possess a fairly larger luminous gland, the luminosity of which is greater than that of the other members of the family Leiognathidae (Haneda, 1950). Dunlop et al. (2007) reported that the light organ of *Secutor megalolepis* harbors the luminescent bacteria, *Photobacterium leiognathi*. Herring (1983) reported that *Secutor ruconius* showed a wavelength of light emission maximum as 500 nm.

3.2.2 ORDER: ACROPOMATIFORMES

3.2.2.1 LANTERNBELLIES (ACROPOMATIDAE)

1. *Acropoma* spp.

Common name(s): Temperate ocean-basses or lanternbellies.

Global distribution: These fishes are native to the Indian Ocean and western Pacific Ocean.

Ecology: These fishes are found to live several hundred meters down from the surface.

Biology: These fishes are generally small. While some species range up to 40 cm, most species are not more than 15 cm. They have two dorsal fins; first with 7–10 spines and the second with a spine in addition to 8 to 10 soft rays. Anal fin has two or three spines, and pelvic fins are with one spine and five soft rays.

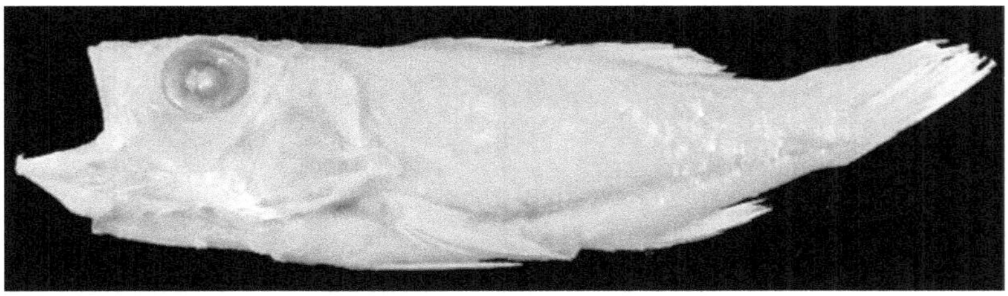

1

2

3

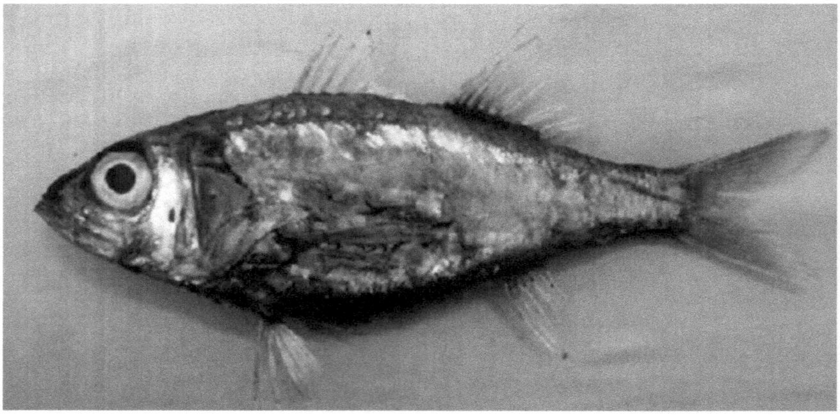

4

Credits: 1: : Museums Victoria, https://fishesofaustralia.net.au/home/species/5503. https://creativecommons.org/licenses/by/3.0/au/; 2. Acropoma argentistigma (Image courtesy Prasanna Yennawar; 3: Image courtesy : Shao, Kwang, Tsao, Taiwan Fish Database; 4. Hamid Badar Osmany. https://creativecommons.org/licenses/by/3.0/deed.en

1. *Acropoma arafurensis*; 2. *Acropoma argentistigma*; 3. *Acropoma hanedai*; 4. *Acropoma japonicum*

Luminous organs and bioluminescence: Acropoma species.

These fishes differ from other luminous fish in possessing unusually large luminous areas which are formed from the half the muscles of their complicated body structure (Haneda, 1950). The family Acropomidae is known for indirect emission type and for its bioluminescent symbiosis with luminous bacteria. The species of *Acropoma* have a ventral luminous organ that has a open type of luminous gland, a lens, and a reflector. The shape and length of the luminous gland help distinguish the species in the genus (Anon., https://en.wikipedia.org/wiki/Acropoma). The luminous gland of *Acropoma* spp. is a whitish-yellow U-shaped filiform body (like a U-tube) which is lying flat and embedded in the muscles of the pectoral and ventral regions. Since this luminous gland is not visible on the surface of the body, these fish have the appearance of ordinary non-luminous fish in daylight. However, in the dark, the ventral area of the fish is brilliantly illuminated when they are caught by a net. The luminosity is believed to be continuous and is never disappearing completely, but it may be dimmed by means of the chromatophores (Haneda, 1950). The description of luminous glands of the different species of *Acropoma* is given below.

Acropoma japonicum: In this species, the luminous organ is not visible externally and the fish looks like a common non- luminous one, but the lower part of the fish body has a lightning-like bluish-white color. It has a short U-shaped luminous gland in its light organ which harbors the luminescent, symbiotic (gram negative) bacteria, *Photobacterium mandapamensis* and *Photobacterium leiognathi* (Okamoto and Golani, 2018; Dunlap et al., 2007). Herring (1983) reported on its wavelength of emission maximum as 486 nm.

The bioluminescent organ of *Acropoma japonicum* is composed of compartmentalized left and right arms, which unite anteriorly and are connected by a medial lumen anterior to the anus. These arms remain, however, separated posteriorly. These arms form the bioluminescent component of the bioluminescent organ and are called the filiform body, U-shaped tube, or glandular organ (Haneda, 1950). Further, the bioluminescent component of the bioluminescent organ is separated into folded chambers or lobules and are lined by a simple cuboidal epithelium where the cells become slightly compressed apically to basally. The lobules are filled with bacterial cells and they open toward the center medial surface of the bioluminescent arms where there is considerable connective tissue (Ghedotti et al., 2018).

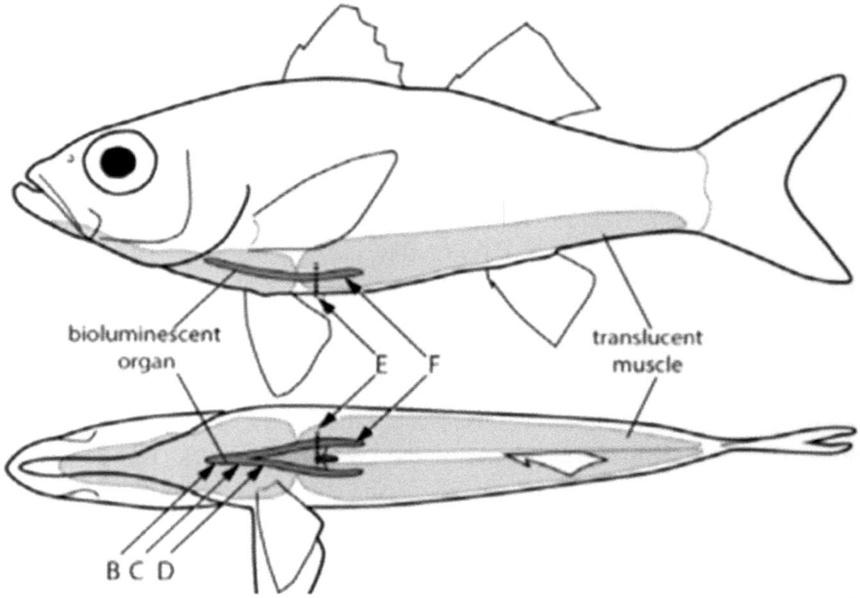

Image courtesy: Michael J. Ghedotti

B,C,D – Anterior bacterial luminescent organ
Bioluminous organs in *Acropoma japonicum*

Acropoma arafurensis: This bioluminescent species has a luminous gland which is U-shaped extending from throat to posterior tip of pectoral fin (Okamoto et al., 2019).

Acropoma argentistigma: It has a short U-shaped and open type of luminous gland containing symbiotic type of luminous bacteria (Haneda, 1980; Okamoto and Ida, 2002).

Acropoma boholensis: In this species, the luminous gland extends from the isthmus to the anal fin base (Anon., https://www.discoverlife.org/20/q?search=Acropoma+boholensis&b=FB59419).

Acropoma hanedai: In this species, two very long light organs run from the isthmus to slightly beyond the origin of the anal fin and are joined with the opposite organ immediately in front of the anal fin (Anon., https://fishbase.mnhn.fr/summary/9232). Tanet et al. (2020) reported that the light organ of this species contains the luminous bacteria, *Photobacterium kishitanii.* According to Ghedotti et al. (2018), the bioluminescent organ of this species is composed of compartmentalized left and right bacteria-containing arms connected by a medial lumen which extends from ventral to the eye near the isthmus to the middle anal fin.

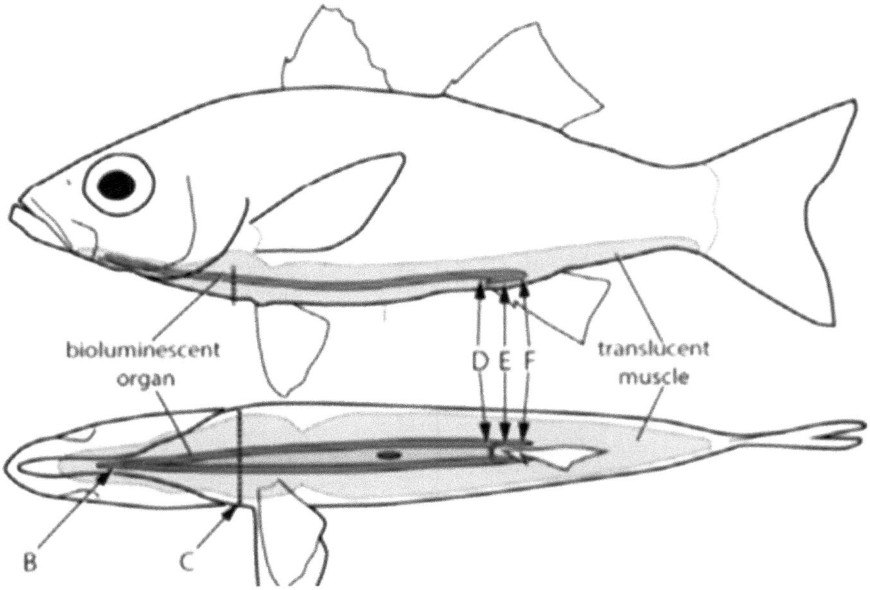

Image courtesy: Michael J. Ghedotti

Here, b – anterior bacterial bioluminescent organ; c – ventral body wall immediately anterior to pelvic – and pectoral – fin origins; d,e,f – left lateral ventral bioluminescent organ in region of anterior anal fin.

Bioluminous organs in *Acropoma hanedai*

Acropoma heemstrai: It has a unique moderate Y-shaped luminous gland which is extending from the throat to midway between the origins of the pelvic and anal fin (Okamoto and Golani, 2018). Haneda (1980) reported that its luminous gland is an open type of gland containing symbiotic type of luminous bacteria

Acropoma lacrima: It has a short U-shaped luminous gland in its light organ (Okamoto and Golani, 2018).

Acropoma lecorneti: It has an O-shaped luminous gland in its light organ (Yamanoue and Toda, 2008).

Acropoma leobergi: It has a short U-shaped luminous organ which is slightly receding behind the anus (Prokofiev, 2019).

Acropoma musorstom: Its luminous gland is short and U-shaped around the anus (Okamoto et al., 2021).

Acropoma neglectum: It has a short U-shaped luminous gland (Okamoto and Golani, 2018).

Acropoma profundum: It has a unique long, U-shaped luminous gland, which extends from the throat to just in front of the anal fin origin (Okamoto, 2014).

3.2.3 ORDER: ANGUILLIFORMES

3.2.3.1 CONGER AND GARDEN EELS (CONGRIDAE)

1. Lumiconger arafura

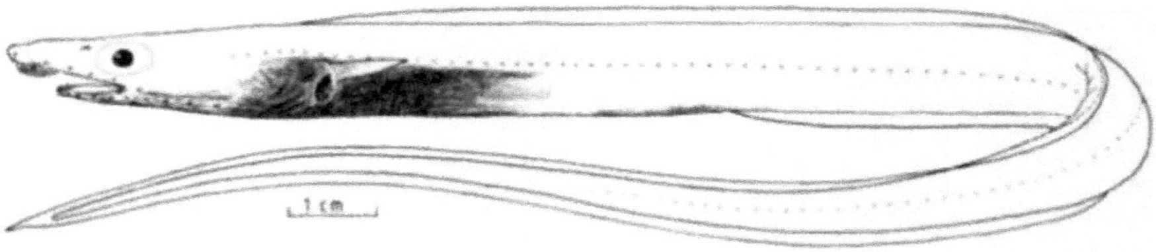

Image credit: Museums Victoria, https://fishesofaustralia.net.au/home/species/3450; https://creativecommons.org/licenses/by/3.0/au/

Common name(s): Luminous conger.

Global distribution: Indo-West Pacific Ocean and eastern Indian Ocean.

Ecology: It is a shallow-living, benthic species and is often found on the continental shelf over soft bottom at depths of 27–152 m.

Biology: This conger eel has a very slender tail and its anus is positioned well before the midlength. It has a few cephalic pores. This species is unique in possessing a deeply curved premaxillary-ethmoid bearing a long tooth patch which is completely exposed on the underside of its long slender snout when the mouth is closed. It grows to a maximum length of 27 cm.

Luminous species: Lumiconger is a monotypic genus (i.e., with single species that is luminescent.

Luminous organs and bioluminescence: It is a bacterially induced bioluminescent species and its luminescent organ is derived from its intestinal tract which houses the bacterial symbionts (Finn, 2020).

3.2.4 ORDER: ARGENTINIFORMES

3.2.4.1 PENCIL SMELTS (MICROSTOMATIDAE)

1. *Nansenia ardesiaca*

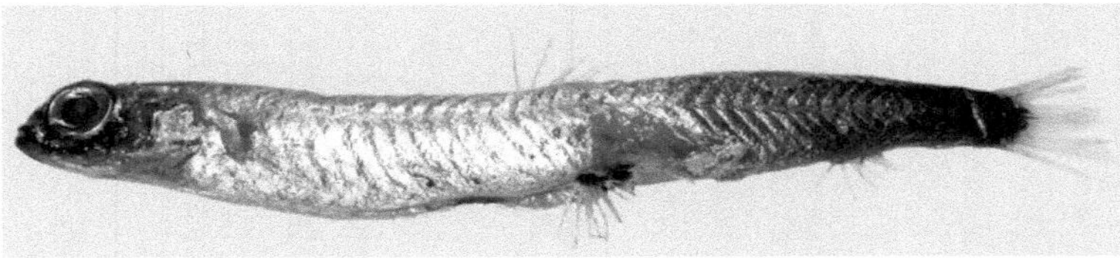

Image credit: Museums Victoria, https://fishesofaustralia.net.au/home/species/3629; https://creativecommons.org/licenses/by/3.0/au/

Common name(s): Robust Smallmouth.

Global distribution: Indo-West Pacific Ocean.

Ecology: This mesopelagic and bathydemersal species has a depth range of 200–1000 m.

Biology: It grows to a maximum length of 20 cm.

2. *Nansenia boreacrassicauda*

Image credit: Agosti and Egloff, W. http://treatment.plazi.org/id/087187E8C16EFFE0FF1E4E1AFA772FD4; https://creativecommons.org/licenses/by/2.0/

Common name(s): Pencil smelt.

Global distribution: Atlantic Ocean and temperate and subarctic waters.

Ecology: This pelagic-oceanic; species has a depth range of 382–400 m.

Biology: It has its characteristic curved vomerine teeth and large eye. The body coloration is golden-brownish and its scales are strongly iridescent. It grows to a maximum length of 26.8 cm.

Luminous organs and bioluminescence: Nansenia spp.

In the luminescent species of *Nansenia,* ventrally positioned, external luminous tissues are present. In *Nansenia boreacrassicauda*, a large ventral strip of bioluminescent tissue which is associated with large scales is running from the isthmus to vent (Poulsen, 2016).

3.2.4.2 BARRELEYES (OPISTHOPROCTIDAE)

1. Dolichopteryx longipes

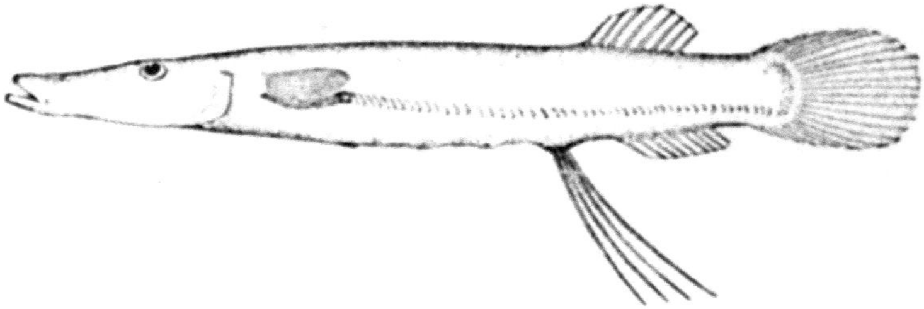

Source: Public domain

Common name(s): Brownsnout spookfish.

Global distribution: It has worldwide tropical and temperate distribution.

Ecology: This mesopelagic and bathypelagic-oceanodromous species has a depth range of 500–2,400 m.

Biology: The body of this species is slender and its eyes are cylindrical facing upward. The body is transparent and is covered by thin scales. Snout is dusky in color. Four pairs of blotches are seen on the peritoneal cavity beneath the gut; a line of chromatophores is seen below the lateral line to the position of the dorsal fin; and two blotches are found on the base of the tail. It grows to a maximum length of 18 cm.

Luminous organs and bioluminescence: It has a luminous organ on the ball of the eye in front of the lens (Haneda, 1950). Poulsen et al. (2016) reported that external ventrally positioned luminous tissues are present in this species.

***2. Monacoa grimaldii** (= Opisthoproctus grimaldii)*

Source: Public domain

Common name(s): Mirrorbelly.

Global distribution: It is found distributed in the tropical to subtropical Atlantic and Pacific oceans.

Ecology: This mesopelagic–bathypelagic species has a depth range of 0–4750 m.

Biology: The body of this species is deep. While its dorsal part of the head is transparent, ventral part and ventral 2/3 of body are peppered with large melanophores. Ventral surface of the rear part of the sole has 4 dark blotches. It grows to a maximum length of 8 cm.

3. Monacoa griseus

Image credit: Museums Victoria, https://fishesofaustralia.net.au/home/species/3631; https://creativecommons.org/licenses/by/3.0/au/

Common name(s): Grey mirrorbelly.

Global distribution: Southwest Pacific Ocean.

Ecology: This bathypelagic species has a depth range of 0–521 m.

Biology: In this species, the anterior part of its sole is greyish and it is abruptly changing, to a dense pigmented posterior part just in front of the pelvic fins. It grows to a maximum length of 6.4 cm.

4. Monacoa niger

Common name(s): Black mirrorbelly.

Global distribution: Eastern Central Pacific Ocean.

Ecology: This bathypelagic species has a depth range of 0–520 m.

Biology: This species is distinguished by the presence of black pigmentation along its entire sole and it has a darker body (scale pockets are darker) compared to its congeners. The black streak on the sole consists of distinct patches posteriorly. It grows to a maximum length of 4.8 cm.

Luminous organs and bioluminescence: Monacoa spp.

In the bacterially induced luminescent mirrorbelly fishes, the ventral space is enlarged and it has developed into a large reflector organ which controls light emission. However, light is produced in the "rectal bulb" (large black light organ) which is a modification of the distal part of the intestinal tract close to the anus, and is channeled through a small opening into an internal reflector-lumen called the hyaloid body (reflector) (Poulsen et al., 2016). In *Monacoa grimaldii,* its light producing rectal bulb is with the luminous bacterium, Photobacterium kishitanii (Tanet et al., 2020). Herring (1983) reported its wavelengths of light emission maximum as 480 and 484 nm.

5. Opisthoproctus soleatus

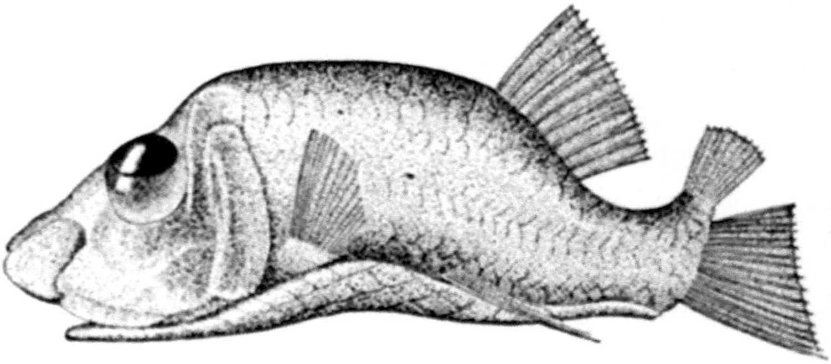

Source: Public domain

Common name(s): Mirrorbelly, barrel-eye or spook fish.

Global distribution: Eastern Atlantic Ocean.

Ecology: This bathypelagic species has a depth range of 300–800 m.

Biology: In this species, the sides of its body are dark. Head below and behind the eye are found sprinkled with large melanophores. Snout is translucent ventrally. It grows to a maximum length of 10.5 cm.

Luminous species: It is a single species in this genus which is luminescent.

Luminous organs and bioluminescence: In this species, its open type light organ is situated near the anus from a gut diverticulum which contains symbiotic luminous bacterial species, *Photobacterium phosphoreum.* This light organ which is a simple rectal bulb without lens of cells or a reflector layer is believed to emit light indirectly (Suntsov et al., 2008). Dunlap et al. (2007), however, reported that the light organ of this species harbors the luminescent, symbiotic bacteria of the species, *Photobacterium kishitanii.*

Its light is emitted along the flat sole of the fish and is used for ventral camouflage.

Poulsen et al. (2016) reported that in these fishes, its ventral space is enlarged and it has evolved into a large reflector organ that controls its light emission. Further, the light produced in the "rectal bulb" is channeled through a small opening into an internal reflector-lumen called the hyaloid body. The entire external surface of the sole which is covered with large, pigmented thin scales function as a light-screen when the reflector is contracted (when there is no light emission) or expanded (when light passes through the thin transparent parts of the scales).

6. Rhynchohyalus natalensis

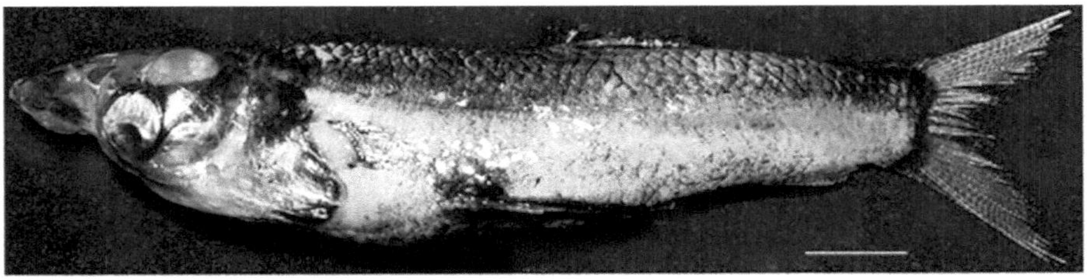

Image credit: Museums Victoria, https://fishesofaustralia.net.au/home/species/2086; https://creativecommons.org/licenses/by/3.0/au/

Common name(s): Glasshead barreleye.

Global distribution: Eastern and Western Atlantic Ocean and Western Pacific Ocean.

Ecology: This mesopelagic and oceanic species has a depth range of 250–1000 m.

Biology: In this species, body compressed and its ventral surface is with scattered melanophores. Pelvic fins and area around anus are black. Head is translucent. It grows to a maximum length of 18 cm.

Luminous organs and bioluminescence: In this species, the light organ is of the open type and is a diverticulum of the gut with symbiotic luminous bacterial species, *Photobacterium phosphoreum*. The light organ which is a simple rectal (anal) bulb is without lens of cells or a reflector layer is believed to emit light directly through the anus (Suntsov et al., 2008). Herring (1983) reported that the emission maximum wavelength of this species was 480 nm.

7. Winteria telescopa

Image credit: Museums Victoria, https://fishesofaustralia.net.au/home/species/3632; https://creativecommons.org/licenses/by/3.0/au/

Common name(s): Binocular fish.

Global distribution: Though it has been reported to occur in all oceans around the world., it is believed to be common in Eastern Atlantic Ocean and Tropical Indian Ocean.

Ecology: This mesopelagic and bathypelagic species has a depth range of 400–2500 m.

Biology: This species is distinguished by its more forward-facing eyes; Body coloration is dark blue-black with silvery reflections on the head. It grows to a maximum length of 15 cm.

Luminous species: It is a monotypic genus.

Luminous organs and bioluminescence: In this species, the light organ is of the open type and is a diverticulum of the gut containing symbiotic luminous bacterial species, *Photobacterium phosphoreum*. The light organ which is a simple rectal bulb is without lens of cells or a reflector layer is believed to emit light indirectly (Suntsov et al., 2008). Paitio et al. (2016) reported that the rectal light organ of this species used mainly for counterillumination to be away from its predators.

3.2.5 ORDER: AULOPIFORMES

3.2.5.1 GREENEYES (CHLOROPHTHALMIDAE)

***1. Chlorophthalmus* spp.**

Common name(s): Greeneyes.

Global distribution: These fishes have circumglobal distribution in tropical and temperate waters.

Ecology: These fishes prefer continental slopes and shelves of depths 50–1000 m. They form schools occasionally.

Biology: These fishes are known for their disproportionately large, iridescent and fluorescent eyes, They are slender fish with slightly compressed bodies. Heads are small with large jaws. The body coloration ranges from a yellowish to blackish brown, and some species may have cryptic blotches. Yet some species may also have iridescent patches covering their heads. Fins are simple and spineless. These fishes feed mainly on benthic invertebrates and pelagic crustaceans such as decapods and mysids. While the largest species of this genus viz. *Chlorophthalmus agassizi* reaches a length of 40 cm, most other species are much smaller.

1

2

Credits: 1: Museums Victoria, https://fishesofaustralia.net.au/home/species/5262. https://creativecommons.org/licenses/by/3.0/au/; 2. Public domain.

1. Chlorophthalmus acutifrons; 2. Chlorophthalmus albatrossis.

Luminous organs and bioluminescence: Chlorophthalmus spp.

In the luminescent species of *Chlorophthalmus*, a small light organ was found around the anus on the ventral part of their body. The light organs of these fishes lie in the rectum and shine out of the anus. The light organ which is a open type of gland in these fishes has been reported to contain luminescent, symbiotic type of bacteria, *Photobacterium kishitanii* (Haneda, 1980; Tanet et al., 2020; Dunlap et al., 2007). Somiya (1977) reported that the luminescence of these fishes is blue-green and continuous and according to him a possible interrelationship exists between the light organ and eyes in the chlorophthalmid deep-sea fish. Herring (1983) reported on the wavelength of light emission maximum in this species as 475 nm. The light of these fishes serve mainly as an intraspecific signal for courtship and it may also play an important role in the formation of schooling.

3.2.5.2 BOMBAY DUCK (SYNODONTIDAE)

1. Harpadon nehereus

Image credit: Hamid Badar Osmany. https://creativecommons.org/licenses/by/3.0/

Common name(s): Bombay duck.

Global distribution: Indo-West Pacific Ocean.

Ecology: This marine and brackish water species is benthopelagic and oceanodromous and occurs at depths of 50 m. It is also inhabiting deep water offshore on sandy mud bottom for most of the year, but it is also known to gather in large shoals in deltas of rivers to feed during the monsoon period.

Biology: The body of this species is elongated and compressed with small eyes and very short snout. Mouth is very wide and is armed with slender, recurved and depressible teeth of unequal size. Scales are mostly restricted to the posterior half of the body. The posterior tip of its pectoral fin is reaching origin of pelvic fin. It grows to a maximum length of 40 cm.

Luminous organs and bioluminescence: Pringgenies et al. (2019) reported that the light organ of this species harbors luminous bacteria of the species, *Alteromonas macleodii.*

3.2.6 ORDER: GADIFORMES

3.2.6.1 GRENADIERS (BATHYGADIDAE)

1. *Gadomus* spp.

Common name(s): Rattails.

Global distribution: Western Central Atlantic Ocean.

Ecology: It is a benthopelagic species.

Biology: These are medium to large sized fishes, ranging in total length from about 30 to 150 cm. They have an elongate, firm body with fairly a large head and short trunk which is tapering into a prolonged, finely pointed tail. Caudal fin is absent. There are two dorsal fins. Anterior dorsal fin which is short-based is comprised of a short first and an elongate second spinous ray. A well-developed chin barbel is present. The body coloration is usually creamish of light brown.

1

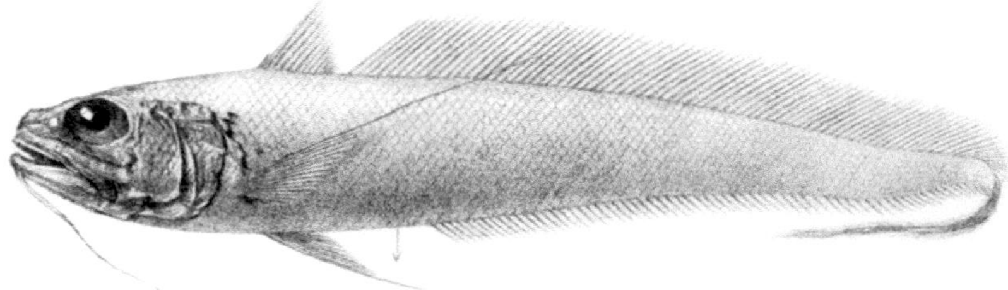

2

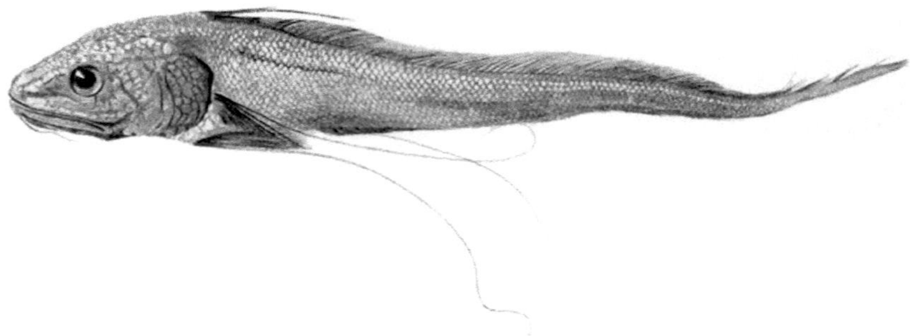

3

Credits: 1: Treatment Bank, https://zenodo.org/record/194743#.ZDVaC-zMJNY. Public domain; 2 & 3: Public domain.

1. Gadomus arcuatus; 2. Gadomus dispar; 3. Gadomus longifilis.

Luminous organs and bioluminescence: Gadomus spp.

In all the luminescent species of *Gadomus*, a ventral abdominal light organ containing luminescent, symbiotic bacteria is seen (Merrett, 2003).

3.2.6.2 RATTAIL FISHES (MACROURIDAE)

In some species of this family, a ventral light organ which is identified by a dermal window (i.e., small to medium-sized black pigmented fossa) is seen between the pelvic fin bases (Merrett, 1986; Anon., https://www.fao.org/3/y4161e/y4161e25.pdf)).

1. Cetonurus crassiceps

Common name(s): Globosehead rattail.

Global distribution: Pacific and Atlantic Oceans.

Ecology: This benthopelagic species which is having a depth range of 290–1490 m inhabits the continental slope waters.

Biology: This species is characterized by their large, soft, globose head; fully scaled including branchiostegal rays; enlarged scales along the dorsal-fin interspace and anterior portion of the second dorsal fin; and interrupted lateral line. These fishes grow to a maximum length of 44 cm.

2. Cetonurus globiceps

Common name(s): Globehead grenadier.

Global distribution: Eastern and Western Atlantic ocean and Western Pacific Ocean.

Ecology: This benthopelagic species lives mostly at depths of 1,000–2,000 m.

Biology: The body of this species is robust, but it is tapering very rapidly behind the anus. Head is large, broad and deep. Snout is high and broad. The body coloration is brown to dark brown and mouth, gill cavity and body cavity are blackish. It feeds mainly on small fishes and planktonic crustacean and it grows to a maximum length of 50 cm.

Luminous organs and bioluminescence: Species of *Cetonurus.*

The light organs of these fishes are bulbous with a glandular lining and they have a single opening just before a naked area around the anus. Anon. (http://species-identification.org/species.php?species_group=fnam&id=1421) reported that the light organ of this species is a short blackish area and it is produced anteriorly into a white or light-colored triangle.

1

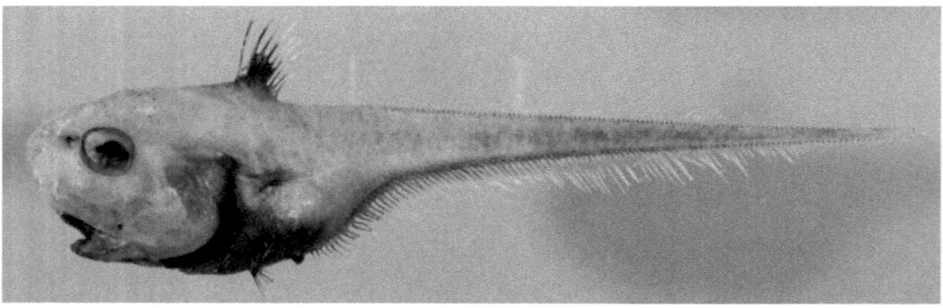

2

Credits: 1: Public domain; 2: Museums Victoria, https://fishesofaustralia.net.au/home/species/3897. https://creativecommons.org/licenses/by/3.0/au/.

1. *Cetonurus crassiceps*; 2. *Cetonurus globiceps*.

3. Coelorinchus spp.

Common name(s): Rattails.

Global distribution: The species of this genus have been reported to occupy all oceans, except the high Arctic.

Ecology: These benthopelagic fishes live on or near the bottom in deep waters at depths of; 200–1500 m.

Biology: Due to their long tapering tails, these fishes are widely named as rattails. Trunk is short and tail is compressed, greatly elongated and is tapering to a slender point. Caudal fin is absent. Head which is small to enormously inflated is compressed, rounded, or cylindrical, with a bluntly rounded to sharply pointed snout. Mouth is large, ending with a small and inferior terminal. Chin barbel is usually present. The eyes are moderate to very large. Rays of the long-based second dorsal fin are shorter than the rays of the anal fin. These fishes feed mainly on epibenthic invertebrates. Adult rattails may reach up to 50 cm of total length.

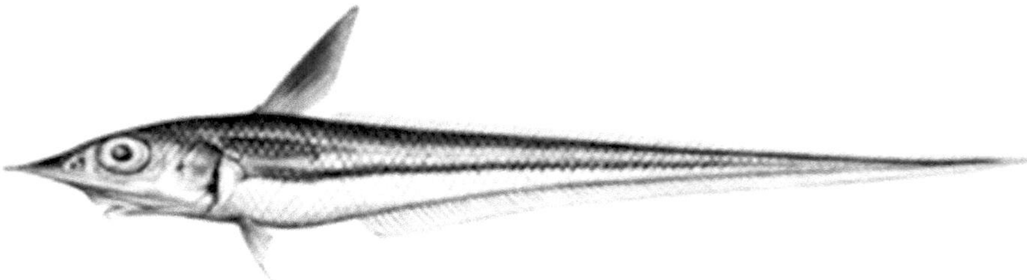

1

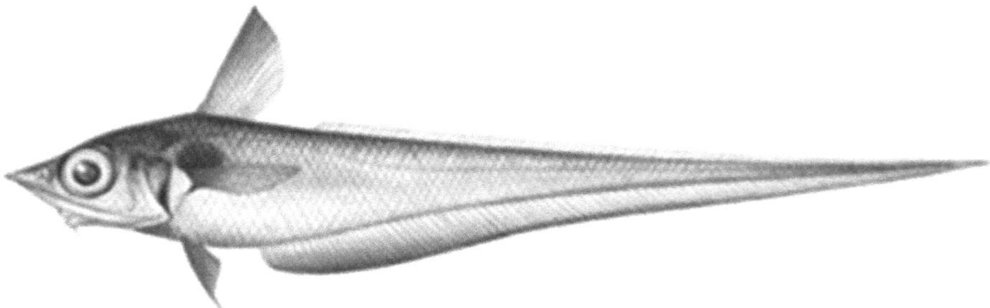

2

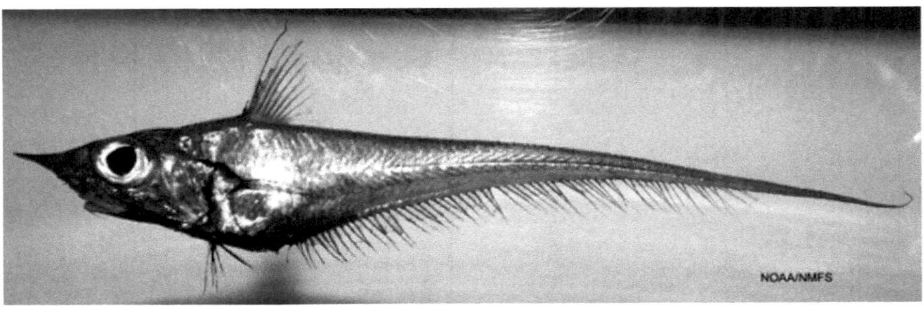

3

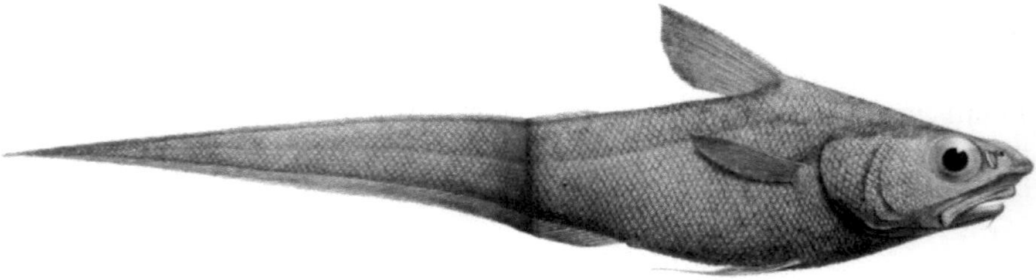

4

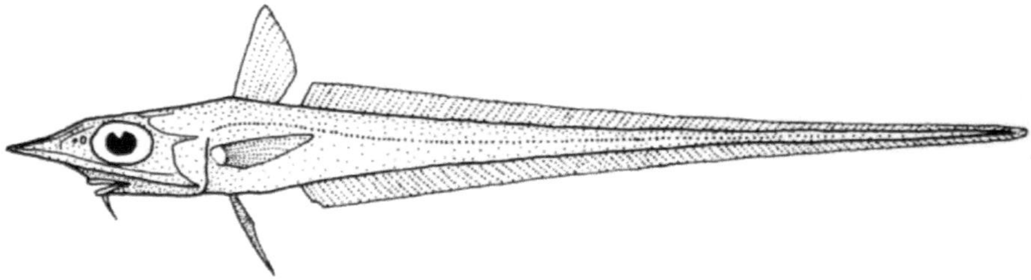

5

6

7

8

9

10

11

12

13

14

Credits: 1 & 2: Kwang-Tsao Shao, Taiwan Fish Database; 3-7, 9, 14: Public domain; 8–13: Museums Victoria, https://fishesofaustralia.net.au/. https://creativecommons.org/licenses/by/3.0/au/.

1. Coelorinchus hubbsi; 2. Coelorinchus kishinouyei; 3. Coelorhinchus caribbaeus;4. Coelorinchus occa; 5. Coelorinchus innotabilis; 6. Coelorinchus fasciatus; 7. Coelorinchus caelorhincus;8. Coelorinchus macrorhynchus; 9. Coelorinchus parallelus; 10. Coelorinchus australis; 11. Coelorinchus maculatus; 12. Coelorinchus mycterismus; 13. Coelorinchus matamua; 14. Coelorinchus carminatus.

Luminous organs and bioluminescence: Coelorhinchus spp.

In the species of the genus, *Caelorinchus,* the shape of the light organ varies from tubular to bulbous with a glandular lining and is with a single opening just before the anus. Iwamoto et al. (2015) reported that in some luminescent species of this genus, the ventral light organ which is not externally visible may develop from a small gland to a large prominent organ with one or two dermal fossae along the ventral midline. Dunlap et al. (2007) reported that the species of this genus viz. *Coelorinchus anatirostris, Coelorinchus denticulatus, Coelorinchus fasciatus, Coelorinchus formosanus, Coelorinchus hubbsi, Coelorinchus japonicus, Coelorinchus kamoharai, Coelorin kishinouyei,* and *Coelorinchus multispinulosus* harbored symbiotic luminescent bacteria, *Photobacterium kishitanii* In their light organs. Tanet et al. (2020) observed the presence of the symbiotic, luminescent bacteria *Aliivibrio fishcheri* in the light organs of *Coelorhinchus formosanus* and *Coelorhinchus multispinulosus*.

The light organs observed in the different species of this genus are given below.

Coelorhinchus kishinouyei: In this species, the light organ is long and its anterior end is found crossing forward of line connecting pelvic fin bases (Chiou et al., 2004).

Caelorinchus sheni: In this species, the light organ is small with a short and slender blackish streak (Chiou et al., 2004).

Coelorhinchus caribbaeus: In this species, the light organ is large and is a broadly lenticular to oval black naked fossa between and slightly anterior to the pelvic fin bases (Anon., https://www.fishbase.se/summary/3102).

Coelorhinchus coelorhynchus: Here, the light organ is large and it looks like a black naked fossa between and somewhat anterior to the pelvic fin bases (Anon., https://www.fishbase.se/summary/1726).

Cohen et al. (1990) recorded the light organs of the different species of this genus as given below.

Coelorinchus aconcagua: In this species, the light organ is fairly short, and is extending in front of anus within body wall as a flattened, elongate diverticulum which is falling well short of pelvic fin bases

Coelorinchus anatirostris: In this species, the light organ is small. The fossa is narrow and is extending forward immediately before anus

Coelorinchus argentatus: In this species, a long black streak of light organ is found extending from before anus to close behind isthmus and is dilated at each end forming shallow fossae

Coelorinchus aspercephalus: In this species, the light organ is lanceolate to narrowly oval and is extending forward to midway between anus and pelvic fin insertions

Coelorinchus australis: In this species, the light organ is elongate to ovoid and is found extending forward to midway between the anus and pelvic fin insertions

Coelorinchus biclinozonali: In this species, the light organ is large and a lens-shaped fossa on abdomen extends forward to midway between anus and pelvic fin insertions

Coelorinchus braueri: In this species, the light organ is shortand narrow and it extends forward to the midpoint between pelvic fin insertions and anal fin origin

Coelorinchus chilensis: In this species, the light organ is very small and is faintly visible externally as a slightly swollen blackish area before the anus

Coelorinchus denticulatus: In this species, the light organ is long and is found extending as a dark streak from anus to chest.

Coelorinchus fasciatus: In this species, the light organ is moderately large and its length is more than half of pupil diameter. The lens-shaped fossa which is located immediately before the anus is extending to more than half distance to pelvic fin bases.

Coelorinchus formosanus: In this species, the light organ is long and is found extending as a black streak from the anus to just behind the isthmus It is expanded at each end but it lacks a scaleless fossa.

Coelorinchus hexfasciatus: In this species, the light organ is small and is externally visible as a narrow blackish streak extending forward of anus to midway between pelvic fin insertions and anal fin origin

Coelorinchus innotabili: In this species, the light organ is small and is not clearly visible from the exterior. It is connected to periproct by a short stalk and is much closer to pelvic fin insertions.

Coelorinchus japonicus: In this species, the light organ is short

Coelorinchus karrerae: In this species, the light organ is small and is externally visible as a blackish streak extending forward of the anus to between the pelvic fins.

Coelorinchus longicephalus: In this species, the light organ is small and is externally visible as a narrow blackish streak. Further, it is extending forward of the anus to about midway between pelvic fin insertions and anal fin origin.

Coelorinchus macrochir: In this species, a naked fossa of light organ is seen immediately anterior to, but is well separated from periproct region

Coelorinchus matamua: In this species, the light organ is a small globular gland and is adjacent to the anus, but it is not externally visible.

Coelorinchus oliverianus: In this species, a black naked fossa of light organ is oval and is located between pelvic fin bases; and is separated from anus by several scale rows.

Coelorinchus parallelus: In this species, the light organ is slightly visible as a black crescent before the anus.

Coelorinchus productus: In this species, the light organ is small and the fossa of light organ is narrow, and is extending forward immediately before the anus.

Coelorinchus scaphopsis: In this species, a large lens-shaped black fossa of light organ is seen.

Coelorinchus ventrilux: In this species, the light organ is large and is visible as a blackish streak leading from anus to a long black naked fossa on the rear half of the chest.

Coelorinchus longissimus: It has a long bioluminescent organ in front of the anus. (Anon., https://en.wikipedia.org/wiki/Coelorinchus_longissimus).

4. *Haplomacrurus nudirostris*

Common name(s): Nakedsnout whiptail.

Global distribution: Southern Atlantic, Indian and Pacific oceans.

Ecology: This bathypelagic species has a depth range of 690–1590 m and is common in continental slope.

Biology: It has no scales on top of its snout or on the underside of head. Scales are small, and those on the head and front of the body are without spinules. Anus is located about midway between the anal and pelvic fins. Spinous ray of the first dorsal fin is stout, flattened, curved and is finely serrated. The body coloration is brownish black with bluish abdomen from behind the head to about one-third of its body length. It attains a maximum length of 58 cm.

Luminous organs and bioluminescence: It has a small light organ between its pelvic fin bases (McMillan et al., 2011).

5. *Hymenocephalus* spp.

Common name(s): Rattails.

Global distribution: These fishes are found distributed in all tropical oceans, except the eastern Pacific ocean.

Ecology: It occurs in the continental slope terrain, chiefly between 300 and 1000 m depths.

Biology: These fishes are characterized by striations on jugular and thorax; an extraordinary development of sensory reception organs; well enlarged eyes; uniquely large and specialized sagittal otoliths; and extremely wide and deep head canals. Total length of these species is about 25 cm.

1

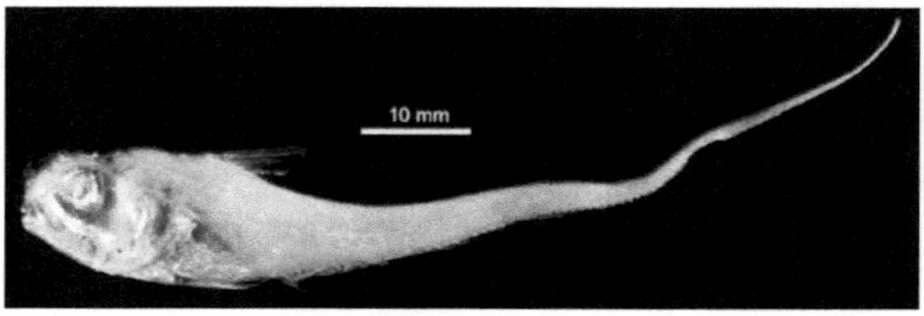

2

3

4

Credits: 1 & 3: Public domain; 2 & 4: Museums Victoria, https://fishesofaustralia.net.au. https://creativecommons.org/licenses/by/3.0/au/.

1. *Hymenocephalus italicus*; 2. *Hymenocephalus iwamotoi*; 3. *Hymenocephalus lethonemus*; 4. *Hymenocephalus longibarbis*.

Luminous organs and bioluminescence: Hymenocephalus spp.

These species are known for the presence of an elongated, tubular light organ with two lenses. While one small lens is present on the chest anterior to the pelvic fin bases, the second one is situated just anterior to the anus. Its blackish striations seen on the silvery chest are also associated with light production (Cohen et al., 1990). The light organ of the species *Hymenocephalus striatissimus* has been reported to harbor the luminescent bacterium, Photobacterium kishitanii (Dunlap et al., 2007). In *Hymenocephalus italicus*, the lens-like swellings of anterior and posterior dermal windows of light organ are small but they are distinct. In *Hymenocephalus iwamotoi*,the silvery luminescent striae extend along both sides below the gill covers and are reaching the pectoral-fin bases and behind the pelvic-fin bases. Its posterior lens of the light organ is rounded (Anon., https://fishesofaustralia.net.au/home/species/5133). In *Hymenocephalus lethonemus*, there is a long luminescent organ with two external lenses just in front of the anus (Anon., https://en.wikipedia.org/wiki/Hymenocephalus_lethonemus).

6. Hymenogadus gracilis

Common name(s): Rattails or grenadier fish.

Global distribution: It is found distributed mostly in tropical and warm temperate seas. worldwide.

Ecology: It is a bathypelagic and benthopelagic species living in depths of about 300–580 m.

Biology: Head of these fishes is large, cavernous, and is covered by thin or membranous skin. Mouth is large and it is sub-terminal to nearly terminal. Gill openings and outer gill slit are wide. There are more than 16 gill rakers on the inner series of first arch. The first dorsal fin is with a weakly serrate spinous long second ray. Anus is located immediately before anal fin origin. Silvery striae are seen on the isthmus, shoulder girdle and chest. These fishes feed mainly on planktonic crustaceans and attain a maximum length of about 40 cm.

Image credit: Museums Victoria, https://fishesofaustralia.net.au/home/species/3788; https://creativecommons.org/licenses/by/3.0/au/

Luminous organs and bioluminescence.: In the species of *Hymenogadus,* the light organs are present underneath skin along the midline of abdomen with opening just before anus. This light organ is tubular with two lens-like windows, one before pelvic fin bases and the other which is round or oblate in outline is situated before the anus. In *Hymenogadus gracilis*, the light organ with its two external lenses is situated underneath the rear end of the body (Chiou et al., 2004). Merrett (1986) reported that its elongated light organ which is situated along its belly has an external circular lens-shaped body at each end.

7. Kumba dentoni

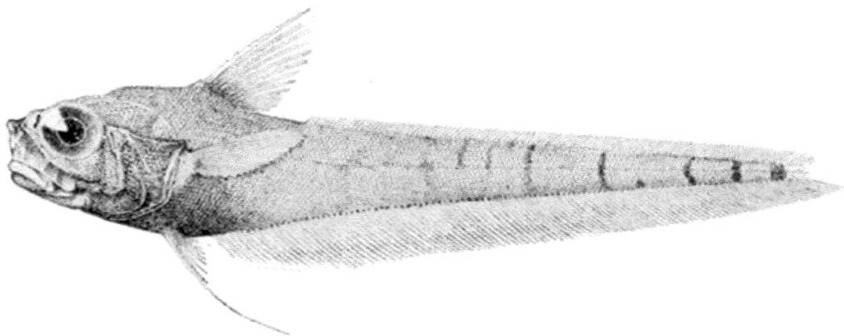

Kumba sp.

Source: Public domain

Common name(s): Grenadier fish or rattail.

Global distribution: Northeast Atlantic Ocean.

Ecology: It is a bathydemersal species living up to a depth of 1207 m.

Biology: It has a large and soft head which is inflated. There is a dorsally scaleless snout which is much longer than orbit is found extending to the level of its anterior nostril. A rudimentary barbel is seen. It grows to a maximum length of 22 cm.

Luminous organs and bioluminescence: In the species of this genus, the light organ is poorly developed and it is situated in the tear-drop shaped black area associated with the abdomen (McEachran and Fechhelm, 2013). In *Kumba detoni*, the light organ is visible only as a small, naked, black projection in front of the anus (Anon., https://www.fishbase.se/summary/Kumba-dentoni).

8. *Kuronezumia bubonis*

Image credit: Museums Victoria, https://fishesofaustralia.net.au/home/species/4361; https://creativecommons.org/licenses/by/3.0/au/

Common name(s): Bulbous whiptail.

Global distribution: Western Atlantic Ocean, Western Indian Ocean, and Pacific Ocean.

Ecology: This benthopelagic and bathydemersal species has a depth range of 700–1200 m. It is commonly found on the continental slope.

Biology: It is a large brownish grenadier fish with an elongate and attenuated body, a blunt snout and a moderately sized subterminal mouth. It has a serrated first dorsal-fin spine. The underside of its head is scaled except for a naked area just above the upper jaw. Fins are black colored. It grows to a maximum length of 80 cm.

Luminous organs and bioluminescence: It has a bulbous, scaly, wartlike and dermal light organ between its pelvic-fin bases (situated anterior to the anus) (Iwamoto et al., 2015).

9. *Lepidorhynchus denticulatus*

Image credit: Museums Victoria, https://fishesofaustralia.net.au/home/species/4355; https://creativecommons.org/licenses/by/3.0/au/

Common name(s): Thorntooth grenadier.

Global distribution: Eastern Indian Ocean and Southwest Pacific Ocean.

Ecology: This bathypelagic and non-migratory species has a depth range of 180–1000 m. It is commonly occurring on the continental slope.

Biology: The head of this species is with scarcely developed ridges. The eyes are large and chin barbel is short. Its first dorsal-fin spine is smooth. Scales are densely covered with lanceolate spinules which are arranged in quincunx or convergent rows. It grows to a maximum length of 55 cm.

Luminous organs and bioluminescence: It has only one small lens-like light organ which is located just anterior to the anus (Cohen et al., 1990).

10. Lucigadus lucifer

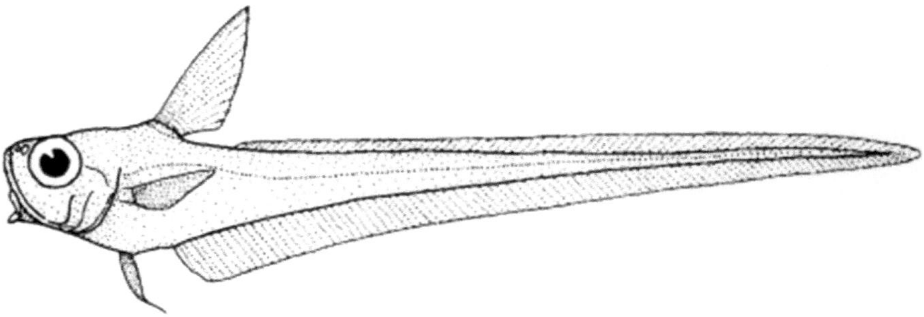

Lucigadus sp.

Image credit: Wikimedia. https://creativecommons.org/licenses/by-sa/1.0/deed.en

Common name(s): Rattail.

Global distribution: Western Pacific Ocean.

Ecology: This benthopelagic species is found up to a depth of 320 m.

Biology: It is a very small rattail with an unusual shape. All the ventral parts of its body seem to have "shifted forward" so that the pelvic fins are located just below the gill openings, and the origin of its long anal fin is just behind the head. Another significant feature of this species is its rounded snout without any coloration. It attains a maxiumum length of 11.0 cm.

Luminous organs and bioluminescence: In this species, there are two lens-like luminous organs. Among them, one is located between the ventral fins and the other is immediately before the anus- (Anon., https://www.etyfish.org/gadiformes1/).

11. Malacocephalus laevis

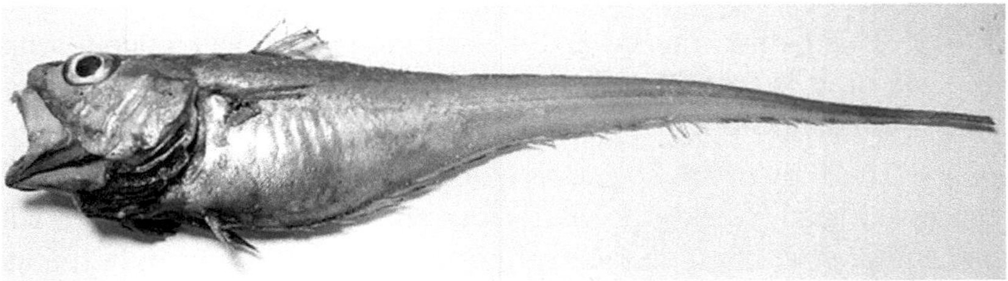

Source: Public domain

Common name(s): Softhead grenadier.

Global distribution: Western Atlantic Ocean and Indian Ocean.

Ecology: This benthopelagic species has a depth range of 200–1,000 m.

Biology: The head of this species is large with a length about 20% of its body length. The trunk is short and is tapering evenly to a pointed tail. Snout is short and blunt, and is less than eye diameter. The mouth is large, and subterminal and its upper jaw is fairly long. Pyloric caeca are branched. Anus is in naked area without scales and is located between pelvic insertion and anal fin origin. Color of the body is greyish or brownish; mouth is white; but ill covers, gular and branchiostegal membranes and abdomen are dark brown to blackish; median fins are greyish (except the tip of first dorsal fin); and paired fins are blackish. It feeds mainly on crustaceans and it grows to a maximum length of 50 cm.

12. Malacocephalus occidentalis

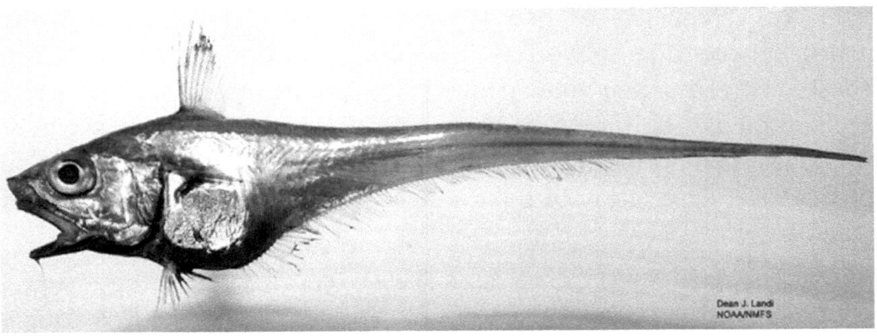

Source: Public domain

Common name(s): Western softhead grenadier.

Global distribution: It is found distributed from the warm temperate to tropical Atlantic Ocean.

Ecology: This bathydemersal and non-migratory fish has a depth range of 140–1945 m. It is found to inhabit commonly in warm to temperate waters on the continental slope.

Biology: The body of this species is elongated and attenuated with a laterally compressed head, weakly pointed snout and large jaws which are subterminal and are extending to the near posterior margin of orbits. Chin barbel is fairly long and is about equal in length to eye diameter. The gap between first and second dorsal fins is wide. Coloration of the body is brown dorsally and silvery ventrally. It feeds mainly on crustaceans and attains a maximum length of 45 cm.

13. Malacocephalus okamurai

Common name(s): Not designated.

Global distribution: Western Central Atlantic Ocean.

Ecology: This bathydemersal species has a depth range of 229–411 m.

Biology: In this species, its snout is high and moderately pointed. Mouth is large and its upper jaw extends to vertical through posterior border of orbits. Orbit is longer than both interorbital width and snout length. Its abdominal area and gill cover are black; gill filaments are pale and gill arches and rakers are dusky. It grows to a maximum length of 28 cm.

Luminous organs and bioluminescence: Malacocephalus spp.

In all the species of this genus, the luminous bacteria containing light organ which consists of two large dermal windows (naked fossae) is present between the pelvic fins. i.e., just before the anus (Anon., https://fishesofaustralia.net.au/home/genus/860). Among these fossae, the anterior one is large and halfmoon or bean-shaped. These light organs which are located just before the anus are bulbous with a glandular lining and a single opening. The description of light organs of the aforementioned species are given below.

Malacocephalus laevis: Cohen et al. (1990) reported that in this species, the light organ has two dermal windows,each associated with a lens. Dunlap et al. (2007) reported that the light organ of this species harbors the luminescent bacteria, *Photobacterium kishitanii*. When the fish is disturbed, it ejects some of the luminous bacteria into the water through its anus. The sudden appearance of this luminescent mass helps the fish disorientate its predator. Anctil (2018) observed that the cloudy liquid of this species exudes a blue luminescence in the dark. This liquid when smeared onto a piece of shark skin, this luminescence was found to last for hours. It is believed that the fishermen used these smeared shark skin pieces as fish bait to attract fishes to the light.

Malacocephalus occidentalis: Its abdomen has a semi-circular shaped naked black area associated with a light organ which is located anterior to the pelvic fin base (McEachran, 2021).

Malacocephalus okamurai: The anterior dermal window and posterior dermal window of the light organ of this species are well developed (Iwamoto and Arai, 1987).

14. Mesovagus antipodum (= *Mesobius antipodum*)

Image credit: Museums Victoria, https://fishesofaustralia.net.au/home/species/3791; https://creativecommons.org/licenses/by/3.0/au/

Common name(s): Bathypelagic rattail.

Global distribution: Indo-West Pacific Ocean.

Ecology: This meso-bathypelagic and benthic species has a depth range of 846–1300 m and it is commonly found on the continental slope.

Biology: The head of this species is laterally compressed and it is entirely covered with scales (except in lips and gular and branchiostegal membranes). Head ridges are poorly developed. There are no prominent tubercles or modified scales on the snout. Chin barbel is absent. Abdomen is short and gas bladder is greatly reduced. The body coloration is mostly blackish and tail is greyish brown. It has a maximum length of 66 cm.

15. Mesovagus berryi *(= Mesobius berryi)*

Common name(s): Berry's grenadier, midwater grenadier.

Global distribution: Eastern Pacific Ocean.

Ecology: This meso-, bathy- and benthopelagic species has a depth range of 0–2700 m. It is commonly found on the continental slope.

Biology: This species largely resembles *Mesovagus antipodum* and it has a maximum length of 39 cm.

Luminous organs and bioluminescence: Mesobius spp.

In the species of *Mesobius*, the light organ is large and is bulbous with a glandular lining. Further, it has a single opening just before the anus (Merrett, 2003).

16. Nezumia aequalis

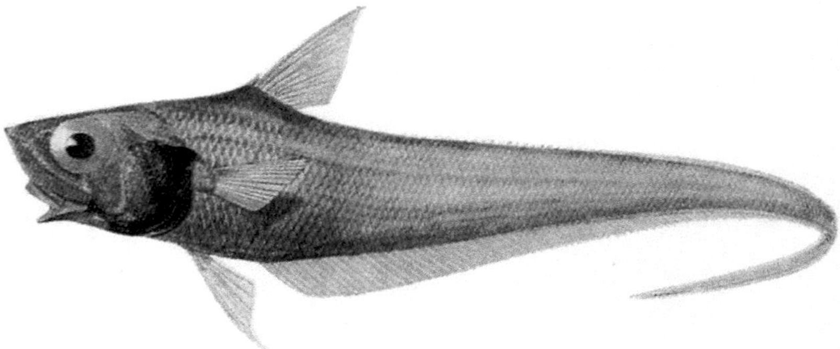

Source: Public domain

Common name(s): Common Atlantic grenadier.

Global distribution: Eastern and Western Atlantic Ocean.

Ecology: This benthopelagic and non-migratory species has a depth range of 200–2320 m.

Biology: The head of this species is short and is completely scaled except for a naked strip on the ventral surface of its snout. The eyes are fairly large with about 35% of head length. Snout is short and pointed. Chin barbel is very short. The body scales have shield-shaped spinules. Overall body color is bluish to violet and the head is slightly tawny to swarthy. It feeds on benthic as well as pelagic animals like polychaete worms, mysids, amphipods, small shrimps, copepods, isopods, and ostracods and it grows to a maximum length of 36 cm.

17. Nezumia stelgidolepis

Common name(s): California grenadier.

Global distribution: Eastern Pacific Ocean.

Ecology: This bathydemersal and non-migratory species has a depth range of 277–909 m.

Biology: It has a large head, large eyes, and a slender black body. Snout is short, narrow and is bluntly pointed with a terminal scute. There are 24–58 pyloric caeca. The body scales are densely covered with conical to narrowly lanceolate spinules which are arranged in short, slightly convergent rows. Overall body coloration is swarthy and it is blackish ventrally, with bluish tinge on abdomen. Fins are dusky to blackish and first dorsal fin is somewhat pale basally. It is an opportunistic feeder and it grows to a maximum length of 45 cm.

Luminous organs and bioluminescence: Nezumia spp: In the species of *Nezumia,* the light organ is bulbous with a glandular lining and is with a single opening just before the anus (Merrett, 2003). While the light organ of *Nezumia aequalis* harbors the luminescent bacteria, *Photobacterium kishitanii* (Dunlap et al., 2007), and *Photobacterium phosphoreum* (Ruby and Morin, 1978), *Nezumia stelgidolepis* has *Photobacterium phosphoreum* in its light organ (Ruby and Morin, 1978).

18. Odontomacrurus murrayi

Image credit: Museums Victoria, https://fishesofaustralia.net.au/home/species/3800; https://creativecommons.org/licenses/by/3.0/au/

Common name(s): Roundhead grenadier.

Global distribution: Atlantic, Indian and Pacific Oceans.

Ecology: This bathypelagic species occurs up to a depth of about 2500 m.

Biology: The body of this species is deep and is well compressed laterally. Snout is broadly rounded and is slightly protruding beyond upper jaw. The mouth is large and terminal with a single row of teeth in its upper jaw. Chin barbel is absent. Swim bladder is reduced. General body coloration is dark brown. It has a maximum length of 64 cm.

Luminous species: It is a single species of this genus.

Luminous organs and bioluminescence: In this species, there is a small dermal window of light organ in front of the anus (Nakayama et al., 2015).

19. Sphagemacrurus hirundo

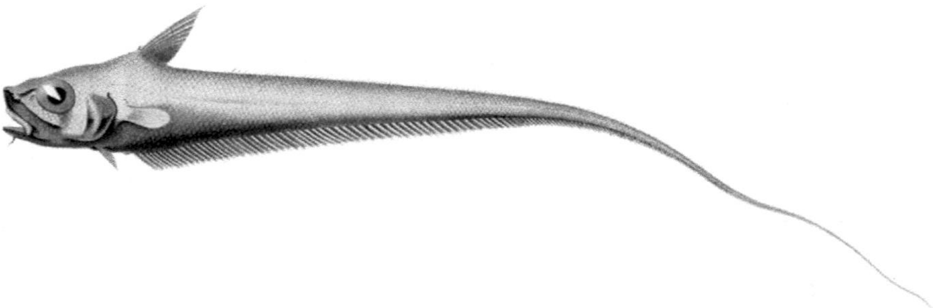

Source: Public domain

Common name(s): Swallow grenadier.

Global distribution: Eastern Central Atlantic Ocean.

Ecology: This bathydemersal species has a depth range of 1266–2330 m.

Biology: Overall body coloration of this species is dark brown. Snout is with spiny scutes at its tip. Mouth is subterminal and chin barbel is short. It is scaleless ventrally and attains a maximum length of 21 cm.

Luminous organs and bioluminescence: In the species of *Sphagemacrurus,* the light organ is bulbous with a glandular lining with a single opening just before the anus (Merrett, 2003). Ruby and Morin (1978) reported that its light organ contains the luminous psychrotrophic bacterium, *Photobacterium phosphoreum.*

20. Spicomacrurus spp.

Common name(s): Rattails.or grenadiers.

Global distribution: Western Pacific Ocean.

Ecology: This benthopelagic species has a depth range of 315–710 m.

Biology: Its diagnostic characters include a protruding snout; a net-like black epithelium covering the gular region; and a membranous flange on the distal portion of its outer pelvic-fin ray. It attains a maximum length of 24 cm.

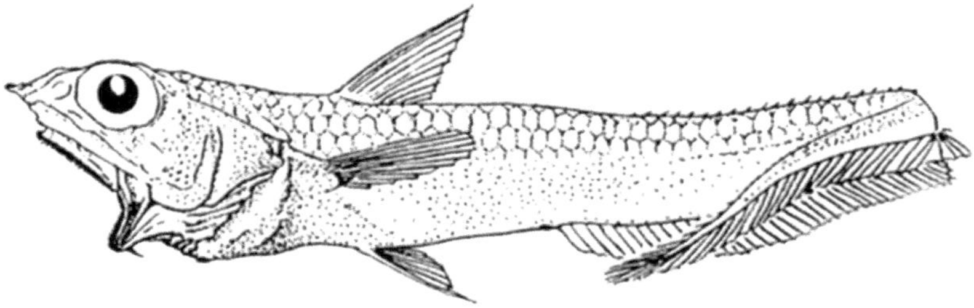

Spicomacrurus adelscotti

Image credit: Museums Victoria, https://fishesofaustralia.net.au/home/species/3783

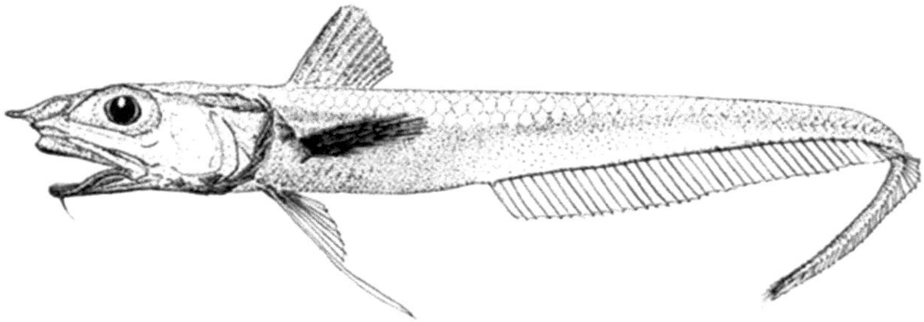

Spicomacrurus kuronumai

Image credit: Museums Victoria, https://fishesofaustralia.net.au/home/species/3789

Luminous organs and bioluminescence: Spicomacrurus spp.

The species of *Spicomacrurus* have been reported to possess an elongate posterior dermal window of the light organ (Anon., https://fishesofaustralia.net.au/home/genus/1724). Schwarzhan (2014) reported that the species of *Spicomacrurus* viz. *Spicomacrurus adelscotti, Spicomacrurus dyctiogadus, Spicomacrurus kuronumai* and *Spicomacrurus mccoskeri* showed ventral luminescent striae on their abdominal wall, parts of chest, and shoulder girdle. These striae which are associated with light-producing function consisted mostly of fine parallel lines of dark, alternating with silvery pigment.

21. Trachonurus sp.

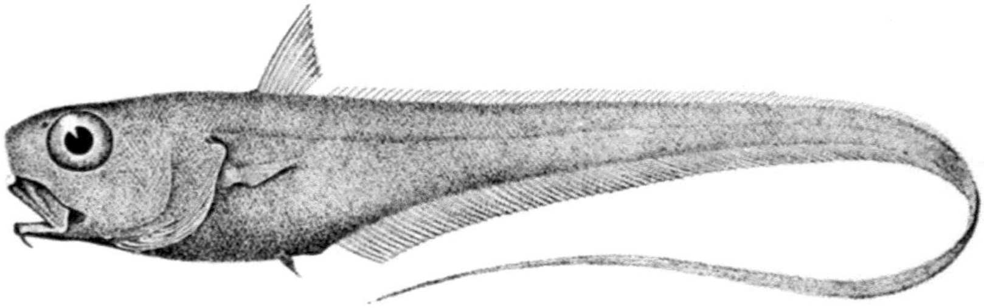

Source: Public domain

Common name(s): Grenadier or rattail.

Global distribution: It has a wide distribution from the Arctic Ocean to the Antarctic Ocean.

Ecology: This benthopelagic species is found at depths of 200–2000 m.

Biology: It has a large head with a projecting snout. A single barbel is seen on its chin. The body is slender and is tapering to a whip-like tail which is without a distinct caudal fin which is often confluent with the posterior dorsal and anal fins.

Luminous organs and bioluminescence: In some species of *Trachonurus*, the light organ is bulbous with a glandular lining with a single opening (dermal window or fossa) just before the anus (Merrett, 2003).

22. Ventrifossa spp.

Common name(s): Grenadiers.

Global distribution: It has worldwide distribution and mostly in tropical to warm-temperate seas.

Ecology: This benthopelagic species has a depth range of 200–1,500 m.

Biology: Snout of these species is slightly pointed to rounded. A small and poorly developed median tubercle is seen on the snout tip in some species. Lateral snout tubercles are not seen. Head is without stout, coarsely scuted ridges. The mouth is subterminal and upper jaw is about 35% of head length. There are 30 to 70 pyloric caeca. These fishes grow to a total length of more than 53 cm.

Luminous organs and bioluminescence: Ventrifossa spp.

In the species of *Ventrifossa,* a small, round dermal window of light organ is present at anterior end of periproct (small region surrounding the anus); and a second window is situated in a shallow fossa between pelvic fin bases, i.e., anterior to periproct (Cohen et al., 1990). Merrett (2003) reported that in the species of *Ventrifossa,* the light organ is bulbous with a glandular lining with a single opening just before the anus as in other species of this family. In *Ventrifossa saikaiensis* and *Ventrifossa rhipidodorsalis*, the ventral luminous organ is rather large and is situated between pelvic bases and is connected with the black periproct by a narrow naked groove. On the other hand, in *Ventrifossa longibarbata*, a small bioluminescent organ is located between the bases of the pelvic fins (Chiou et al., 2004). In *Ventrifossa macropogon,* its ventral fossa is bean-shaped (Anon., http://www.oads.org.br/livros/23.pdf). Tanetet al. (2020) reported that the light organs of *Ventrifossa garmani, Ventrifossa longibarbata, Ventrifossa macropogon, Ventrifossa mucocephalus,* and *Ventrifossa rhipidodorsalis* harbor the luminescent bacteria, *Photobacterium kishitanii.*

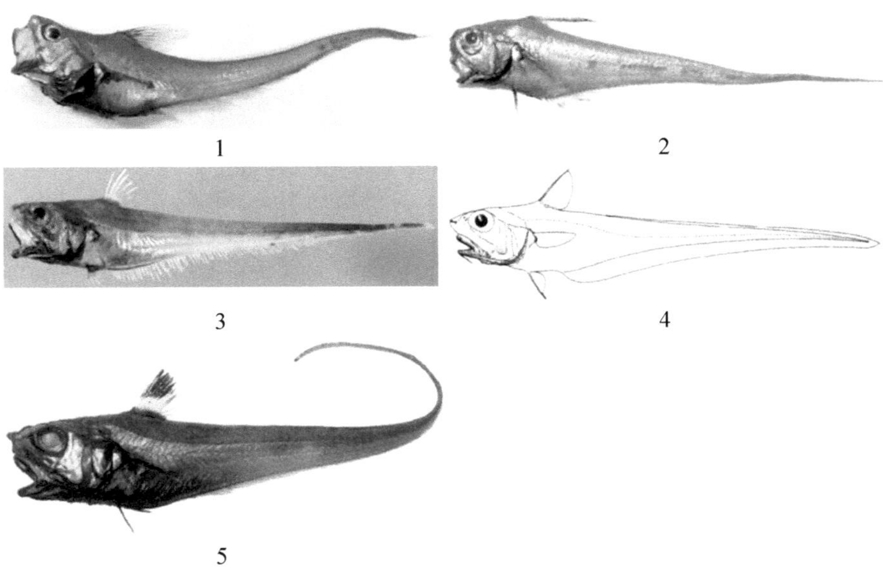

Credits: 1 & 2: Kwang-Tsao Shao, Taiwan Fish Database; 3 & 5: Museums Victoria, https://fishesofaustralia.net.au. https://creativecommons.org/licenses/by/3.0/au/; 4. https://www.inaturalist.org/taxa/92782-Ventrifoss. https://creativecommons.org/licenses/by-sa/3.0/.

1. *Ventrifossa longibarbata*; 2. *Ventrifossa rhipidodorsalis*; 3. *Ventrifossa macropogon*; 4. *Ventrifossa mucocephalus*; 5. *Ventrifossa nigrodorsali.*

3.2.6.3 BLUE GRENADIERS (MERLUCCIIDAE)

1. *Steindachneria argentea*

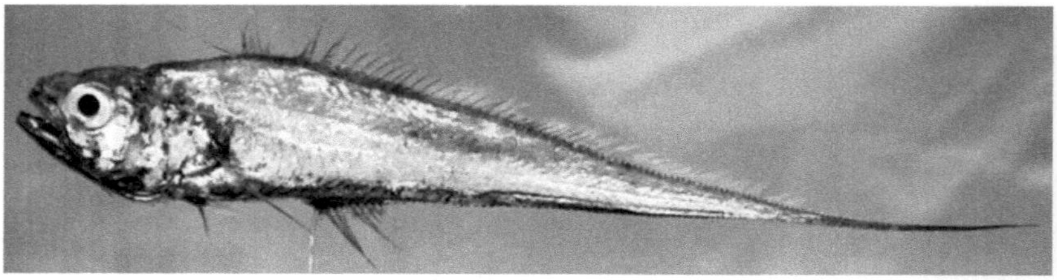

Source: Public domain

Common name(s): Luminous hake.

Global distribution: Western Atlantic Ocean.

Ecology: This bathydemersal species has a depth range of 400–500 m and is commonly inhabiting the outer shelf and upper slopes and on soft bottoms.

Biology: The body of this species is long and compressed and is tapering to a fine point posteriorly. The body coloration is silvery; dorsal portions are slightly brownish; purplish ventrally; and oral cavity is darker. It grows to a maximum length of 30 cm.

Luminous organs and bioluminescence: The light organ of this species is an open type, doughnut- shaped gland surrounding the anus and is containing symbiotic type of bacteria (Haneda, 1980). In this species, the bacterially produced light is emitted through the striated skin and hypaxial abdominal musculature (Finn, 2020).

3.2.6.4 MORA CODS (MIRIDAE)

Among the 20 genera of the family Moridae, 4 genera viz. *Gadella, Physiculus, Salilota,* and *Tripterophycis* have been reported to possess light organs. In the first three genera, the light organ consists of a pigmented gland and is located in the body wall behind the ventral fin bases.It seems to be a scaleless, circular dermal fossa and is connected to the anus by an internal duct (Paulin, 1989).

1. *Gadella* spp.

Common name(s): Morid cods.

Global distribution: These species are found distributed in warm temperate and tropical seas.

Ecology: This benthopelagic species lives generally at depths of 40–1500 m.

Biology: These are morid fishes with an elongate compressed body. Snout is broad and obtusely rounded and is not projecting beyond the mouth. Chin barbel is absent. There are two dorsal fins and one anal fin, and a slightly rounded caudal fin. These fishes grow to a maximum length of 30 cm.

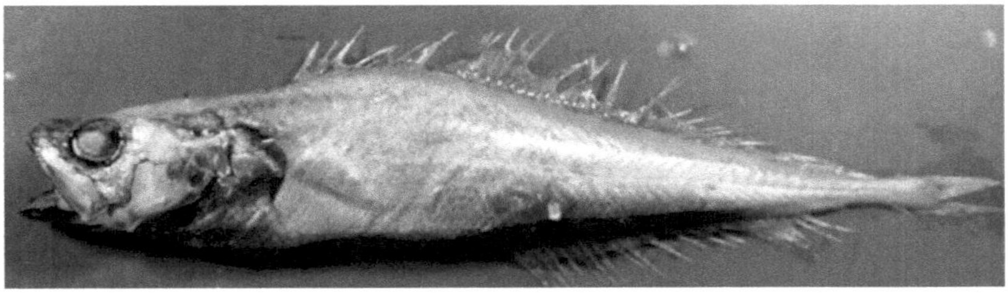

Gadella imberbis

Source: Public domain.

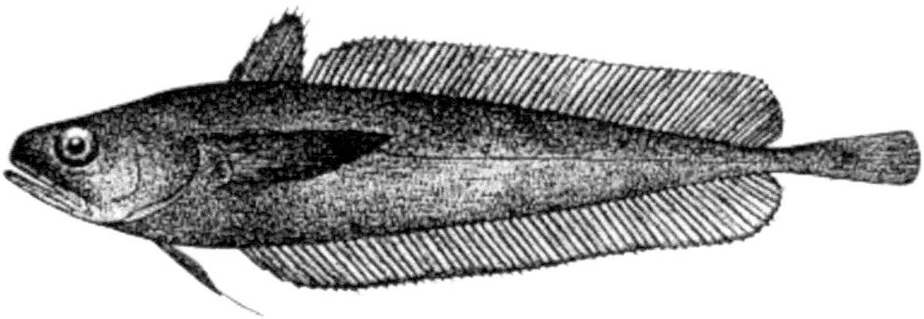

Gadella maraldi

Source: Public domain

Luminous organs and bioluminescence: Gadella spp.

In these species, a small circular luminescent organ which is a scaleless, dermal fossa is seen on midline of belly and is connected to anus by duct. Barnes (2003) reported that the light organ is a bulbous gland containing bacteria and is opening into the rectum in the species of *Gadella norops, Gadella molokaiensis, Gadella imberbis,* and *Gadella jordani.* Tanet et al. (2020) reported that the light organ of *Gadella maraldi* contains the symbiotic luminous bacterium, *Photobacterium mandpamensis*. The location of the light organ in the different species of *Gadella* as a measure of the distance (between a transverse line) between the anterior edges of the ventral fin insertions and the anal fin origin is given below as % InV-af (Paulin, 1989).

Species of *Gadella*	% InV-af
Gadella filifer	8.4
Gadella obscurus	4.6
Gadella edelmanni	5.7
Gadella norops	4.0
Gadella molokaiensis	5.7
Gadella imberbis	8.6
Gadella jordani	3.7
Gadella maraldi	9.1

Source: Paulin (1989).

2. *Physiculus* spp.

Common name(s): Morid cods.

Global distribution: These species are found distributed in tropical and temperate waters of all oceans.

Ecology: These benthopelagic species which may prefer a rough, rocky bottom are occurring at depths ranging from 40 to 1500 m.

Biology: These morid fishes have an elongated and somewhat compressed body. Chin barbel is present. There are two dorsal and one anal fin of uniform height. Snout is broad and obtusely rounded, and is not projecting beyond mouth; Coloration of body in the various species varies from black or blue-black to pale pink or scarlet.

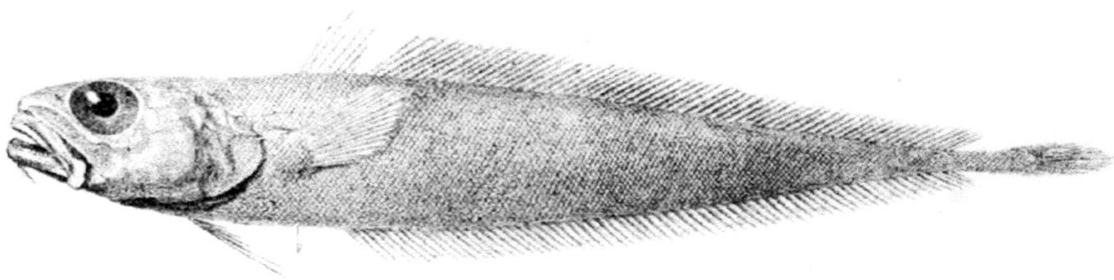

1

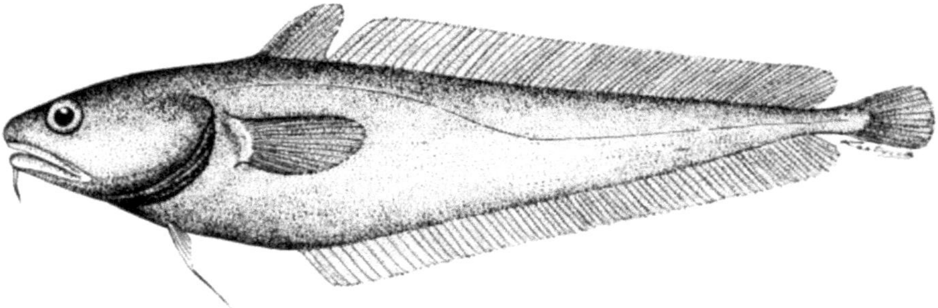

2

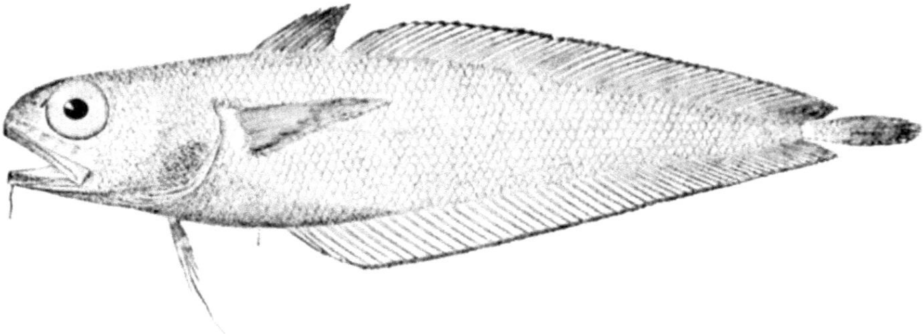

3

Credits: 1 & 3: Public domain; 2: https://thewebsiteofeverything.com/animals/fish/Gadiformes/Moridae/Physiculus-japonicus. https://creativecommons.org/licenses/by-nc/3.0/.

1. *Physiculus dalwigkii;* 2. *Physiculus japonica;* 3. *Physiculus fulvus.*

Luminous organs and bioluminescence: Physiculus spp.: In the species of *Physiculus*, a ventral light organ with luminescent symbiotic bacteria is present. This light organ seems to be a circular, scaleless dermal fossa located on the midline of its belly in front of the anus. This light organ is also consisting of a pigmented gland in the body wall behind the ventral fin base and is connected to the anus by an internal duct (Paulin, 1989). While the species *Physiculus japonicus* harbors the luminous symbiotic bacterium, *Photobacterium kishitanii*, it is *Photobacterium phosphoreum* in the case of *Physiculus rastrelliger* (Tanet et al., 2020). Dunlap et al. (2007), however, reported that the light organ of *Physiculus japonicus* harbors two species of luminescent, symbiotic bacterial species viz. *Vibrio fischeri* and *Photobacterium kishitanii*.

Paulin (1989) reported on the light organ of the different species of the *Physiculus* as follows (Paulin, 1989). (The location of the light organ in the different species of *Physiculus* as a measure of the distance (between a transverse line) between the anterior edges of the ventral fin insertions and the anal fin origin is given as % InV-af).

Physiculus argyropastus: Light organ is moderately small; 7.0% InV-af; and is placed approximately midway between interventral line and anus.

Pbyslculus capensis: Light organ is moderately small; 7.3% InV-af; and is placed approximately midway between interventral line and anus.

Physiculus chigodarana: Light organ is moderately small; 8.4% InVoaf; and is placed closer to interventral line than anus.

Physiculus coheni: Light organ is small; 7.4% InV-af; and is placed closer to anus than interventral line.

Physiculus cynodont: Light organ is moderately small; 7.7% InV-af; and is placed closer to anus than interventral line.

Physiculus dalwigkii: Light organ is very large; 18.8% InV-af; and is placed immediately behind interventral line.

Physiculus fulvus: Light organ is moderately large; 11.7% InV-af; and is placed midway between interventral line and anus.

Physiculus grinnelli: Light organ is moderately large; 10.5% InV-af; and is placed slightly closer to interventral line than anus.

Physiculus helenaensis: Light organ is moderately large; S.S% InV-af; and is placed approximately midway between interventral line and anus.

Physiculus huloti: Light organ is moderately large; 9.5% InY-af; and is placed approximately midway between interventral line and anus.

Physiculus japonica: Light organ is moderately small; 9.0% InV-af; and is placed approximately midway between interventral line and anus.

Physiculus karrerae: Light organ is large; 17.1% InV-af; and is placed closer to interventral line than anus.

Pbysiculus kaupi: Light organ is moderately large; 10.1% InV-af; and is placed slightly closer to interventral line than anus.

Physiculus longicavis: Light organ is moderately large; 13.6–14.8% InV-af; and is placed closer to anus than interventral line.

Physiculus longifllis: Light organ is moderately small; 8.3% InV-af; and is placed slightly closerto anus than interventral line.

Physiculus luminosa: Light organ is large; 15.9% InV-af; and is placed closer to interventral line than anus.

Physiculus marisrubri: Light organ is moderately large; 8.0–8.9% InV-af; and is placed closer to anus than interventral line.

Physiculus natalensis: Light organ is large; 14.2% InV-af; and is placed close to interventral line.

Physiculus nematopus: Light organ is moderately small; 7.8% InV-af; and is placed slightly closer to interventral line than anus.

Physiculus nigrescens: Light organ is small; 7.1% InV-af; and is placed closer to anus than interventral line.

Physiculus nigripinnis: Light organ is moderately small; 8.4% InV-af; and is placed closer to anus than interventral line.

Physiculus normani: Light organ is moderately small; 8.3% InV-af; and is placed closer to anus than interventral line.

Physiculus peregrinus: Light organ is moderately small 8.5% InV-af; and is placed closer to the anus than interventral line.

Physiculus rastrelliger: Light organ is large 16.2% Inv-af; and is placed close to interventral line.

Physiculus rhodopinnis: Light organ is moderately large; 11.6% InV-af; and is placed slightly closer to interventral line than anus.

Physiculus roseus: Light organ is moderately large; 10.7% InV-af; and is placed closer to interventral line than anus.

Physiculus slerops: Light organ is small; 8.0% InV-af; and is placed closer to anus than interventral line.

Physiculus sudanensis: Light organ is large 17.2% InVoaf; and is placed close to interventral line.

Physiculus talarea: Light organ is moderately small; 8.9% InV-af; and placed closer to interventral line than anus.

Physiculus therosideros: Light organ is large; 13.7% InVoaf; and is placed closer to anus than interventral line.

Physkulus yoshidae: Light organ is small; 7.1% InV-af; and is placed closer to anus than interventral line.

3. Salilota australis

Source: Public domain

Common name(s): Tadpole codling.

Global distribution: Southeast Pacific and Southwest Atlantic Oceans. It is found however only in the cool temperate waters of South America.

Ecology: This demersal and oceanodromous species is found at depths ranging from 40 to 1500 m.

Biology: These are morid fishes with an elongate compressed body. Snout is broad, and obtusely rounded but is not projecting beyond the mouth. Chin barbel is present Two dorsal and one anal fin of uniform height are seen. A slightly rounded caudal fin is present. Coloration of body is uniformly brown and fins are dark-edged. Fishes of this species grow to a maximum length of 50 cm.

Luminous species: It is a monotypic genus (genus with a single species).

Luminous organs and bioluminescence: It has a ventral, circular light organ which seems to be a dermal fossa on the midline of its belly in front of the anus. This light organ is a pigmented gland in the body wall behind the ventral fin base and is connected to the anus by an internal duct (Paulin, 1989).

4. Tripterophycis gilchristi

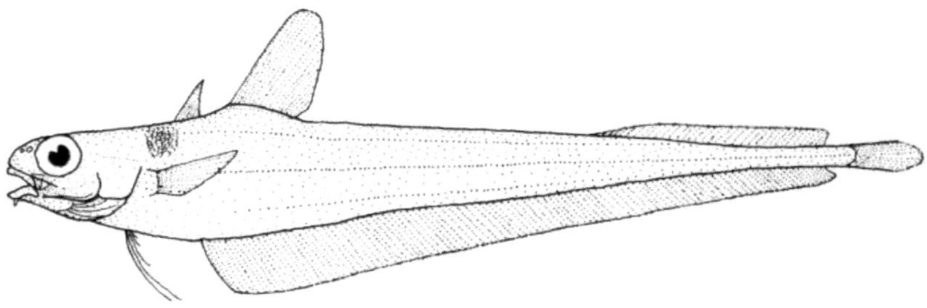

Image credit: Dr Tony Ayling, https://creativecommons.org/licenses/by-sa/1.0/deed.en

Common name(s): Chiseltooth grenadier cod.

Global distribution: Indo-Pacific ocean and mid-south Atlantic ocean.

Ecology: It is a deep sea benthopelagic species occurring at depths from 300 to 750 m. It seems to be common on the continental slope areas at depths more than 1,000 m.

Biology: It is a pale brown to pink colored cod and is growing to a maximum length of 33 cm.

5. Tripterophycis svetovidovi

Common name(s): Brown grenadier cod.

Global distribution: South Atlantic, Indian and Southeast Pacific Oceans.

Ecology: It is a deep sea bentho – bathypelagic species having a depth range of 385–950 m.

Biology: It is a pale tan grenadier cod with a pale tan head, darker brown snout and orbit, a bluish colored abdomen with a silvery sheen, and dark bluish-brown to black fins. It grows to a maximum length of 24 cm.

Luminous organs and bioluminescence: Species of *Tripterophycis* spp.

In the species of *Tripterophycis*, their ventral light organs are found restricted to an area around the anus. These light organs are bulbous glands containing luminous bacteria and they open into the rectum (Paulin, 1989).

3.2.7 ORDER: LOPHIFORMES

Anglerfish (Lophiformes) are mainly found in the deep-ocean where the absence of sunlight, and extremely high pressures and low temperatures prevail. They live mostly solitary lives. They are strange looking deep sea fish with sharp teeth. Among these fish, only female anglerfish are bioluminescent. Protruding from the dorsal spine of the females is a bulb of flesh that contains photophores (light-producing glands or organs). This appendage resembles a fishing pole and lure that hangs above the animal's mouth. Adult female anglerfishes of most species have a lure (called 'esca') which is present at the end of a "fishing rod" (called 'illicium,' a modified dorsal fin ray;). These escae contain a single species of bioluminescent bacteria; These bacteria produce the necessary chemicals to emit light. Among the different families of the angler fishes, the adult female fishes of the family Linophrynidae, also have a chin barbel, the luminescence of which is provided by the fish itself and not bacteria (Anon., https://

oceanexplorer.noaa.gov/explorations/19biolum/background/midnight-zone/midnight-zone.html). The luminescent bulb of these anglerfish lights up and attracts prey in the dark aquatic environment to the large open mouth of these fishes. The lure is also believed to serve as a means to attract male anglerfish. A unique symbiosis is existing between these fishes and microbes. In this mutualistic relationship, the bacteria receive protection and a place to live and grow; and the anglerfish benefit from the relationship by gaining a means of attracting food (Anon., https://www.thoughtco.com/amazing-bioluminescent-organisms-373898).

3.2.7.1 PRICKLY SEADEVILS (CENTROPHRYNIDAE)

1. Centrophryne spinulosa

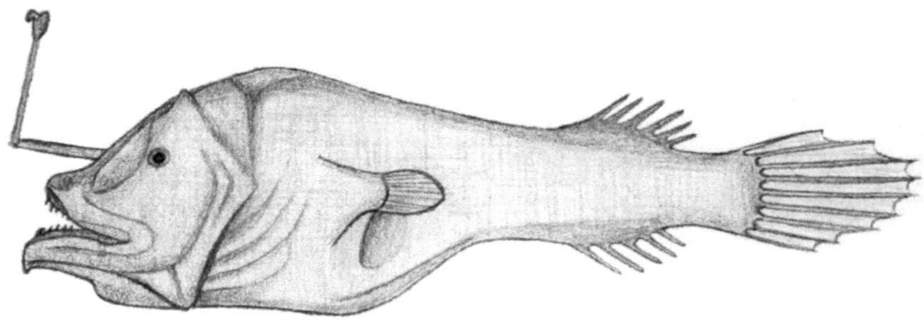

Image credit: Tambja, https://creativecommons.org/licenses/by-sa/3.0/deed.en

Common name(s): Horned lantern fish.

Global distribution: This species is found distributed in the tropical and subtropical parts of all oceans.

Ecology: This bathypelagic and mesopelagic fish has a depth range of 590–2325 m.

Biology: The body of this fish is long and slender, with a large head. Its jaws are of equal length and are filled with slender, recurved, depressible teeth of both large and small sizes. There is a large oval pit in front of each eye in fishes larger than 4 cm in length. The eye itself lies beneath the skin and it appears through a translucent patch. The body coloration is reddish brown to black. its skin is covered with numerous close-set spines. The illicium ("fishing rod") and its esca (lure) are attached to the snout. Esca has a fan-shaped appendage in front and another short appendage on the back, and the tip is white with scattered large melanophores. A small hyoid (chin) barbel is also present in both the sexes. It grows to a maximum length of only 1.3 cm.

Luminous organs and bioluminescence: Its illicium and esca (fishing rod and lure like body modification) are provided with bioluminescent symbionts (Freed et al., 2019). It uses its luminescent lure to attract its prey (Anon., https://sites.google.com/pingree.org/deepsea/species/vertebrates). Though a short hyoid (chin) barbel which is believed to possess intrinsic luminescence is present in both the sexes of this species, the function of this barbel is still unknown (Anon., https://en.wikipedia.org/wiki/Horned_lantern_fish).

3.2.7.2 SEADEVILS OR ANGLERFISHES (CERATIIDAE)

The ceratioid anglerfishes spend most of their lives in the dark, bathypelagic zone which is below 2000 m. The genus *Ceratias* differs from *Cryptopsaras,* the only other genus of the Ceratiidae, by having nine caudal-fin rays against 8 caudal-fin rays of the latter. In the species of *Ceratias,* the ninth or ventral-most ray is reduced to a small remnant. While the female fishes of *Ceratias* are larger and luminescent, the males are much smaller and are obligatory sexual parasites as adults. The females are characterized by an amazing skeletal adaptation. That is their first dorsal spine, the illicium is modified to carry a lure (esca) which is filled with bioluminescent bacteria. The pterygiophore (stem) of the illicium emerges anteriorly well behind the tip of the snout and posteriorly on the back, near the soft dorsal-fin origin; (Mincarone et al., 2021). The esca may be solidly planted right above the fish mouth as a lamppost or it may dangle at the end of a long line which may be equivalent to many times the length of the fish. In all the case, the purpose of the luminescent lure is only to attract the swimming prey (Wilson, 2013). In some species, the females of ceratioid fishes may have two or 3 club-shaped, bioluminescent caruncles (low fleshy appendages) on the dorsal midline just anterior to the soft dorsal fin. Further, female fishes of the genera *Ceratias* and *Cryptopsaras* are known to discharge a luminous secretion from 2–3 sacs on the back (Matsui and Rosenblatt, 1984).

2. Ceratias holboelli

Common name(s): Krøyer's deep sea angler fish.

Global distribution: It has circumglobal distribution in tropical to temperate waters of all oceans.

Ecology: The species of *Ceratias* are the largest known deep-water angler-fishes. and are found to live in depths of 0–4,400 m.

Biology: The bioluminescent females of this species have fairly long, compressed body, small vertical mouth, and single distal escal appendage. As in other species of this genus, mature males are found living parasitically on the females. Pietsch (1986a) reported that the females of this species are the largest deep-sea angler, reaching 145 cm TL (weighting around 10 kg) and males are parasitic and are reaching a total length of 118 mm (about 150 g).

3. Ceratias tentaculatus

Common name(s): Southern seadevil.

Global distribution: It is endemic to the Southern Ocean.

Ecology: This bathydemersal species has a depth range of 100–2900 m.

Biology: Metamorphosed female fish of this species has the following characteristic features: a pair of distal esca appendages; darkly pigmented proximal portion of escal bulb; proximal portion tapering into stem of illicium (pterygiophore); two slender escal appendages which are bifurcate, trifurcate or multibranched; and presence of escal pore at apex of bulb. The females grow to a maximum length of 88 cm.

4. Ceratias uranoscopus

Common name(s): Stargazing seadevil.

Global distribution: It has circumglobal distribution in tropical waters and is seen in the Atlantic, Pacific, and Indian Oceans.

Ecology: This bathypelagic and mesopelagic species has a depth range of 500–1,000 m.

Biology: Female fishes of this species have simple esca but without distal escal appendages. Escal bulb is darkly pigmented except its distal tip. While the distal portion of the escal bulb is oval shaped,its proximal portion is tapering into the stem of illicium. These female fishes grow to a maximum length of 24 cm.

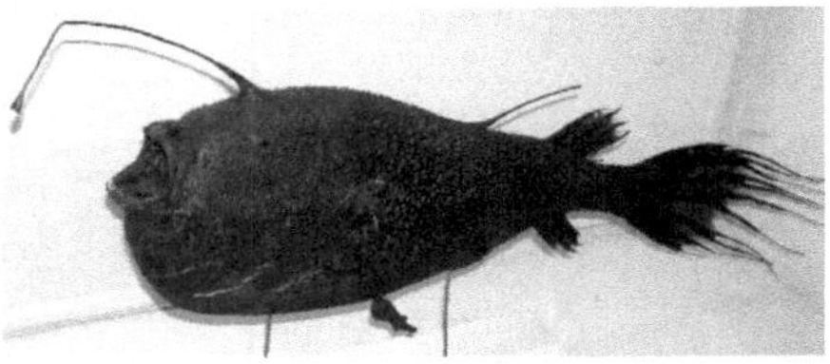

1

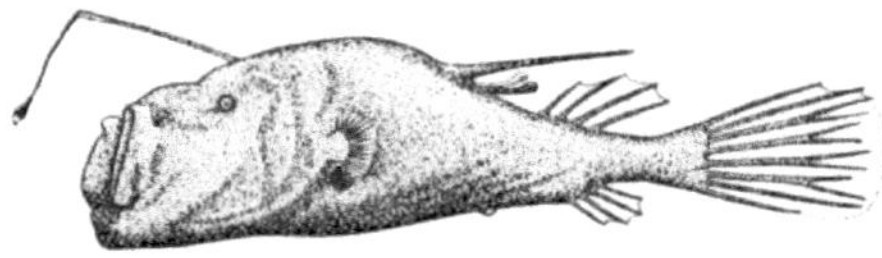

2

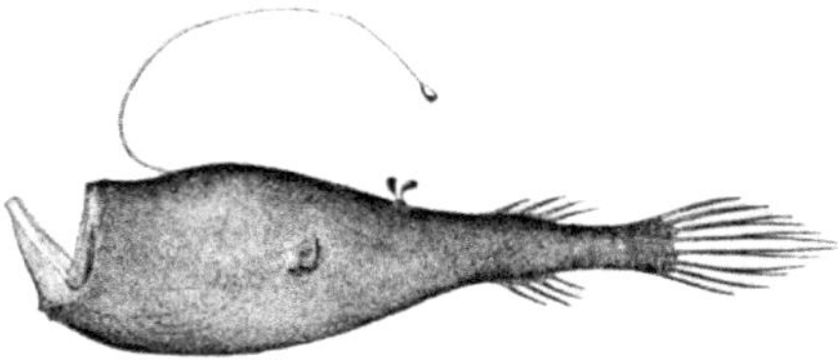

3

Credits: 1: Andrew Butko. https://creativecommons.org/licenses/by-sa/3.0/deed.en; 2: Institute of Oceanographic Sciences (Great Britain). https://creativecommons.org/publicdomain/zero/1.0/deed.en; 3. Public domain.

1. *Ceratias holboelli;* 2. *Ceratias tentaculatus;* 3. *Ceratias uranoscopus.*

Luminous organs and bioluminescence: Ceratias spp.

Adult females of *Ceratias* possesses a long "fishing rod" (illicium) and a bioluminescent lure (esca) on the snout. This lure is fully packed with the luminous bacteria, *Candidatus Enterovibrio escacola* (Tanet et al., 2020; Maia et al., 2021). The bacterial luciferase in the escae of species of *Ceratias* might have been produced by the luminous symbiotic bacterial species of *Photobacterium*. In addition to these luminescent structures, two club shaped (soft fleshy wart-like) bioluminescent appendages (caruncles, additional bacterial light organs) are seen in the species of *Ceratias* on the dorsal midline of the trunk just anterior to the origin of the soft dorsal fin (Freed et al., 2019; Anderson and Leslie, 2001). These caruncles have been reported to degenerate in the species of *Ceratias* at about 40 cm SL, but are retained by the species of *Cryptopsaras*. Caruncles secrete a whitish-yellow luminescent fluid which appears to be an anti-predator, startle substance.

It is also reported that the species of *Ceratias viz. Ceratias holboelli, Ceratias uranoscopus,* and *Ceratias tentaculatus* possess a hitherto unknown escal gland which is not associated with the escal light gland of these species. This accessory escal gland (AEG) is located in the connective tissue around the illicium and forms part of the existing escal light gland. The presence of ducts opening on the surface of the esca suggests that the AEG, like the already existing escal light gland, might have been derived from the epidermis. Though the function of the AEG is unknown, its location suggests that this gland may supplement the escal light gland, by secreting substances which may attract the prey animals (Munk, 1992).

Herring (1983) reported that light emitted by *Ceratias holboelli* showed the maximum emission with a wavelength of 470 nm.

5. Cryptopsaras couesii *(= Ceratias carunculatus)*

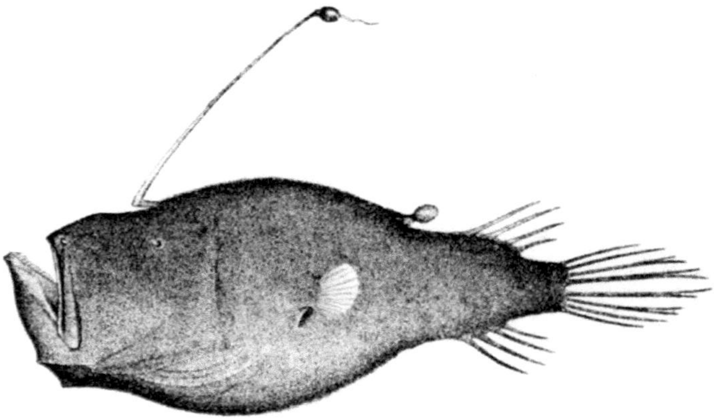

1

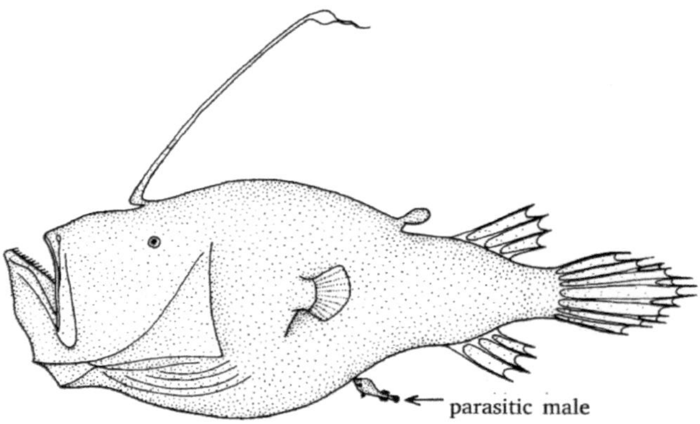

2

Credit: Public domain.

1. *Female;* 2. *Female with parasitic male attached (arrow).*

Common name(s): Triplewart seadevil.

Global distribution: It is found distributed in tropical to subpolar waters of all three major oceans.

Ecology: Adults are mesopelagic and are found at depths of 75–4000 m. It is known to get dispersed from the deep ocean to shallower water due to its weak swimming power.

Biology: Female fishes of this species have a laterally compressed and elongated body with a large and massive head and a mouth which is almost vertical when it is closed. The eye is minute, and subcutaneous. The body is covered by deeply embedded hollow spines. Adult female fishes possess jet-black pigmentation. Adult females of this species are characterized by the presence of a highly reduced illicium. While adult females reach a length of 30 cm, the parasitic males attain a length of only 3 cm.

Luminous organs and bioluminescence: Its luminescent lure consists of a minute esca , a retractable illicial pterygiophore, and a remnant illicium. Esca is with a distal appendage which develops as a central filament with few to several side filamentous branches.

Females of the species of *Cryptopsaras* have also been reported to secrete and discharge a slime containing luminous granules from three club-shaped caruncles (bioluminescent appendages) situated on the back, i.e., just anterior to origin of dorsal-fin (Matsui and Rosenblatt, 1984). Among the caruncles inserted in a triangular pattern; the median is the largest. (Anderson and Leslie, 2001). Histological study and genome sequence analyses of *Cryptopsaras couesii* caruncle symbionts showed that, like the esca, these caruncles contain dense populations of luminous bacteria that can be expelled through a distal pore (Freed et al., 2019).

The bacterial luciferase in the terminal esca of the species of *Cryptopsaras* may be due to the luminous symbiotic species of bacteria, *Photobacterium.* Tanet et al. (2020) and Freed et al. (2019) reported that this species hosted a novel bacterial species named '*Candidatus Enterovibrio luxaltus*."

The terminal esca containing bioluminescent bacterial symbionts is creating a glowing lure to attract r their prey (Anon., https://en.wikipedia.org/wiki/Triplewart_seadevil).

Herring (1983) reported on the wavelength of its light emission maximum at 470 nm.

3.2.7.3 DOUBLE ANGLERS (DICERATIIDAE)

1. *Bufoceratias microcephalus*

Common name(s): Double angler or Small-head toady seadevil.

Global distribution: Indian Ocean.

Ecology: It was collected at depths of 755–770 m.

Biology: This species is characterized by a slightly elongated and somewhat compressed body; oval profile when viewed from the lateral side; small and narrow head; single tube-like nostril about midway between eye and mouth; small mouth with shorter jaws; and small and numerous teeth on both jaws. The body coloration is uniformly blackish, except for grayish rays of dorsal, anal and caudal fins. It reaches a maximum length of about 15 cm.

2. *Bufoceratias shaoi*

Common name(s): Double angler.

Global distribution: Western Indian Ocean and Western Pacific Ocean.

Ecology: This bathypelagic species has a depth range of 0–1200 m.

Biology: The body of this species is short and globular with rounded to oval profile in lateral view. Mouth is large and oblique. It grows to a maximum length of 10 cm.

3. *Bufoceratias thele*

Common name(s): Double angler.

Global distribution: Western Pacific Ocean.

Ecology: This bathypelagic species occurs up toa depths of 1500 m.

Biology: The body of this species is short and globular with a rounded profile in lateral view. Mouth is large and oblique. It grows to a maximum length of 16 cm.

4. *Bufoceratias wedli*

Common name(s): No common name.

Global distribution: Western and Eastern Atlantic Ocean and Indian Ocean.

Ecology: This mesopelagic and bathypelagic species has a depth range of 300–1750 m. It may also occur near or on continental shelf.

Biology: The body of this species is short and globular and is rounded in lateral view. Mouth is large and oblique. It grows to a maximum length of 25 cm.

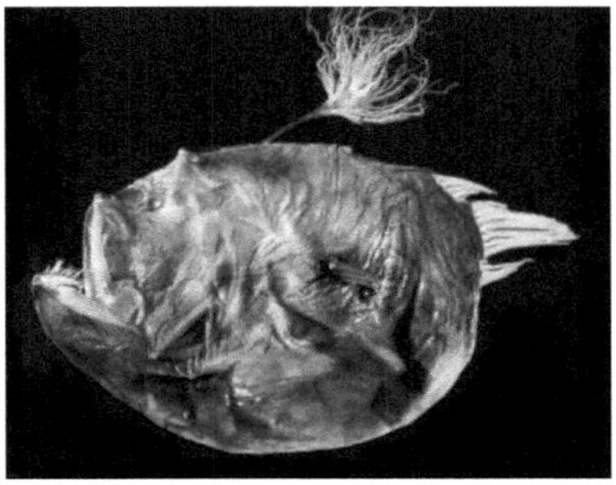

1

2

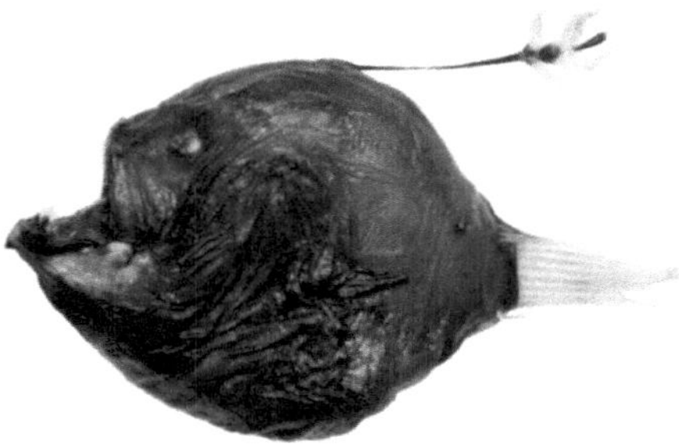

3

Credits: 1 & 3: Masaki Miya et al. https://creativecommons.org/licenses/by/2.0/deed.en; 2: Kwang-Tsao Shao, Taiwan Fish Database.

1. *Bufoceratias shaoi;* 2. *Bufoceratias thele;* 3. *Bufoceratias wedii.*

Luminous organs and bioluminescence: Bufoceratias spp.

In the fishes of Diceratiidae, there are no caruncles (bioluminescent appendages) on dorsal midline. However, these fishes possess a second cephalic ray which is present immediately posterior to the base of illicium is with a distal luminous gland (Anderson and Leslie, 2001). The luminous organs in the different species of *Bufoceratias* are given below.

Bufoceratias microcephalus: In this species, a moderately long illicium is present. Its large esca is bearing large bulb-like terminal papilla, simple filament-like posterior appendage and pair of short simple filament-like lateral appendages. Anterior appendage is absent in the esca (Ho et al., 2016).

Bufoceratias shaoi: In this species, illicium is fairly short and its terminal esca is small (Ho et al., 2016). In the species of *Bufoceratias shaoi, Bufoceratias thele* and *Bufoceratias wedli*, the Illicium has been reported to emerge from dorsal surface of head behind sphenotic spines.

Bufoceratias thele: In this species, Illicium is very long., and the esca has a simple bulb (Ho et al., 2016).

Bufoceratias wedli: In this species, a luminescent terminal papilla is seen between the anterior and posterior appendages of the esca (Anderson and Leslie, 2001). Illicium is fairly short. Esca is moderate in size (Ho et al., 2016). Anon. (https://www.frogfish.ch/deepsea-anglerfish.html) reported that this species has two lures (a smaller lure behind the larger one) which are visible on its back.

5. *Diceratias pileatus*

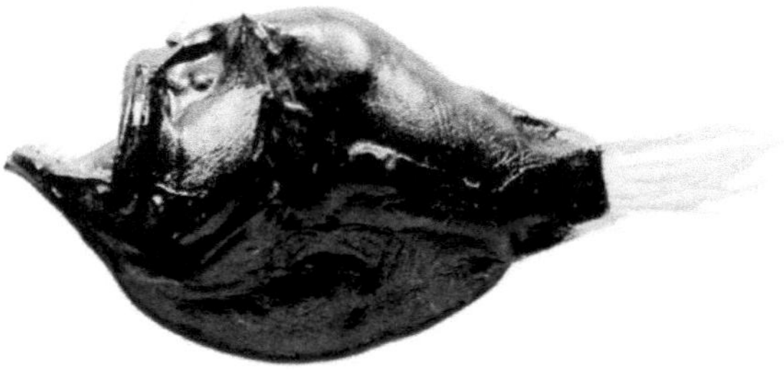

Image credit: Keiichi Matsuura.

Common name(s): Double anglerfish.

Global distribution: Eastern and Western Atlantic Ocean.

Ecology: This bentho-, meso- and bathypelagic species has a depth range of 640–1430 m. It is commonly found near or on continental shelf with adults which are probably benthic.

Biology: In this species, females are bioluminescent and males are free-living. The body of the fishes is solid black and is covered in minute spinules. It grows to a maximum length of 23.5 cm.

6. *Diceratias trilobus*

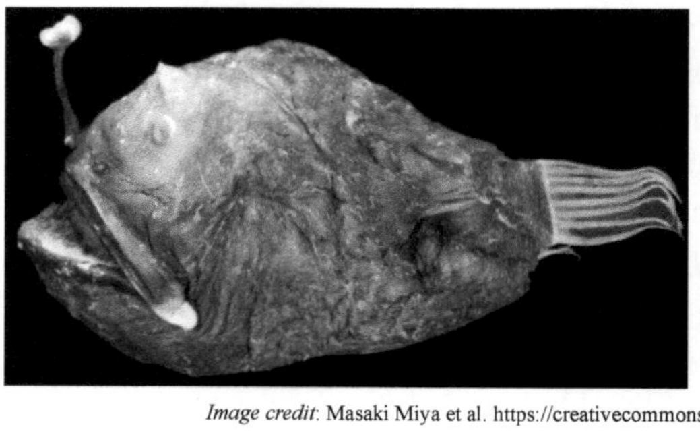

Image credit: Masaki Miya et al. https://creativecommons.org/licenses/by/2.0/deed.en

Common name(s): Double angler.

Global distribution: Northwest Pacific Ocean.

Ecology: This bathypelagic species inhabits depths up to 1216 m.

Biology: The body of this species is short and globular. Jaw is large and oblique. It has a maximum length of 14 cm.

Luminous organs and bioluminescence: Diceratias spp.

Diceratias pileatus: In female fishes, there is a small esca and the width of which is almost equal to its length. A bulbous and posteriorly oriented terminal escal papilla is seen. Posterior and anterior escal appendages are absent. Escal pore is at the posterobasal margin of the terminal papilla (Anon., https://www.fishbase.in/summary/Diceratias-pileatus.html).

Diceratias trilobus: Esca of this species is unusually large and is laterally compressed and its greatest width is more than 1.5 times its length. Terminal escal papilla is rounded. Well developed posterior and anterior escal appendages bearing 1 or more tiny, slender terminal filaments are seen. Illicium is short (Pietsch et al., 2006; Anon., https://www.fishbase.se/summary/Diceratias-trilobus.html).

3.2.7.4 WHIPNOSE ANGLERS (GIGANTACTINIDAE)

The metamorphosed females of the Gigantactinid fishes are distinguished by their elongated and laterally compressed body; and a long slender illicium arising from the anterior most tip of the snout and its highly variable lengths (less than standard length to nearly five times standard length) (Maia et al., 2021).

1. *Gigantactis* spp.

Common name(s): Whipnose anglers.

Global distribution: These fishes are distributed in all oceans.

Ecology: It usually inhabits well above the bottom of the deep oceanic waters at depths of 1,000–2,500 m well above the bottom.

Biology: All the species of this genus exhibit extreme sexual dimorphism (i.e., differences between males and females). While the largest female grows to about 40 cm in length, the largest male only grow to 2.2 cm. Males use their highly developed sense organs to find their counterparts mainly.

1

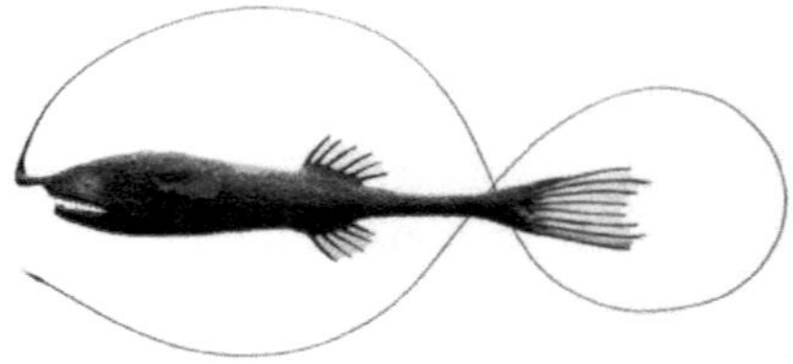

2

3

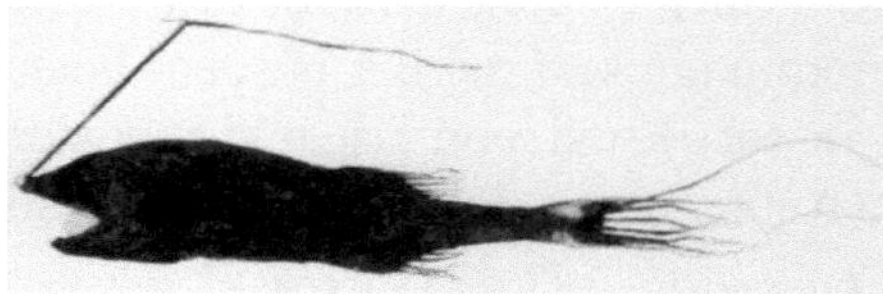

4

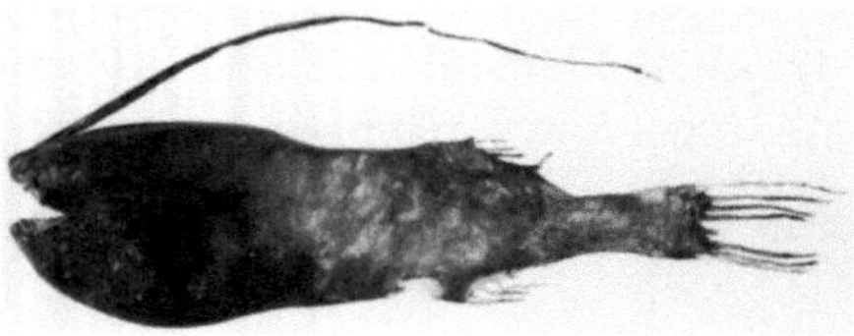

5

6

1. *Gigantactis elsmani*; 2. *Gigantactis macronema*; 3. *Gigantactis kreffti*; 4. *Gigantactis gargantua*; 5. *Gigantactis vanhoeffeni*; 6. *Gigantactis paxtoni.*

Luminous organs and bioluminescence: Gigantactis spp.

In the species of the genus, *Gigantactis,* the extremely enlarged first filament of dorsal fin, called the illicium has its origin at the tip of the snout. The esca (bioluminescent photophore containing luminous bacteria) which is an expanded luminous bulb present at the end of the illicium (Anon., https://en.wikipedia.org/wiki/Gigantactis). In some cases, the lure may be whip-like and four or five times longer than the fish itself. The description of the lure (illicium and esca) of the metamorphosed females of some species of this genus are given below.

Gigantactis watermani: In this species, the characteristic features of the illicium and esca are: length of the illicium is 130–490% of its standard length; distal escal filaments are simple; distal escal filaments are heavily pigmented for more than one-half their length; proximal escal filaments are restricted to the anterior margin of the escal bulb; presence of a group of anterior filaments arising from the base of esca; distal part of escal bulb is bearing four or five pairs of stout filaments along posterior margin; presence 10 filaments at esca base; presence of 4 or 5 pair of filaments on the distal part of escal bulb and each of these filaments is with a pigmented swollen base (Maia et al., 2021).

Gigantactis perlatus: In this species, the characteristic features of the illicium and esca are: presence of an extremely large distal prolongation of escal bulb; entire esca is densely covered with slightly elongated papillae; length of illicium is less than 120% standard length; and presence of a pair of posterior escal appendages (Anon., https://www.fishbase.in/Summary/SpeciesSummary.php?id=23260&lang=italian).

Gigantactis golovani: In this species, the characteristic features of the illicium and esca are: presence of branched distal escal filaments; 3 slender filaments are seen on and below the anterior margin of escal bulb, and many such filaments are present on and below the posterior margin of escal bulb; and length of illicium is about 2000% standard length (Anon., https://www.fishbase.se/summary/16986).

Gigantactis gargantua: In this species, the characteristic features of the illicium and esca are: length of illicium is between 134- and 54% standard length; presence of 4 or 5 pairs of large, distal escal filaments; and presence of 30–50 proximal filaments are found centered on an anterior margin of escal bulb (Anon., https://www.fishbase.se/summary/Gigantactis-gargantua).

Gigantactis ios: In this species, the characteristic features of the illicium and esca are: Esca has a dense cluster of about 15 short, lanceolate appendages; length of illicium is 247–256% standard length; and escal bulb is without distal prolongation (Anon., https://www.fishbase.se/summary/16990).

Gigantactis longicauda: In this species, the characteristic features of the illicium and esca are: length of illicium is 330% standard length; escal bulb is nearly spherical and is with a short, truncated distal prolongation; escal bulb is bearing two pairs of short, spatulate appendages;and stem of illicium is without a filament (Anon., https://www.fishbase.in/summary/Gigantactis-longicauda.html).

Gigantactis microdontis: In this species, the characteristic features of the illicium and esca are: Length of illicium is 216–240% standard length; escal bulb is nearly spherical and is with a short, truncated distal prolongation; escal bulb is with 8–10 distal appendages; and escal bulb is with short distal prolongation which is pigmented on distal surface (Anon., https://www.fishbase.in/summary/52854)

Gigantactis savagei: In this species, the characteristic features of the illicium and esca are: length of illicium is 165–273% standard length; and distal prolongation of esca is tapering to a point which bears 10–18 long, unpaired distal appendages (Anon., https://www.fishbase.se/summary/52855).

Gigantactis herwigi: In this species, the characteristic features of the illicium and esca are: length of illicium is 373% standard length; presence of four pairs of large distal escal filaments; and 7 proximal filaments are found centered on anterior margin of escal bulb (Anon., https://www.fishbase.se/summary/52856).

Gigantactis gibbsi: In this species, the characteristic features of the illicium and esca are: length of illicium is less than 120% standard length; escal bulb is with a short conical, spinulose, and darkly pigmented distal prolongation which is slightly constricted at base; and escal bulb is with distally flattened papillae, short distal filaments and slender proximal filaments (Anon., https://www.fishbase.in/summary/Gigantactis-gibbsi.html).

Gigantactis gracilicauda: In this species, the characteristic features of the illicium and esca are: length of illicium is less than 120% standard length; escal bulb is with an elongate, darkly pigmented and spinulose distal prolongation; and presence of short distal filaments and slender proximal escal filaments (Anon., https://www.fishbase.se/summary/Gigantactis-sexfilis.html).

Gigantactis meadi: In this species, the characteristic features of the illicium and esca are: presence of filaments on dorsal surface of head just behind illicium' base; length of illicium is less than 120% standard length; presence of short filaments along the entire posterior margin of the illicium; escal bulb is with an elongate spinulose and black pigmented distal prolongation which is bearing distally flattened papillae; presence of slender proximal and short distal escal filaments (Anon., https://fishbase.mnhn.fr/summary/Gigantactis-meadi.html).

Gigantactis elsmani: In this species, the characteristic features of the illicium and esca are: length of illicium is less than 130% standard length; presence of a single proximal lateral pair and two distal pairs of large escal filaments and absence of distal prolongation of the escal bulb; (Anon., https://www.fishbase.se/summary/16980).

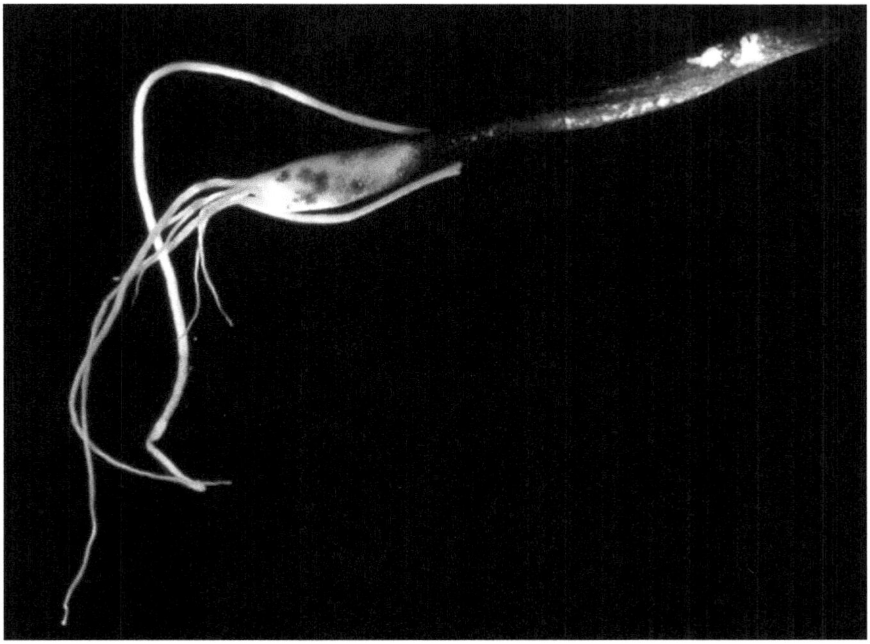

Esca or lure of *Gigantactis elsmani*

Image courtesy: Stuart Humphries / Australian Museum. https://creativecommons.org/licenses/by/3.0/au/

Gigantactis kreffti: In this species, the characteristic features of the illicium and esca are: length of illicium is less than 120% standard length; distal escal prolongation has spinules at base and many digitiform filaments on lateral margins; absence of escal papillae; and presence of a posteromedial papilla at the base of escal bulb and a second papilla on illicium below escal bulb; (Anon., https://www.fishbase.se/summary/Gigantactis-kreffti.html).

Gigantactis macronema: In this species, the characteristic features of the illicium and esca are: length of illicium is 340–447% standard length; escal bulb is with lightly pigmented and truncated distal prolongation; and presence of 6–20 long distal filaments; (Anon., https://www.fishbase.de/summary/28245).

Gigantactis meadi: In this species, the length of illicium is less than 120% of its standard length. Short filaments are seen along the entire posterior margin of illicium. Escal bulb has an elongate, spinulose and black pigmented distal prolongation which bears distally flattened papillae. Short distal and slender proximal escal filaments are present (Anon., https://www.fishbase.se/summary/52849).

Gigantactis paxtoni: In this species, the length of illicium is 168–198% of its standard length. Short filaments are seen on the base of illicium. Escal bulb is gradually tapering into a conical, spinulose black pigmented distal prolongation. Both escal bulb and distal prolongation bear low unpigmented papillae. Short filaments are present on the distal prolongation of escal bulb (Anon., https://www.fishbase.se/summary/Gigantactis-paxtoni.html).

Gigantactis vanhoeffeni: In this species, esca is covered with minute spines and bears distally-flattened papillae. Escal bulb has an elongate, darkly pigmented distal prolongation. Short, black filaments are found concentrated around the tip of this distal prolongation. Escal bulb also has 2–3 unpigmented filaments; and close-set pair of small appendages are seen on the posterior margin of illicium just below the escal bulb (Anon., https://www.fishbase.se/summary/16995).

3.2.7.5 FOOTBALLFISHES (HIMANTOLOPHIDAE)

***1. Himantolophus* spp.**

Common name(s): Footballfish.

Global distribution: These fishes are found distributed in the tropical and subtropical waters of the Atlantic, Indian, and Pacific Oceans.

Ecology: These are mesopelagic and demersal species.

Biology: Female fishes of the family Himantolophidae are distinguished by the following characteristic features: presence of a short, deep, globular body; lower jaw is unusually blunt, and is extending anteriorly beyond the upper jaw; illicium is thick and stout and the pterygiophore (stem) of the illicium is fully embedded in the dermis of the head; and esca is unusually large and complex (Maia et al., 2021. The largest females may attain a length of 60 cm and are globose in shape. On the other hand, adult males do not exceed 4 cm and are fusiform. Both male and female are reddish brown to black in life and are with a shortened, blunt snout.

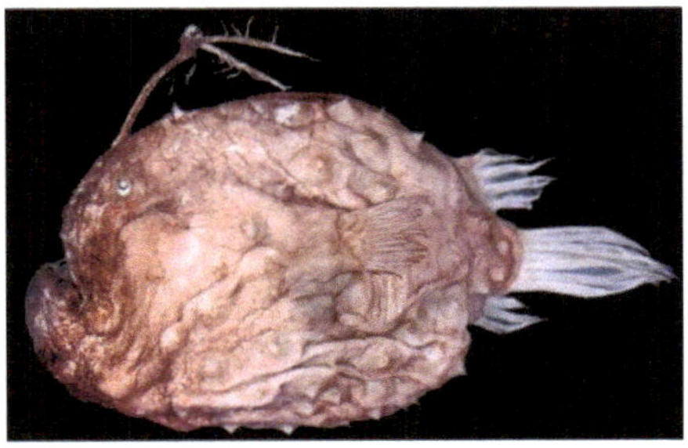

1

2

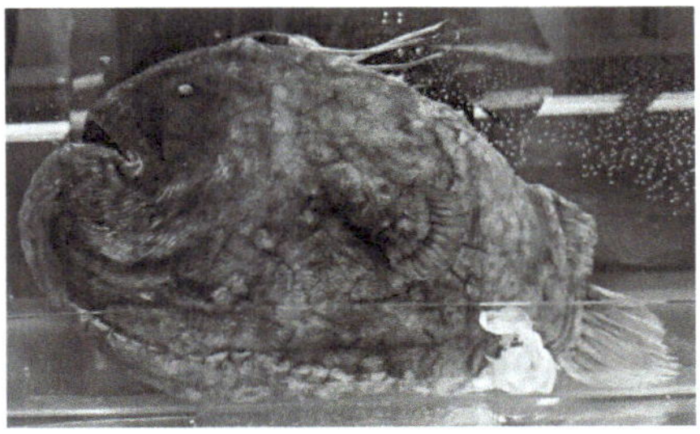

3

4

5

6

Credits: 1: Masaki Miya et al.. https://creativecommons.org/licenses/by/2.0/deed.en; 3: OpenCage. https://creativecommons.org/licenses/by-sa/2.5/; 2, 4, 5 & 6. From Ho, et al. 2016. https://creativecommons.org/licenses/by/4.0/.

1. *Himantolophus appelii*; 2. *Himantolophus danae*; 3. *Himantolophus groenlandicus*; 4. *Himantolophus macroceratoides*; 5. *Himantolophus nigricornis*; 6. *Himantolophus sagamias*.

Luminous organs and bioluminescence: Himantolophus spp.

In the fishes of the genus *Himantolophus,* the illicium is originating above or slightly in advance of the small eye and at its end, a bioluminescent and bulbous esca is seen. The escal morphology varies between species, and it may or may not have denticles or accessory appendages. When present, the escal appendages may be either branched or unbranched. The pterygiophore (stem) of the illicium is not protruding from the snout (Anon., https://en.wikipedia.org/wiki/Footballfish). The description of the luminous organ of the different species of *Himantolophus* is given below:

Himantolophus albinares: In the females of this species, Illicium is short and its base is above its eyes. There are two pairs of swellings in the distal end of esca, and of these swellings, the length of the anterior pair is only half of the posterior pair. Other distinguishing characteristics of metamorphosed females are: length of illicium is 36–54% standard length especially in fishes of more than 3 cm length; deeply cleft distal escal appendage; length of posterior escal appendage is 8.5–21% standard length in fishes of 3–19 cm; and illicial stem (pterygiophore) is with 2–9 posterolateral appendages which are simple or branched at tip (Anon., https://www.fishbase.se/summary/16888).

Himantolophus appelii: In this species, a series of 2–7 filaments are found emerging from the anterior margin of the two primary branches of paired posterior escal appendage. Posterolateral appendages are simple or bifurcated at tip. Small dermal spinules are seen on stem of illicium, escal bulb and escal appendages especially in fishes of greater than 3 cm length (Anon., https://www.fishbase.in/summary/Himantolophus-appelii.html).

Himantolophus danae: Distinguishing characteristics of the females of this species are: anterior escal appendage is 29% of standard length; distal escal appendage is only 2.5% of standard length; presence of pointed distal escal lobes; presence of 3 pairs of posterolateral appendages on or near the base of escal bulb and among these, distal-most and longest pair is 36% of standard length and is bifurcated distally (Anon., https://www.fishbase.se/summary/58775).

Himantolophus groenlandicus: In this species, Illicium is short and there are two pairs of swellings in the distal end of esca. These swellings are almost equal in size and shape (Anon., https://www.fishbase.in/summary/3099). Haneda (1980) reported that this species produces luminous secretion and its luminescence is extracellular.

Himantolophus macroceratoides: Distinguishing characteristics of the metamorphosed females of this species are: length of illicium is 33–38% standard length; each primary branch of distal escal appendage has a posterior side branch which is emerging 42–49% standard length from base; total length of distal escal appendage is 92–100% standard length; the length of side branchesis is 9–10% standard length; all side branches are fusiform; and small dermal spinules on distal appendage are either absent or restricted to base (Anon., https://www.fishbase.se/summary/60774).

Himantolophus mauli: Distinguishing characteristics of the metamorphosed female of this species are: length of illicium is 56–87% standard length; distal escal appendage is bifurcated and is located some distance from base; each primary branch is simple and black pigmented especially in small fishes and is faintly pigmented in large fishes; length of primary branch is 35–40% standard length in fishes of 3–3.8 cm size, and 78–83% in fishes of 8.6–15.5 cm size; unpigmented posterior escal appendage and its length is 3.5–13% standard length; 0–5 posterolateral appendage on illicium and among them longest appendage is 0–6.3% standard length; and escal bulb and appendages are without dermal spinules (Anon., https://www.fishbase.se/summary/27745).

Himantolophus nigricornis: Distinguishing characteristics of the metamorphosed female of this species are: length of illicium is 70–87% standard length; each primary branch of posterior escal appendage is simple; and stem of illicium is with a single pair of posterolateral appendage (Anon., https://www.fishbase.se/summary/Himantolophus-nigricornis.html).

Himantolophus pseudalbinares: Distinguishing characteristics of the metamorphosed female of this species are: length of illicium is 52–66.7% standard length; proximal part of distal escal appendage is undivided; each tip of primary branch is bifurcated; primary branch is black pigmented except on tip; posterior escal appendage is 13% standard length; distal part of each primary branch is bifurcated; five posterolateral appendages on stem of illicium; escal bulb and appendages lack dermal spinules (Anon., https://www.fishbase.se/summary/58781).

Himantolophus sagamius: Distinguishing characteristics of adult female of this species are: length of anterior escal appendage is 11–38% standard length; distal escal appendage is 6.2–11% standard length; distal lobes of esca are blunt; posterior escal appendage are 12–41% standard length; and length of distal and longest pair is 30–45% standard length (Anon., https://www.fishbase.se/summary/Himantolophus- sagamius.html).

3.2.7.6 NETDEVILS (LINOPHRYNIDAE)

1. Haplophryne mollis (= *Edriolychnus schmidti)*

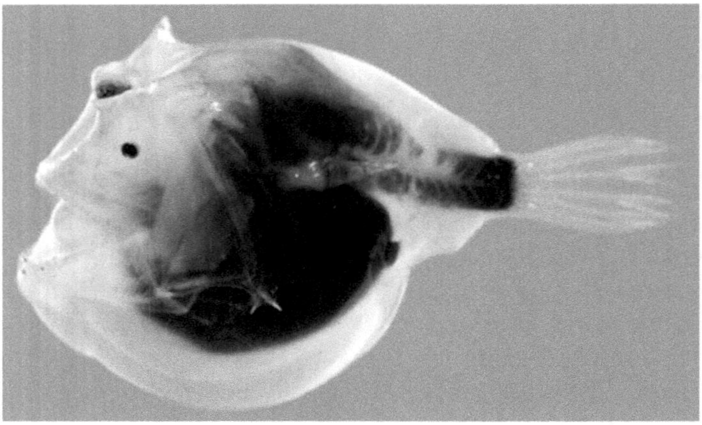

Image credit: Museums Victoria, https://fishesofaustralia.net.au/home/species/2847; https://creativecommons.org/licenses/by/3.0/au/

Common name(s): Soft leftvent angler, phantoma anglerfish or chostly seadevil.

Global distribution: It is found distributed in all tropical and subtropical parts of all oceans.

Ecology: This mesopelagic and bathypelagic has a depth range of 220–2250 m.

Biology: Adult luminescent females of this species possess a short, deep body with totally transparent skin and a large head which is with prominent spines on top and a large spine on the lower edge of the gill cover. Unlike other species of this family, it lacks a chin barbel. Dwarf male fishes are obligatory sexual parasites on their larger female partners. Amazingly both males and females reach sexual maturity only when they are in such an association.

Luminous organs and bioluminescence: As in other anglerfish, the front dorsal fin is replaced by a small bioluminescent lure (illicium or esca) that protrudes forward and over the mouth. The illicium consists of a rounded flap without a pole (Anon., https://en.wikipedia.org/wiki/Haplophryne). Herring and Munk (1994) reported that its esca (light gland) shows a central cavity, with diverging branched ducts which ramify into several tightly-packed radial tubules. In larger specimens, all glandular lumina contained symbiotic, luminescent bacteria. It is also reported that its light emission may be rapidly modulated from within the esca. Herring (1983) reported its wavelengths of light emission maxima as 470 and 475 nm. The emission spectra from the esca of this species is almost identical to that of the luminescent strain of *Photopbacterium phosphoreum* isolated from the species of *Opisthoproctus*.

2. *Linophryne* spp.

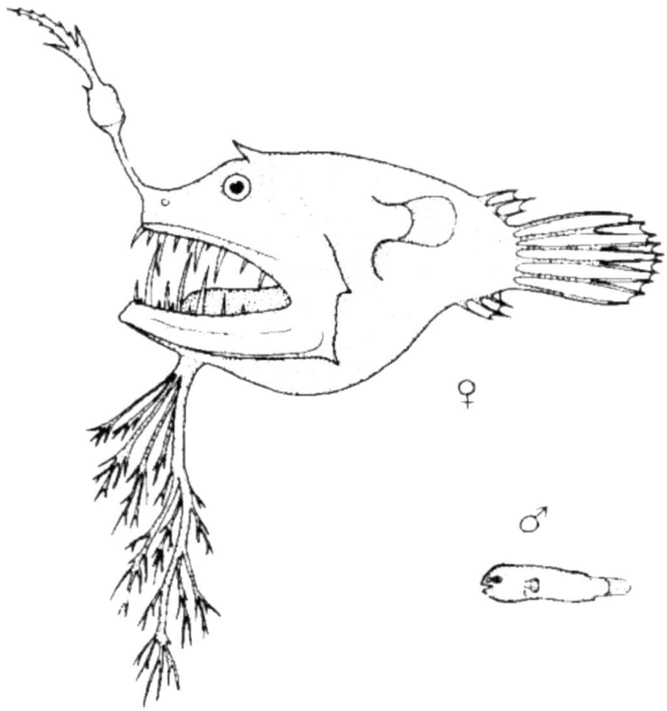

A species of *Linophryne* showing female and male animals

Source: Drawing by Dr Tony Ayling. https://creativecommons.org/licenses/by-sa/1.0/deed.en

Common name(s): Bearded seadevils or leftvents.

Global distribution: These fishes are found distributed throughout tropical to subtropical waters of all oceans.

Ecology: It is a solitary, meso- and bathypelagic species.

Biology: Luminescent female fishes of this family have a short, globose or tapering body. Head is massive and teeth are fang-like. Skin is black and without scales (naked). Sphenotic spines are present. The body coloration of these linophrynid fishes are dark brown to black over entire surface of head, body, and fins (except for the distal portion of the esca. These fishes are characterized by extreme sexual dimorphism in which male fishes are dwarfed, parasitic on females, and attain only a fraction of the size of the females. While females have a maximum length of 23 cm, males are only with 2.9 cm. They are carnivores and sexual reproduction prevails in these fishes.

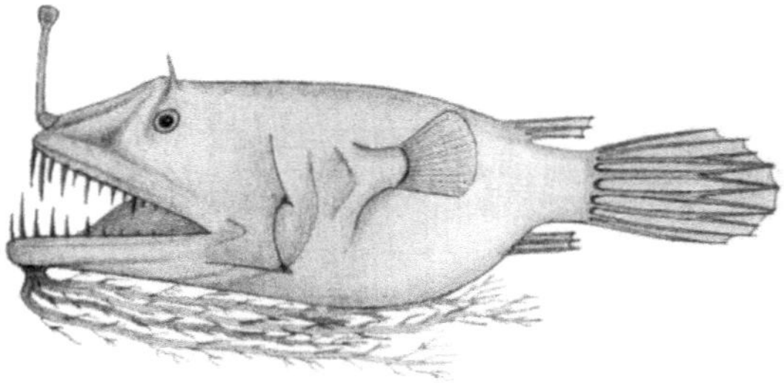

1

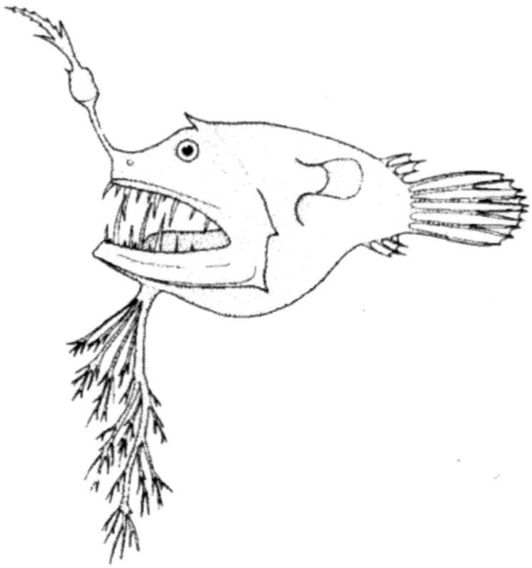

2

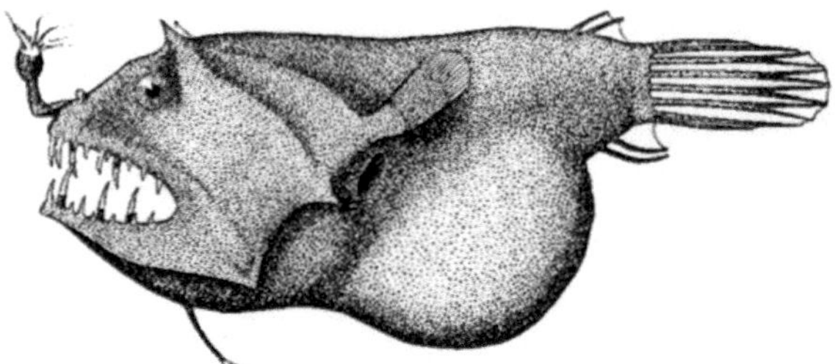

3

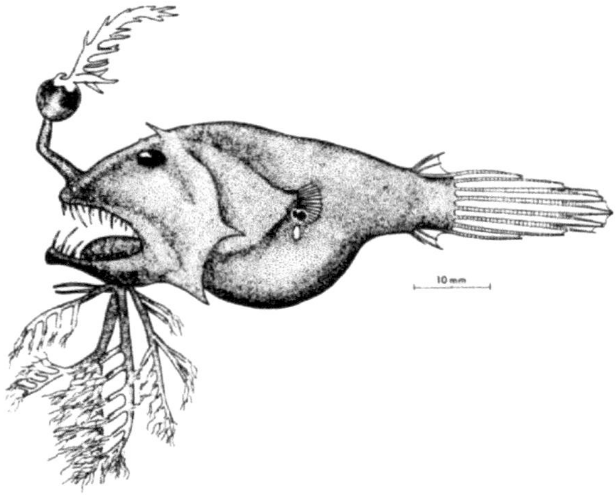

4

5

Credits: 1: Tambja. https://creativecommons.org/licenses/by-sa/3.0/deed.en; 2: Drawing by Dr Tony Ayling. https://creativecommons.org/licenses/by-sa/1.0/; 3: https://fishesofaustralia.net.au/home/species/2848. https://creativecommons.org/licenses/by/3.0/au/; 4. Museums Victoria, https://fishesofaustralia.net.au/home/species/3426. Wikipedia. https://creativecommons.org/licenses/by-sa/1.0/deed.en; 5. Tambja. https://creativecommons.org/licenses/by-sa/2.5/deed.en.

1. *Linophryne algibarbata;* 2. *Linophryne arborifera;* 3. *Linophryne indica;* 4. *Linophryne densiramus;* 5. *Linophryne maderensis.*

Luminous organs and bioluminescence: Linophryne spp.

The species of *Linophryne* are known for their dual bioluminescent systems. Davis et al. (2016) reported that the female fishes of the different species of this genus have developed an intrinsic bioluminescent hyoid (chin) barbel in addition to their bacterially illuminated escal lure on the head (Finn, 2020; Davis et al., 2016). The elongate, multibranched barbel which bears numerous, small, globular light organs. On the other hand, the bacterial luminous system is its lure which is consisting of an illicium (modified first dorsal-fin spine which is short and is less than the length of head and body) which bears a distinct terminal bioluminescent bait (esca). The barbel light organ has been reported to possess complex paracrystalline photogenic granules. While the esca is ectodermal in origin, the barbel organ has been derived from the mesoderm. (Hansen and Herring, 1977; Anderson and Leslie, 2001). The luminous organs of the different species of *Linophryne* are given below.

Linophryne algibarbata: Metamorphosed females possess the following characteristics: globular escal bulb which is surrounded distally and laterally by a blunt ridge in small fishes; lack of escal filaments or other appendages; and hyoid barbel is divided at base into four primary branches of almost equal length. Each of these primary branches has a varying number of secondary and tertiary branches with prolonged into filaments, and each filament is with a distal photophore (Anon., https://www.fishbase.se/summary/27743).

Linophryne arborifera: In this species, its escal light organ is very much similar to a pearl onion and is containing luminous bacteria. On the other hand, its hyoid barbels, which resemble seaweed fronds, do not contain luminous bacteria but care with complex paracrystalline photogenic granules (Anon., https://amp.en.google-info.in/4466297/1/linophryne-arborifera.html).

Linophryne bicornis: Metamorphosed females possess the following characteristics: escal bulb is without distal prolongation or filaments; absence of posterior and anterior appendages; presence of a stout, tapering filament on each side of the upper part of escal bulb; presence of 0–2 minute conical papillae below each lateral filament; primary stem of chin barbel is undivided and distal end is divided into a pair of short branches, each with a lateral series of 3–4 long filaments which are decreasing in length distally and gradually continuing as a cluster of stalked photophores on tip and medial side of each branch; each filament is with a distal series of internal photophores; and small branch is with internal photophores in cleft between pair of primary branches (Anon., https://www.fishbase.in/summary/Linophryne-bicornis.html).

Linophryne densiramus: In this species, its hyoid barbel has five main branches near base. Esca has long and slender appendages with 6–10 filaments. Its posterior escal appendage is small, papilliform or digitiform (Hansen and Herring, 1977).

Linophryne digitopogon: The self luminescent, hyoid barbel of this species is elongated and unbranched at the base. Further, it is longer than the total length of the fish g and is bifurcated at its tip (Hansen and Herring, 1977).

Linophryne indica (= Linophrynx corymbifera): In this species, the illicium which is located on the snout is short and is broadened into a larger esca at its tip. This has a number of short filaments and a long trailing appendage. There is a barbel on the chin with a single stem which is branching into a cluster of short branches (Anon., https://en.wikipedia.org/wiki/Linophryne_indica). Haneda (1980 (reported that its escal light organ which is a open type of gland contains symbiotic type of bacteria. Herring (1983) reported on its wavelength of light emission maximum as 489nm for esca and 500 nm for barbel.

Linophryne maderensis: In this species, esca is without distal appendage and it is with two simple filaments on the upper part of escal bulb. Further, there are no appendage behind the escal pore. Hyoid barbel of this species has four primary branches. Of these branches, the proximal is simple with few photophores at its tip. On the other hand, other branches are with 2–4 terminal, slender, unpigmented filaments each with an internal series of photophores (Anon., http://species-identification.org/species.php?species_group=fnam&id=1521). Tanet et al (2020) reported that its escal bulb has the luminescent bacteria, *Candidatus Enterovibrio escacola.*

Linophryne parini: In this species, its hyoid barbel has three main branches near its base. Its distal escal prolongation is short and papilliform without any filaments. Its posterolateral escal appendages are paired and are long, and digitiform, with highly filamentous tips (Hansen and Herring, 1977).

3. *Photoorynus spiniceps*

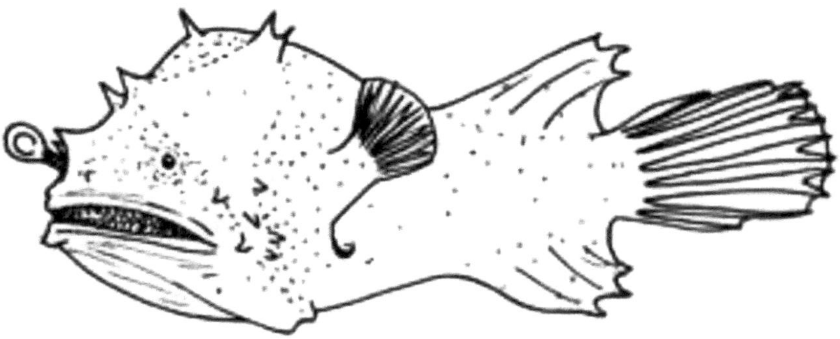

Image credit: Apokryltaros. https://creativecommons.org/licenses/by-sa/3.0/deed.en

Common name(s): Leftvent.

Global distribution: It is found distributed in the tropical and subtropical waters of the western Atlantic Ocean. It is also known from both sides of the Pacific Ocean.

Ecology: This mesopelagic and bathypelagic species has a depth range of 990–1420 m.

Biology: Male fish of this species which is 6.2 mm (less than a quarter of an inch) long is the world's smallest known, sexually mature vertebrate (Anon., https://www.livescience.com/17896-smallest-vertebrate-controversy-frog-angler-fish.html). This dwarfed male is parasitic and it looks like a small nub in the middle of the female 's back. Adult males are characterized by having skin which is very faintly pigmented, and olfactory organs and eyes (to a lesser extent) are degenerated. Metamorphosed females are distinguished by the following characteristics: moderately strong maxillae; jaws with numerous short teeth which are arranged in several series; and absence of vomerine teeth. While the male has a maximum length of 1 cm, female attains a maximum length of 4.6 cm.

Luminous organs and bioluminescence: Female fishes of this species possess bioluminescent lures which are growing from their heads (Anon., https://www.washington.edu/news/2006/01/27/flap-over-fishes-whos-the-smallest-of-them-all/). The lure consists of an illicium which is the highly modified first ray of the dorsal fin and a bioluminescent sac (esca) is located at the end of an illicium. Length of illicium is less than 10% standard length and esca is almost sessile on snout, without appendages. Hyoid barbel is absent (Anon., https://www.fishbase.se/summary/Photocorynus-spiniceps.html). The females wave the lures to attract prey to their mouths.

3.2.7.7 GOOSEFISHES (LOPHIIDAE)

1. Lophiodes caulinaris

Common name(s): Spottedtail angler.

Global distribution: Eastern Pacific Ocean.

Ecology: This demersal species has a depth range of 15–311 m.

Biology: In this species, head and anterior part of the body are depressed; posterior part of body is gradually narrowing; and back and sides of head have numerous spines. Coloration of body is Back brown or dark brown, belly is lighter, and fins are darker distally. It grows to a maximum length of 40 cm.

2. *Lophiodes infrabrunneus*

Common name(s): Shortspine goosefish or shortspine monkfish.

Global distribution: Eastern Indian Ocean.

Ecology: This mesopelagic species has a depth range of 208–1412 m.

Biology: The body of this species is olive-shaped and is tapering posteriorly. Tendrils are present along the edges of head, trunk, and tail. Head is strongly depressed with a large mouth. Lower jaw extends beyond the upper jaws anteriorly. Alcohol preserved specimens are light to dark brown in dorsal surface, paler in ventral surface and all fins are much darker. It grows to a maximum length of 40 cm.

Lophiodes sp.

Source: Public domain

Luminous organs and bioluminescence: Lophiodes spp.

Lophiodes caulinaris: In this species, its first dorsal fin spine is modified into a fishing pole with complex "lure" which is bearing a pennant-like flap (Anon., https://biogeodb.stri.si.edu/sftep/en/thefishes/species/761).

Lophiodes infrabrunneus: In this species, the illicium is fairly short and its terminal esca a leaf-like transparent flap (Ho et al., 2009).

3. Lophiomus setigerus

Source: Public domain

Common name(s): Blackmouth angler,

Global distribution: Indian and western Pacific Oceans.

Ecology: This demersal species has a depth range of 30–800 m.

Biology: It is a goosefish with a flattened, and flabby body with a well-developed skin flaps on its head and sides. A very broad disc-shaped head with a very wide mouth and a gradually tapering trunk are seen. This species grows to a maximum length of 40 cm.

Luminous species: It is a monotypic genus.

Luminous organs and bioluminescence: In this species, the first dorsal-fin spine is modified into a fishing lure on the snout. (Anon., https://fishesofaustralia.net.au/home/species/2841). Illicium is long, slender and is without any tendrils (Caruso, 1983).

4. Lophius spp.

Common name(s): Monkfish, fishing-frogs, frog-fish or sea-devils.

Global distribution: Atlantic and Indian Oceans.

Ecology: These fishes live on the bottom in the bathyal zone which is the zone between 1000 and 4000 m. They are very lethargic as they lie quietly or moving slowly. They have the slowest ventilatory cycles (change from inspiration to expiration) with a typical ventilatory cycle which is lasting more than 90s. That is when individual fishes were disturbed from the sediment, they were found to breathe more rapidly by reducing the duration of the cycle to 18–30s (Farina and Bemis, 2016).

Biology: Head of these fishes is large, broad, flat, and depressed and the rest of the body appears merely like an appendage. Their wide mouth extends all around the anterior circumference of the head, and both jaws are found armed with bands of long, pointed teeth. Skin bears fringed appendages all around its head and also along the body These structures help to change the color of the body of the fish to match its surroundings and to assist the fish largely in concealing itself in its hiding places. These carnivorous fishes grow to a maximum length of about 150 cm.

1

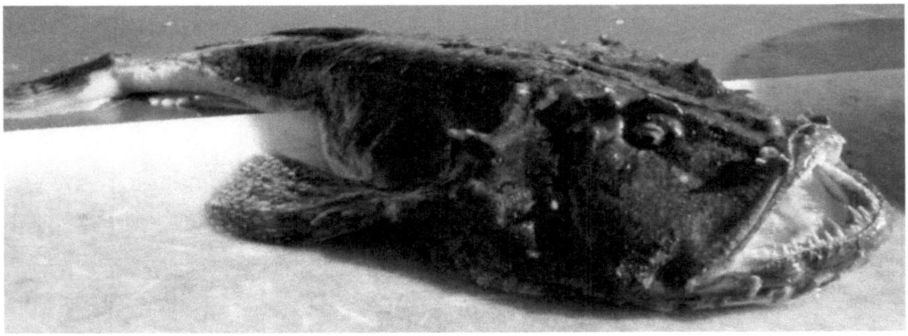

2

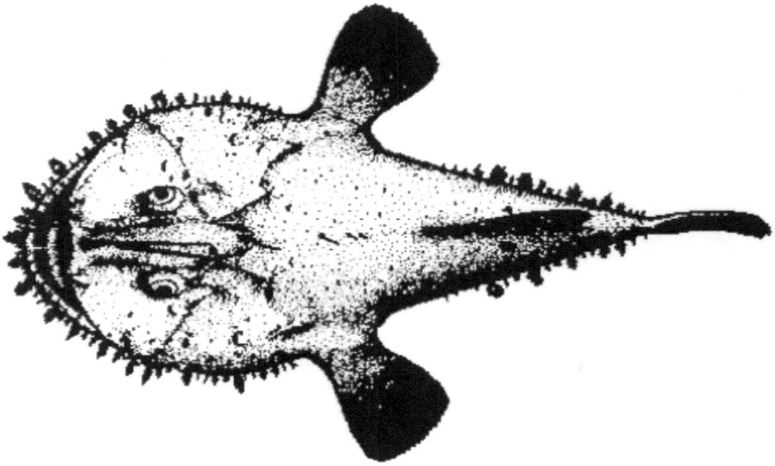

3

Credits: 1 & 2: Public domain; 3: Food and Agriculture Organization of the United Nations, Original Scientific Illustrations Archive.

1. *Lophius americanus*; 2. *Lophius piscatorius*; 3. *Lophius budegassa.*

Luminous organs and bioluminescence: Species of *Lophius:*

In all the species of *Lophius,* a bioluminescent fishing rod (illicium and esca) is seen to attract the prey. This modified fin ray is movable in all directions and is used as a light to attract other fish (Anon., https://en.wikipedia.org/wiki/Lophius), The luminous organs of the major species of this genus are given below.

Lophius americanus: In this species, illicium is long and is 19.5–26.9% of standard length. Esca is a simple pennant-like (long, narrow, triangular flag-like) flap (Caruso, 1983; Anon., https://www.fao.org/fishery/species/3380/en). Anon. (http://gma.org/fogm/Lophius_americanus.htm) reported that the first spine of its dorsal fin bears an irregular leaflike flap of skin at its tip and this plays an important role in the daily life of this fish as a lure for its prey (Anon., http://gma.org/fogm/Lophius_americanus.htm).

Lophius budegassa: In this species, there is a fleshy appendage at the end of the first dorsal spine and this is the esca, a simple pennant-like flap (Anon., faoadriamed.org/html/Species/LophiusBudegassa.html). This esca has a dark spot at its base. The illicium of this species is long and pigmented (Caruso, 1983).

Lophius gastrophysus: In this species, esca is a simple pennant-like flap and illicium is pigmented and is devoid of tendrils (Caruso, 1983).

Lophius litulon: In this species also, the esca is a simple pennant-like flag-like) flap and illicium is short (Caruso, 1983).

Lophius piscatorius: It has a long spiny protrusion(illicium) emerging out of its forehead with a bioluminescent growth on the end called an esca which is bifid and is with black spot at base This illicium is of a moderate length and is pigmented. This fish can move this protrusion with its brightly lit esca to lure other sea creatures (Caruso, 1983; Anon., https://www.animalvogue.com/2015/02/bioluminescence-aggressive-mimicry-and-dating/).

Lophius upsicephalus: In this species, the esca is a simple pennant -like flap with black pigment at its base. Its illicium is devoid of tendrils (Caruso, 1983).

Lophius vaillanti: In this species, a fleshy esca is absent. Illicium is terminating with a simple, rounded, darkly pigmented tip (Caruso, 1983).

Lophius vomerinus: In this species, esca is a simple pennant-like flap as in other related species (Caruso, 1983).

3.2.7.8 BLACKDEVILS (MELANOCETIDAE)

1. Melanocetus spp.

Common name(s): Black seadevils.

Global distribution: These fishes are found distributed in the tropical to temperate waters of the Atlantic, Indian, and Pacific Oceans.

Ecology: This mesopelagic and bathypelagic species has a depth range of 250–2000 m.

Biology: Females of melanocetid fishes are short-bodied and globular with large pendulous and greatly distending bellies. Skin is dark brown and velvety black with baleful appearance. These gelatinous and scaleless fishes have a large head with an oblique mouth. Jaws are with large, sharp, glassy, fang-like teeth. These fishes are cavernous Males of these species are much smaller and are with enormous nostrils which they use to sniff out their mates in the dark. If and when a male finds a female, it attaches itself to the female's underside with its teeth and it remains there as a parasite. This male feeds from the female until it is needed to fertilize the eggs.

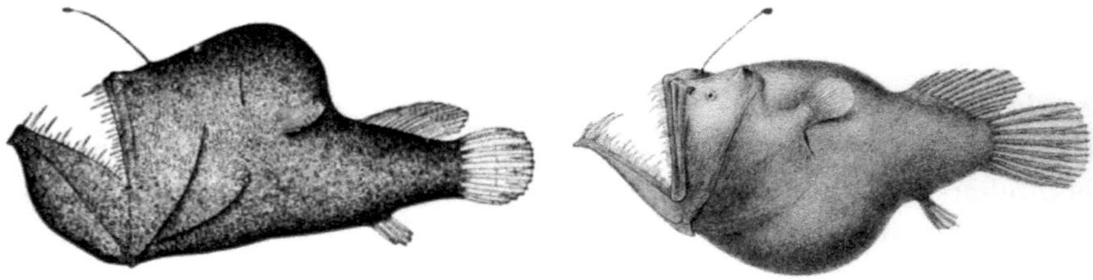

1. *Melanocetus johnseni;* 2. *Melanocetus murrayi.*
Source: Public domain.

Luminous organs and bioluminescence: Melanocetus spp.

Like other anglerfishes, the female black seadevils have a light organ on a stalk between their eyes. This light organ consists of an illicium which is a modified dorsal spine—the "fishing rod"; and an esca which is a bulbous, bioluminescent "fishing lure" situated at the tip of the illicium. The Illicium is emerging on snout and its supporting pterygiophore is fully embedded in skin of head. In black seadevils, the esca takes on a simple form without elongated filaments or appendages (a conical terminus or anterior and posterior ridges may be seen in the esca of some species). Symbiotic, luminous bacteria are housed in the numerous tubular, light glands of the esca. These bacteria are believed to gain access to the light gland through the escal pore and duct. The light emitted by these species has shown peaking at 470–490 nm which is said to correspond to the range emitted by most luminous bacteria at 475–510 nm (Anon., https://sites.evergreen.edu/animatingthesea-s20/black-seadevil-melanocetidae/; Anon., https://en.wikipedia.org/wiki/Black_seadevil). The bacterial luciferase has been identified recently from the species of Melanocetus.

The luminous organs of the well known species of this genus are given below.

Melanocetus eustalus: In this species, the escal bulb is large and slightly compressed with a low conical distal prolongation. Width of escal bulb is 11.3% of standard length; esca is without anterior or posterior crest; and illicium length is 30.6% of standard length. When the live fish of this species was maintained in the laboratory for several minutes, the escal bulb was found to glow continuously with a bright, golden- orange light (Anon., 1980; Anon., https://www.fishbase.se/summary/Melanocetus-eustalus.html).

Melanocetus johnsonii: This is the largest species of this genus and the size of the female's esca, has been reported to be 140 mm. Its escal bulb harbors the luminescent bacteria, *Candidatus Enterovibrio escacola* (Tanet et al., 2020).

Melanocetus murrayi: In this species. The escal bulb is not compressed and is with a low, rounded distal prolongation. Further, the posterior and anterior crests of the escal bulb are either minute or absent (Anon., 1980). Its escal bulb has been reported to house a novel luminescent bacterial species, *Candidatus Enterovibrio escacola* which are believed to enter the light organ from the sea (Anderson and Leslie, 2001; Tanet et al., 2020).

Melanocetus niger: In this species, the escal bulb is not compressed and is with a low, rounded or conical distal prolongation which is almost pigmented on tip. The anterior and posterior crests of the escal bulb are however absent. Unlike other species of this genus, this species has been reported to emit both short-wave, blue, bioluminescence and long-wave red light. It is also believed that by emitting red light, this species can also visualize it, thereby it is able to light up potential prey (Anon., http://lowlightintensity.weebly.com/bioluminescence-aiding-predation.html).

Melanocetus polyactis: In this species, its escal bulb is not compressed and is with conical, distal prolongation which is almost slightly constricted at the base. Further, this prolongation is often longer than the length of the escal bulb and is pigmented on the tip. In some specimens, the anterior and posterior crests of the escal bulb were found to be absent.

Melanocetus rossi: In this species, the esca bears a medial crest, but its top is not pigmented (Orlov et al., 2015).

3.2.7.9 DREAMERS (ONEIRODIDAE)

1. Bertella idiomorpha

Common name(s): Not designated.

Global distribution: Northern Pacific Ocean.

Ecology: This bathydémersal species has a depth range of 580–3475 m.

Biology: Females of this species are distinguished by the following characteristic features: presence of a short, moderately elongate and laterally compressed body with a large head; oblique and horizontal mouth; and anterior jaws. Skin is scaleless and covers the caudal fin base. Numerous small, rounded, darkly pigmented papillae are seen on the head, body, and tail. The body coloration of the fish is dark brown to black. It feeds on amphipods, krill, and small decapods. While the females reach a maximum length of 8.4 cm, the male which is attached to the female is only 1.1 cm.

Luminous organs and bioluminescence: In this species, the illicium has a bulbous, distal light organ (esca bulb). The pterygiophore (stalk) of illicium is cylindrical throughout its length and is emerging on the snout and extending posteriorly on the back behind the head (Anon., https://www.fishbase.in/summary/23258). End of esca is clear and is not pigmented.

2. Chaenophryne spp.

Common name(s): Dreamers.

Global distribution: These fishes are found distributed worldwide.

Ecology: These are solitary and meso- bathypelagic fishes and are having a depth range of 500–1,000 m.

Biology: The fishes of this genus are characterized by extreme sexual dimorphism in which males are dwarfed. While the females are attaining a total length of 28 cm, males are only 1.8 cm. Metamorphosed females are highly variable in shape. The body of these females is short, compressed, globose to elongate and slender. Unlike the fishes of other genera of this family, male fishes of this genus are free-living and are not becoming parasitic on females. The eyes of these fishes are large and are directed laterally. The body coloration is dark brown to black over the entire surface of the head, body, fins and oral cavity.

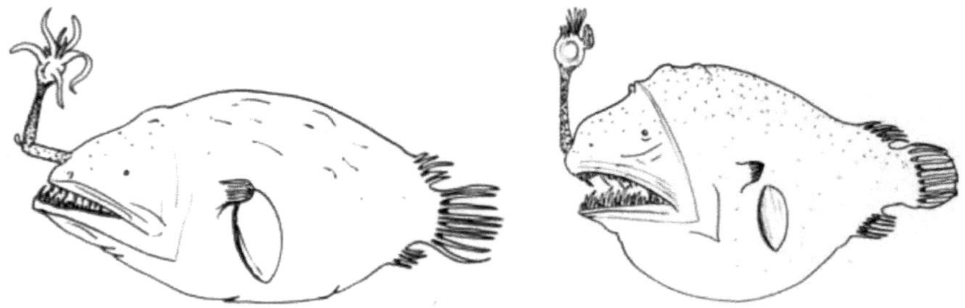

1 2

1. *Chaenophryne draco;* 2. *Chaenophryne longiceps.*

Luminous organs and bioluminescence: Chaenophryne spp.

The metamorphosed females of the species of this family have an illicium which is sharply bent and is positioned directly in front of the mouth. (Anon., https://www.frogfish.ch/deepsea-anglerfish.html). In the different species of this genus), Anon. (https://www.fao.org/3/y4161e/y4161e30.

pdf) reported that the modified first dorsal-fin spine (illicium) may be extremely short with a nearly sessile bait (esca) on snout in some species; or longer illicium (than total length of fish) with a conspicuous terminal bioluminescent bait (esca) in some other species (Anon., https://www.fao.org/3/y4161e/y4161e30.pdf). The light organs of the different species of this genus are described below.

Chaenophryne draco: In the females of this species, a slightly shorter illicium is present. Esca is with an unpaired anterior appendage but is without anterolateral appendages (Maia et al., 2021). This species possessed luminous symbiotic bacteria and the light generated by these bacteria in the main lure is conducted along an anterior light tube and is emitted finally from its trip. Pringgenies et al. (2001) observed that its luminous gland has a central lumen surrounded by an epithelium and is organized as a branched tubular gland. This gland is completely enclosed within an inner reflecting and an outer pigmented layer. Luminou bacteria are present in luminous sacs (Pringgenies et al., 2001).

Chaenophryne longiceps: The esca of this species has 1–3 filamentous, anterolateral appendages on each side; a pair of internally pigmented anterior appendages; and medial appendage or appendages (Maia et al., 2021). Tanet et al. (2020) reported that the bulb of the retractile esca (lure) of this species harbored luminescent, symbiotic bacteria, *Candidatus Enterovibrio escacola*. The emission spectra from this species were nearly identical to that of Photobacterium phosphoreum isolated form the esca of *Opisthoproctus* sp.

Chaenophryne melanorhabdus: In this species, its esca has a single elongate internally pigmented anterior appendage, the length of which is less than one-fourth to almost one-third length of the escal bulb. Medial escal appendages are absent. Posterior escal appendages is with a swollen basal portion and a compressed distal crest (Anon., https://www.fishbase.de/summary/27427).

Chaenophryne quasiramifera: In this species, there are 1 or 2 elongate illicial filaments which are originating laterally just below the base of escal bulb (Maia et al., 2021).

Chaenophryne ramifera: In this species, a fairly longer illicium is present. Esca has 1–3 filamentous, anterolateral appendages on each side; and a single, elongate, anterior escal appendage. Medial escal appendages are absent (Maia et al., 2021).

3. Danaphryne nigrifilis

Credit: Apokryltaros. https://creativecommons.org/licenses/by-sa/3.0/deed.en

Common name(s): Not designated.

Global distribution: It is found distributed in the tropical to temperate parts of all oceans.

Ecology: This mesopelagic and bathypelagic species has a depth range of 0–2000 m.

Biology: Metamorphosed females of this species are distinguished by the following characteristics: slightly elongate, oblong and compressed body, a narrow snout, large and slightly oblique mouth, and very deep caudal peduncle. It may attain a maximum length of 8.2 cm.

Luminous organs and bioluminescence: In this species, the illicium is longer than the length of escal bulb. The pterygiophore (stalk) of illicium is cylindrical throughout its length and it is emerging on the snout (Anon., https://www.fishbase.se/summary/16901).

4. Leptacanthichthys gracilispinis

Image credit: Poulsen, 2019. https://creativecommons.org/licenses/by/4.0/

Common name(s): Plainchin dreamarm.

Global distribution: North and north-eastern Atlantic Ocean and Pacific Ocean.

Ecology: This bathypelagic has a depth range of 0–2000 m.

Biology: Females of this species are dark colored and are growing to a maximum length of 6.9 cm while its parasitic males grow only to a length of 0.8 cm.

Luminous species: It is a monotypic genus.

Luminous organs and bioluminescence: It has a slender illicium which has a dark streak on the dorsal margin including a large flattened appendage (Poulsen, 2019). Anon. (https://fishbase.mnhn.fr/summary/Leptacanthichthys-gracilispinis.html) reported that the Illicium of this species is longer than the length of the escal bulb. The pterygoiphore (stalk) of illicium is cylindrical throughout its length and is emerging on the snout from between frontal bones. While the anterior end of the pterygiophore is slightly exposed, the posterior end is concealed beneath the skin.

5. Lophodolos acanthognathus

Common name(s): Whalehead dreamer or pugnose dreamer.

Global distribution: It is found distributed in the tropical to temperate parts of all oceans.

Ecology: This bathypelagic species has a depth range of 650–1500 m.

Biology: Female fish of this species is fairly long and slender, and its forehead is thrust forward to have an extremely short snout and a concave face. Owing to this character, this fish is also called as "pugnose dreamer." Further, the mouth is large and the opening extends posteriorly past its eyes and jaws are equal anteriorly. Metamorphosed male fishes of *Lophodolos* are unknown (Pietsch, 2009). The female grows to a maximum length of 7.9 cm.

6. Lophodolos indicus

Common name(s): Not designated.

Global distribution: This species is found distributed in the tropical parts of all oceans.

Ecology: This bathypelagic fish has a depth range of 750–1625 m.

Biology: It largely resembles *Lophodolos acanthognathus* and female reaches a maximum length of 7.7 cm.

Luminous organs and bioluminescence: Lophodolos spp.

Lophodolos acanthognathus: The length of illicium of this species is greater than 25% of its standard length (SL) in fishes of size more than 30. Its escal bulb's width is 5.5% SL in fishes of more than 2.5 cm; and it has long escal appendages with a range of length 10.2–20.9% SL in fishes of more than 2.5 cm (Anon., https://www.fishbase.se/summary/16302). It is known to lure its prey to its mouth with the light emitted from its esca containing luminescent bacterial community (Anon., https://species.nbnatlas.org/species/NBNSYS0020420397). Herring (1983) reported about its wavelength of light emission maximum as 475 nm.

Lophodolos indicus: The length of illicium of this species is above 25% SL in fishes of more than 3.0 cm size; its escal bulb's width is 2.1–4.0% SL in fishes of more than 2.5 cm size; and it has short escal appendages the length of which is only 1.2–5.0% SL in fishes of more than 2.5 cm length (Anon., https://www.fishbase.de/summary/16923).

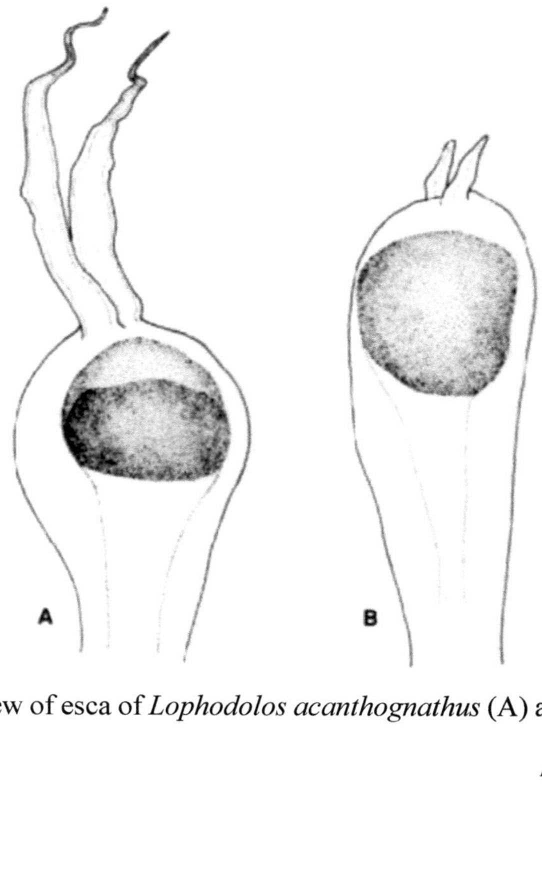

Left lateral view of esca of *Lophodolos acanthognathus* (A) and *Lophodolos indicus*

Image courtesy: Ted Pietsch

7. *Microlophichthys andracanthus*

Common name(s): Dreamer.

Global distribution: Caribbean Sea and Eastern Pacific.

Ecology: It is a bathypelagic species.

Biology: The metamorphosed females are identified by the presence of a lateral series of spines on each side of the upper part of the upper denticular. Adult male of this species grows to a maximum length of 2.4 cm.

8. *Microlophichthys microlophus*

Common name(s): Dreamer.

Global distribution: It is found distributed in the tropical and subtropical parts of all oceans.

Ecology: This mesopelagic and bathypelagic species has a depth range of 800–2200 m.

Biology: The skin of female fish is smooth and naked without dermal spinules. Its black pigmented skin of caudal peduncle extends well past the caudal fin base. It grows to a maximum length of 11.8 cm.

Luminous organs and bioluminescence: Microlophichthys spp.

In both the luminescent species of *Microlophichthys*, their illicium is extremely short and is more or less equal to the length of the escal bulb. The pterygiophore of illicium which is emerging from the snout is cylindrical (Anon., https://www.fishbase.in/FieldGuide/FieldGuideSummary.php?genusname=Microlophichthys&speciesname=microlophus&c_code=608&print=1).

9. *Oneirodes* spp.

Common name(s): Dreamers.

Global distribution: These fishes are found distributed in all three major oceans of the world.

Ecology: It is a oceanic species inhabiting meso- and bathypelagic zones.

Biology: Metamorphosed female fishes of this genus are highly variable in their body form which is ranging from short and deep-bodied to more-or-less globular. While the skin between nostrils and medial surface of the subopercle are unpigmented, caudal peduncle lacks subdermal pigment. The color of females in alcohol preserved specimens is dark brown to black over the entire external surface of the body (except escal appendages and distal part of the escal bulb). Oral cavity and viscera, except the outer surface of the stomach wall are unpigmented. Metamorphosed males are also with outer pigmentation like the females, except its nasal area which is often unpigmented. Female fishes grow to a maximum length of 18 cm.

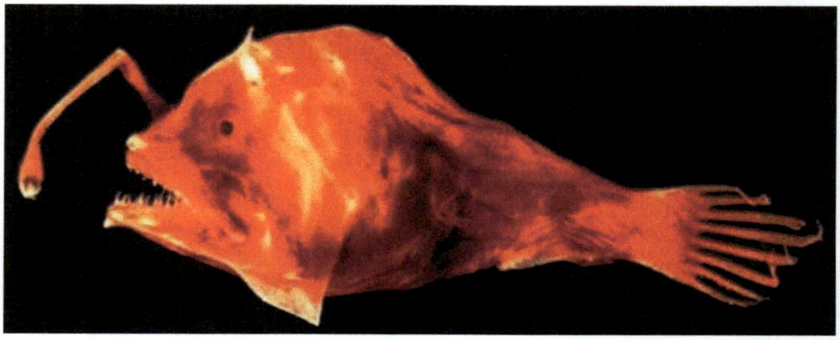

Image credit: C. P. Kenaley. https://creativecommons.org/licenses/by/2.0/deed.en

Oneirodes sp.

Luminous organs and bioluminescence: Species of *Oneirodes.*

Metamorphosed females of the species of the genus *Oneirodes* differ from those of all other genera by the nature of the posterior end of the pterygiophore of the illicium. This structure is protruding from the dorsal midline of the trunk behind the head (Maia et al., 2021). The presence of bacterial luciferase has been recently identified from the esca of the species of this genus. It is further stated that the emission spectra from the escas of these species were largely identical to that of the luminous bacterial strain of *Photopbacterium phosphoreum* isolated from *Opisthoproctus.* Tanet et al. (2020) and Haneda (1980) however reported that certain species of this genus harbor the symbiotic, luminescent bacteria, *Candidatus Enterovibrio escacola* in their escal bulbs. The luminous organs observed in the different species of this genus are given below (Anon., 1974).

Oneirodes acanthias: Esca of this species is with a stout, internally pigmented, anterior appendage which is not longer than escal bulb. This appendage is bearing distally a large, compressed papilla, at the anterior base of which there are two or three short filaments. The posterior margin of the anterior appendage is bearing one to five stout papillae. Filaments and papillae of anterior appendage are darkly pigmented. A minute pair of filamentous medial appendages in most fishes smaller than 110 mm. Each medial appendage which is represented as a pair of minute, dark spots of pigment in most fishes larger than 110 mm. The posterior appendage of the esca is shorter than the anterior appendage and is bearing 3–6 short filaments or papillae which are darkly pigmented. Lateral appendages are absent. A minute, unpigmented, filamentous anterolateral appendage is seen in most fishes of 60 mm size. Widder et al. (1983) reported its emission maximum wavelength as 480nm. Luminous bacteria are present in the deep invaginations of the cells of the tubules of its esca.

Oneirodes amaokai: In this species, the Illicium is fairly long (56.3% standard length) compared to most. On the other hand its escal bulb is small.

Oneirodes anisacanthus: The significant characteristics of the esca of this species are: presence of a well-developed lateral escal appendage; posterior escal appendage is one-third the length of escal bulb; anterior escal appendage is anterodorsally directed with numerous short filaments, and two unpigmented tapering filaments on the anterior margin near the distal tip; and esca is without elongate medial appendages (Maia et al., 2021).

Oneirodes basili: Esca of this species has a branched, filamentous anterior appendage which is two times the length of escal bulb. Spots of dark pigment are present on most branches of the anterior appendage. Medial appendage is absent. A rounded or conical terminal papilla with two, bilaterally placed distal spots of pigment is seen. A posterior appendage which is as long as escal bulb has an antero-posteriorly compressed distal end, the posterior surface of which is fairly concave and darkly pigmented. Lateral appendages are absent. There are two pairs of filamentous anterolateral appendages in which the inner pair is of tapering, branched appendages flanking the anterior appendage; and outer pair of appendages is equal to or less than the length of the inner pair.

Oneirodes bradburyae: It has a rather long, fine filament which is arising from left side of distal end of esca. A short, club-tipped prolongation is seen on the right side, and a short, delicate cluster of filaments are present centrally. All filaments are colorless.

Oneirodes bulbosus: Esca of this species is with a stout, internally pigmented anterior appendage which is bearing distally several tapering filaments, some of which may be twice the length of escal bulb and darkly pigmented distally. There is pair of highly branched, unpigmented, tapering medial filaments which are 2–5 times the length of escal bulb. A short, rounded terminal papilla with a distal spot of pigment is seen in fishes of more than 57 mm size. The unpigmented, tapering posterior appendage is almost twice the length of escal bulb, and is usually bearing 1–2 small, unpigmented filaments at the distal tip. Lateral and anterolateral appendages are absent. A truncated terminal papilla is present in larger fishes. An unpigmented posterior appendage which is nearly half length of escal bulb, and is bearing 2–5 short, unpigmented filaments is present near distal tip. An unpigmented, tapering, unbranched lateral appendage which is less than one-third length of escal bulb is seen in fishes of 47 mm and smaller, unbranched and nearly 6 times length of escal bulb is present in 78 mm fish, and branched and nearly 7 times length of escal bulb is in 173 mm fish. Anterolateral appendage is absent in the esca of this species.

Oneirodes carlsbergi: Esca of this species is with a tapering, internally pigmented, anterior appendage which is 2–3 times as long as escal bulb. This appendage is bearing too short, unpigmented filaments on anterior margin near its distal tip. There is a medial, unpaired, unpigmented appendage which is usually consisting of numerous branched filaments, flanked on each side with a similar filamentous medial appendage. A truncated terminal papilla, with a distal spot of pigment may be present in the esca of some fishes. A laterally compressed, crescent-shaped posterior appendage which is pigmented on distal margin is seen in somefishes. An unpigmented, filamentous lateral appendage is present on each side. Anterolateral appendages are absent in the esca of this species.

Oneirodes cristatus: In this species esca has a large, laterally compressed, internally pigmented anterior appendage which is with a membranous, scalloped distal margin. A pair of filamentous medial appendages which are lightly pigmented at the distal tips especially in 53 mm fishes. A conical terminal papilla is present. There is a short, compressed posterior appendage which has a single, small branch on posterior margin, and a similar branch on each side. Lateral and anterolateral appendages are absent in the esca of this species.

Oneirodes dicromischus: In this species, esca has a stout, internally pigmented anterior appendage and its bifurcated distal end bears numerous, unpigmented, tapering filaments. There are three pairs of filamentous medial appendages in which most anterior pair is the longest, almost as long as the escal bulb. A rounded terminal papilla with a distal streak of pigment is present. There is an unpigmented tapering posterior appendage which is slightly expanded and trifurcated at its distal tip. Lateral appendages absent. However, a stout, unpigmented, anterolateral appendage with a few small filaments is present in the esca of this species).

Oneirodes eschrichtii: Esca of this species has a stout, internally pigmented, anterior appendage which is not longer than escal bulb but is bearing distally a large, compressed papilla. At the anterior base of this anterior appendage, there is a pair of filaments which are shorter than the length of this appendage. One or two additional, shorter filaments are seen along the anterior margin and several along the posterior margin of the anterior appendage. Papilla and distal ends of some filaments of anterior appendage are darkly pigmented in fishes of about 100 mm length and larger. There is a pair of filamentous, medial appendages which are half as long as escal bulb in smaller fishes of about 15 mm size to more than twice the length of escal bulb in fishes of intermediate size (ca.70 mm), and less than half as long as escal bulb in the largest fish of 150 mm size. Tips of tapering filaments of medial appendages are darkly pigmented especially in large fishes of 134–188 mm size. A terminal papilla, which is truncated with a distal streak of pigment, or conical with a distal spot of pigment is seen. An unpigmented, unbranched, tapering posterior appendage which is 1–3 times the length of escal bulb is seen in fishes s less than 100 mm in size.

Oneirodes flagellifer: Esca of this species is with an anterior group of short, unpigmented filaments emerging from a common base in smaller fishes of 12.5 mm size and larger specimens on the other hand, possess a stout, internally pigmented anterior appendage which is slightly more than half length of escal bulb, and is bearing numerous unpigmented filaments on distal end. Medial appendages are absent. A truncated or rounded terminal papilla with a distal streak of pigment is seen in 22 mm sized fishes. An unpigmented, tapering posterior appendage which is almost twice as long as escal bulb is seen in 12.5 mm fishes and the appendage of 3–4 times length of escal bulb is seen in 22.0 mm fishes. Lateral and anterolateral appendages are absent.

Oneirodes heteronema: Esca of this species has an anterior appendage bearing several filaments. Two or three of such filaments are stouter than others and are darkly pigmented on the distal one- third of length in fishes of about 65 mm size and larger. Anterior appendage is internally pigmented. There is a pair of medial appendages and each of them is consisting of three main branches which give rise to numerous, unpigmented, tapering filaments of six times length of escal bulb. A truncated terminal papilla with a distal streak of pigment is seen in fishes of 44 mm and larger. There is a stout posterior appendage which is bearing 4–9 tapering filaments, the distal tips of which are lightly pigmented in some fishes. Lateral appendages are absent. On each side, there is a filamentous anterolateral appendage with two bifurcated, unpigmented filaments.

Oneirodes kreffti: Esca of this species has a stout, internally pigmented anterior appendage which is not longer than escal bulb. Filaments of anterior appendage are lightly pigmented. There is a pair of branched, tapering, unpigmented medial filaments which are more than six times the length of escal bulb. A truncated terminal papilla with a distal streak of pigment is present. An unbranched, tapering posterior appendage with more than two times the length of escal bulb is present on each side. An unpigmented, tapering lateral appendage which is unbranched, bifurcated or trifurcated, and considerably longer than medial filaments is seen. Anterolateral appendage is absent.

Oneirodes luetkeni: In this species, esca has a lateral appendage which may be minute or absent Maia et al., 2021). Esca is with a stout anterior appendage, the distal tip of which is slightly swollen and pigmented in most specimens larger than approximately 60 mm; anterior appendage without internal pigment; a truncated terminal papilla with a distal spot of pigment in some specimens, flanked on each side by a medial appendage consisting of two or three short, unpigmented filaments; an unpigmented,tapering, posterior appendage less than length of escal bulb, becoming proportionately longer with growth; usually a pair of lateral appendages, each consisting of three or four short, unpigmented filaments; and a broad, fringed, membranous anterolateral flap.

Oneirodes macronema: The esca of this species has a short anterior appendage. Its pair of medial appendages which are as long as illicium are bearing minute hair-like filaments along entire length, and one of which bears a single short branch. A truncated terminal papilla is seen. An unpigmented posterior appendage less than one-fourth length of escal bulb is seen. Lateral and anterolateral appendages of the esca are absent.

Oneirodes macrosteus: Esca of this species is with a stout, internally pigmented anterior appendage which is less than half length of escal bulb to nearly twice length of bulb in fairly larger specimens; and is bearing numerous tapering filaments at distal tip and two or three filaments along anterior margin. Medial appendages are absent. A truncated terminal papilla with two bilaterally placed, distal, pigmented spots is present in fishes of 20.0 mm and larger. A posterior appendage, which is approximately one-half as long as to longer than escal bulb is seen in largest fishes. Distal one-third to one-half of the posterior appendage is split into numerous filaments. Lateral appendages of the esca are absent. A filamentous, anterolateral appendage which is less than the length of the anterior appendage is present. Distal tips of filaments of anterior and posterior appendages and the entire length of filaments of anterolateral appendage are found with dark melanophores in larger fishes.

Oneirodes mirus: Esca of this species has a stout anterior appendage which is shorter than escal bulb, and is without internal pigment but is with a single short branch. There is a pair of unpigmented, filamentous medial appendages which are slightly longer than anterior appendage. A rounded terminal papilla with a distal spot of pigment is seen. There is an antero- posteriorly compressed, unpigmented posterior appendage which is as long as escal bulb and is with two short branches at mid-length. Lateral appendages are absent. Of its two pairs of filamentous anterolateral appendages, inner pair of stout, tapering appendages with numerous, short filaments are flanking anterior appendage; and each of the outer pair of unpigmented, bifurcated appendages bears two short filaments.

Oneirodes myrionemus: Esca of this species has a stout, internally pigmented anterior appendage which is approximately twice the length of escal bulb, and is bearing distally 2–3 unpigmented, tapering filaments. This anterior appendage has also one or two branches or papillae which are darkly pigmented in 121 mm sized fishes. There are numerous, highly filamentous, medial appendages in the 43 mm sized fishes. An unpigmented, tapering posterior appendage which is about the length of escal bulb bears two small filaments in 43 mm sized fishes. Lateral appendages are absent. Four highly branched, unpigmented, anterolateral appendages are seen on each side of the esca.

Oneirodes notius: Esca of this species is with a stout, internally pigmented anterior appendage which is shorter than escal bulb especially in smaller fishes (60 mm) and is slightly longer than length of bulb in largest fish (150.0 mm). This appendage is usually bearing a compressed papilla and two tapering filaments on its distal tip. Several small, tapering filaments are seen along the posterior margin of an anterior appendage of 60 mm sized fishes. Papilla and filaments of anterior appendage are unpigmented. A pair of unpigmented, filamentous medial appendages present in almost all fishes. A rounded or truncated terminal papilla with a distal streak of pigment is seen. There is an unpigmented, unbranched, tapering posterior appendage which is as long as, or shorter than the escal bulb. Lateral and anterolateral appendages are absent.

Oneirodes plagionema: In this species, esca has a lateral appendage which may be minute or absent. While the posterior escal appendage is minute, its anterior appendage which is with a single short distal filament is narrow, elongate, and anteroventrally directed (Maia et al., 2021).

Oneirodes posti: In this species, esca has a lateral appendage which may be minute or absent. Two or three medial filaments of this esca are more than twice the length of the escal bulb (Maia et al., 2021).

Oneirodes quadrinema: In this species, the illicium is relatively short and is only 15.6% of standard length. The posterior end of the pterygiophore (stalk) of the illicium is protruding from the dorsal midline of the trunk. Esca bulb is small (Maia et al., 2021.

Oneirodes rosenblatti: Esca of this species is with a stout, unpigmented anterior appendage which is without internal pigment. There is a pair of medial appendages, and each of them is consisting of two or three branched filaments, the distal tips of which are pigmented in fishes of more than 48.0 mm size. A truncated terminal papilla with a distal spot of pigment is seen in somefishes. The posterior appendage is stout and unpigmented. L lateral appendages are absent. On each side, there is a large, membranous, oval or crescent-shaped, anterolateral flap, the distal half of which is darkly pigmented.

Oneirodes schmidti: Esca of this species has a filamentous anterior appendage which is shorter than the length of escal bulb and is without internal pigment. A pair of unpigmented, filamentous, medial appendages which are shorter than the length of the anterior appendage is seen. A rounded terminal papilla with a distal streak of pigment is present. There is a short, unpigmented, unbranched posterior appendage. Lateral appendages are absent. Of the two pairs of filamentous anterolateral appendages, an inner pair is of stout, tapering appendages and the outer pair is of branched appendages which are less than half the length of escal bulb.

Oneirodes theodoritissi: Esca of this species is with a branched, filamentous anterior appendage which is as long as escal bulb and is without internal pigment. A pair of filamentous medial appendages which are as long as the escal bulb is seen. A conical terminal papilla with two distal spots of pigment is seen. An unpigmented, antero-posteriorly compressed posterior appendage which is as long as escal bulb, is seen and the distal half of this appendage bears six short branches in fishes of 58 mm size. Lateral appendages are absent. There is a single pair of unpigmented, bifurcated, anterolateral appendages in which each fork is highly branched and longer than escal bulb.

Oneirodes thompsoni: In the esca of this species, there is a stout, internally pigmented anterior appendage which is shorter than the escal bulb in smaller fishes and is more than three times the length of escal bulb in largest fishes. This appendage is also bearing a large, compressed distal papilla which is closely associated with one to several smaller papillae on the posterior margin and two on anterior margin. Papillae of anterior appendage are darkly pigmented in fishes of about 40 mm size. Medial appendages are absent. There is a truncated or conical terminal papilla with a distal streak of pigment. An unpigmented, tapering and unbranched posterior appendage which is less than the length of escal bulb is seen in smaller fishes and this appendage with more than twice length of escal bulb is seen in the largest fishes. Lateral and anterolateral appendages are absent in the esca of this species.

10. Phyllorhinichthys balushkini

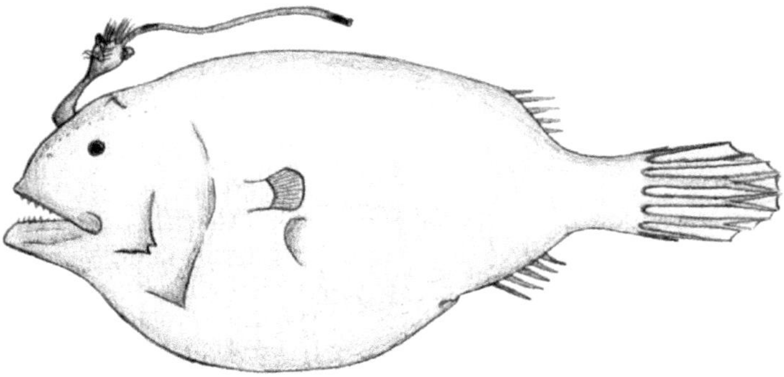

Image credit: Tambja. https://creativecommons.org/licenses/by-sa/3.0/deed.en

Common name(s): Dreamer.

Global distribution: Eastern and Western North Atlantic; and Eastern South Atlantic Ocean.

Ecology: This bathypelagic species has a depth range of 0–3200 m.

Biology: Adult female *Phyllorhinichthys* has a short and somewhat globulose body. Mouth is small and oblique, with slender and depressible teeth which are slightly larger but fewer in the lower jaw. In smaller fishes of 9–18 mm size, there is a distinct band of melanophores on the caudal peduncle. The body coloration is dark brown to black. It grows to a maximum length of 14.5 cm.

11. Phyllorhinichthys micractis

Common name(s): Dreamer.

Global distribution: Atlantic, Indian and Pacific Oceans.

Ecology: This bathypelagic species has a depth range of 0–3500 m.

Biology: It largely resembles *Phyllorhinichthys balushkini.*

Luminous organs and bioluminescence: Phyllorhinichthys spp.

In the species of *Phyllorhinichthys*, the illicium ("fishing rod") is short. The escal bulb ("lure") is internally pigmented on the upper half and unpigmented on the lower half and it bears internally pigmented appendages with silvery tips. Anon. (https://en.wikipedia.org/wiki/Phyllorhinichthys) reported that the most distinctive feature of the genus is the presence of a pair of translucent leaf-like flaps on the snout which are located in front of the eyes and are flanking the esca. Each of these flaps has a central opaque nerve fiber running from the base to the tip. In some fishes, there is an additional smaller pair of similar flaps may also be present above the main pair (Anon., https://en.wikipedia.org/wiki/Phyllorhinichthys). The description of the light organs of the two species of this genus is given below.

Phyllorhinichthys balushkini: In this species, the illicium is fairly longer than that of *Phyllorhinichthys micractis*. Further its esca has two forward appendages at the tip and the rear appendage is longer and slender compared to the other species (Anon., http://species-identification.org/species.php?species_group=fnam&id=1544).

Phyllorhinichthys micractis: In this species, the esca has a bifurcated or paired anterior appendage, a pair of medial filamentous appendages, a cylindrical distal appendage, a pair of antero- lateral filamentous appendages and a small trilobed posterior appendage. All these escal appendages have tubular light guides with an axial light – transmitting core which is enclosed by a tubular wall with a window through which light is transmitted (Munk, 1998).

12. Spiniphryne duhameli

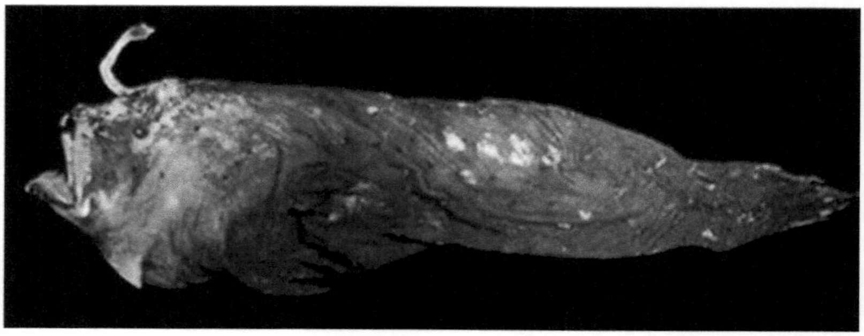

Image credit: Masaki Miya et al. 2010. Courtesy of the American Society of Ichthyologists and Herpetologists. https://creativecommons.org/licenses/by/2.0/deed.en

Common name(s): Not designated.

Global distribution: Central Pacific Ocean.

Ecology: This bathypelagic species has a depth range of 0–2500 m.

Biology: Body of metamorphosed females of this species is relatively long and slender and is not globular. The mouth is large and the cleft is extending well past the eye. Jaws are equal anteriorly. It has a greater number of jaw teeth, particularly in its lower jaw. The female fishes attain a maximum length of 11.7 cm.

13. Spiniphryne gladisfenae

Common name(s): Prickly dreamer.

Global distribution: Eastern central and western Atlantic Ocean.

Ecology: This bathypelagic species lives up to a depth of 1955 m.

Biology: Adult females of this species possess numerous close set skin spines which cover the entire body and fins. These fishes feed on other fishes and crustaceans, and they attain a maximum length of 10.5 cm. In this species males are hitherto unknown.

Luminous organs and bioluminescence: Spiniphryne spp.

In the females of these species, the first dorsal fin ray is modified into an illicium (fishing rod) which ends with a bioluminescent, escal bulb at the tip. The two species of the genus *Spiniphryne* are mainly distinguished from each other by the variations of their esca ("lure) as detailed below.

Spiniphryne duhameli: In this species, there are three pairs of long, slender, lateral escal filaments on each side of the esca (Anon., https://www.fishbase.se/summary/62981).

Spiniphryne gladisfenae: In this species, the esca has two bulbous appendages on the tip, These appendages are covered with tiny papillae and they possess clusters of tiny filaments around the base. A large appendage which is seen on the back of the esca is divided into three lobes or many filaments at the end. The number of lateral escal filaments may vary from nil to three (Anon., https://en.wikipedia.org/wiki/Spiniphryne_gladisfenae).

***14. Tyrannophryne pugnax* /**

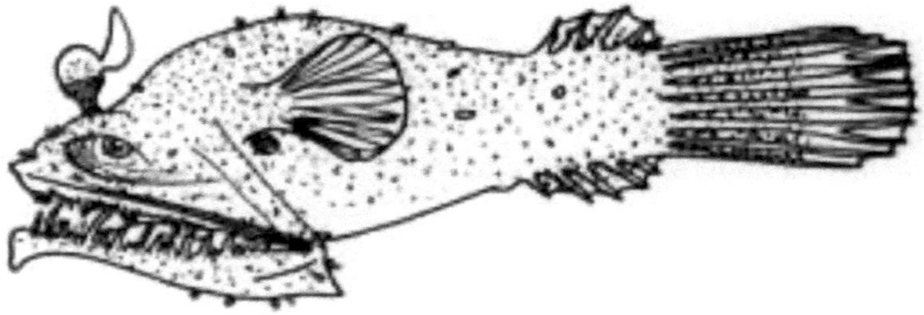

Image credit: Apokryltaros. https://creativecommons.org/licenses/by-sa/3.0/deed.en

Common name(s): Tyrant devil or pugnaceous dreamer.

Global distribution: Eastern, western, south and central Pacific Ocean.

Ecology: This bathypelagic species has a depth range of 400–2100 m.

Biology: The body of the fish is fairly slender, naked, and entirely black in color. It has an extremely large oblique mouth and a long lower jaw, the joint of which is found extending backwards well past the pectoral fin base. Darkly pigmented skin of its caudal peduncle extends well past the caudal fin base.

Luminous organs and bioluminescence: In this species, the fishing rod, illicium is extremely short. Both the illicium and bioluminescent lure are present above the eyes (Anon., https://www.fishbase.se/summary/Tyrannophryne-pugnax.html).

3.2.7.10 WOLFTRAP ANGLERS (THAUMATICHTHYIDAE)

The metamorphosed female fishes of the family Thaumatichthyidae are distinguished mainly by its elongate body; esca with 1–3 large toothlike denticles (bony hooks); and extension of upper jaw forward far beyond the lower jaw. On the other hand, the males are unusually slender and elongate and are never parasitic on females (Maia et al., 2021).

***1. Lasiognathus* spp.**

Common name(s): Wolftrap anglers or compleat anglers.

Global distribution: Atlantic and Pacific Oceans.

Ecology: This mesopelagic and abyssal species is occurring up to a depth of 4,000 m.

Biology: These are small fishes possessing a slender body with a large, slender head which is more than 60% of standard length. The mouth is huge, with the premaxillaries of the upper jaw which are enlarged and extending far beyond the short lower jaw. Skin is completely naked, without spines or denticles. Coloration of the body is a deep chocolate brown. *Lasiognathus amphirhamphus* is the largest species of this genus with a total length 15.7 cm and *Lasiognathus beebei* attains a maximum length of 11.5 cm, *Lasiognathus dinema,* 9.5 cm, *Lasiognathus intermedius* 12.9 cm, *Lasiognathus saccostoma* 7.7 cm, and *Lasiognathus waltoni* 9.4 cm.

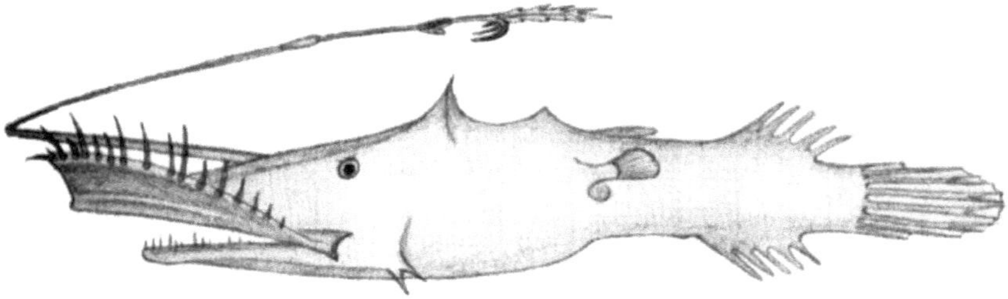

1

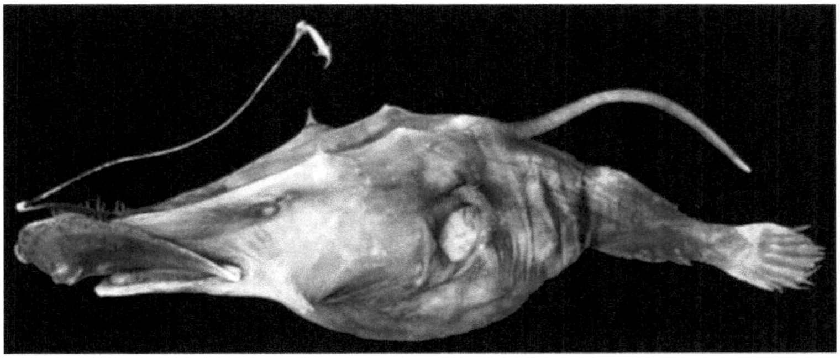

2

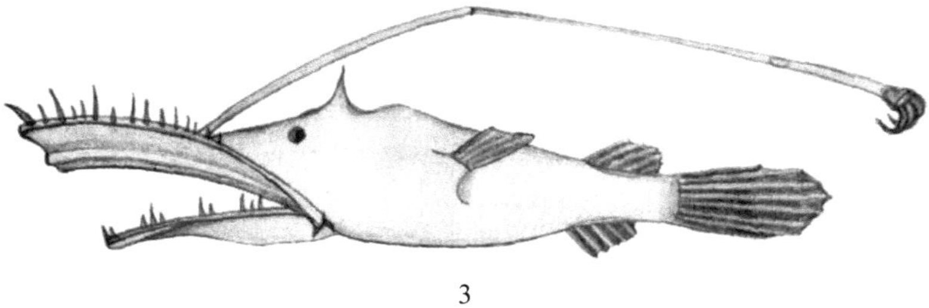

3

1. Lasiognathus saccostoma; 2. Lasiognathus amphirhamphus; 3. Lasiognathus beebei.

Luminous organs and bioluminescence: Lasiognathus spp.

The species of this genus possess a characteristic lure apparatus consisting of a fishing rod (projecting basal bone or pteropterygium), a fishing line (illicium, a modified dorsal fin ray), bait bioluminescent esca), and bony hooks (large dermal denticles) (Anon., https://www.frogfish.ch/deepsea-anglerfish.html). The description of the luminous organs of the different species of *Lasiognathus* is given below.

Lasiognathus amphirhamphus: This species is characterized by having only two lightly pigmented bony hooks on its esca, Distal appendage of the esca is elongated and cylindrical with a long, compressed prolongation at the tip. The prolongation has 6 tiny filaments at the tip and they have no lateral serrations. The posterior appendage of the esca is broad and laterally compressed. (Anon., https://en.wikipedia.org/wiki/Lasiognathus_amphirhamphus).

Lasiognathus beebei: In this species, the escal bulb is without a membranous anterior crest. Distal escal appendage is compressed and is transversely fan-shaped with three bony escal hooks along its posterior margin. Posterior escal appendage is cylindrical, and is tapering to a point. Length of illicium is 43–44% of standard length (SL) in small fishes and 47–50% of SL in large fishes of 8–11 cm size. Posterior escal appendage is 0.5–2.5% of SL, and is emerging from below escal pore. The thin edge of the distal appendage behind and between the bases of escal hooks are more or less folded. Length of distal appendage (from surface of escal bulb to bases of escal hooks) is 1.3–2.8% of SL (Anon., https://www.fishbase.de/summary/50780).

Lasiognathus dinema: In this species, the escal bulb has an elongate, cylindrical and internally pigmented anterior appendage. Distal escal appendage which is a short cylindrical stalk is with an expanded distal end and a pair of tapering, filamentous prolongations arising anteriorly from bases of the escal hooks. Distal escal appendage is bearing three, lightly pigmented, bony, hook-like denticles. Posterior escal appendage is membranous and is antero-posteriorly compressed, with a rounded distal margin (Pietsch and Sutton, 2015).

Lasiognathus intermedius: This species has an elongated, cylindrical distal appendage with a short, cylindrical prolongation at the tip. Posterior escal appendage is cylindrical in shape (Anon., https://en.wikipedia.org/wiki/Lasiognathus_intermedius).

Lasiognathus saccostoma: The illicium of this species is having a tentacle with a luminous bulb (Antonio, 2017). This species has an illuminated lure with hooklike denticles. The distal escal appendage is elongated and cylindrical with a long, compressed prolongation at the tip.

Lasiognathus waltoni: In this species, the escal bulb has a membranous anterior crest. Distal escal appendage is an elongate, cylindrical stalk which is flattened and expanded distally and is with slender filaments on the periphery. Posterior escal appendage is broad and is laterally compressed. Length of illicium is 38% of standard length (SL). Anterior crest of escal bulb has a rounded outer margin and broad base. Posterior escal appendage is a compressed flange of 2.1% SL in length and is with irregularly rounded edge and broad base (Anon., https://www.fishbase.in/summary/Lasiognathus-waltoni.html).

2. Thaumatichthys **spp.**

Common name(s): Anglerfish.

Global distribution: It is found distributed in tropical oceanic waters worldwide.

Ecology: Benthic species inhabiting near ocean floor at a depth range of 1000–3600 m.

Biology: The fishes of this genus show extreme sexual dimorphism. Females are much larger with difference in morphology compared to the males. Female fishes possess the following characteristic features: body is fairly slender and depressed with a fairly well developed dorsal fin; presence of a long, broad, flattened head with enlarged premaxillaries on the upper jaw which overhangs the short lower jaw; eyes are very small and are placed close to the corners of the mouth; and skin is velvet black or dark brown and lower part of the head and body is with numerous small spines. The maximum length of females of the different species of this genus may vary from 6 to 37 cm. Male fishes are small and slender, with highly enlarged olfactory organs; long hooked denticles are seen at the tips of the jaws; and esca is absent externally. The skin of male fishes is uniform dark brown with small spines throughout. When male matures, its body and head become less deep and jaws become more prolongated. The largest known male fish had a total length of only 5 mm.

1

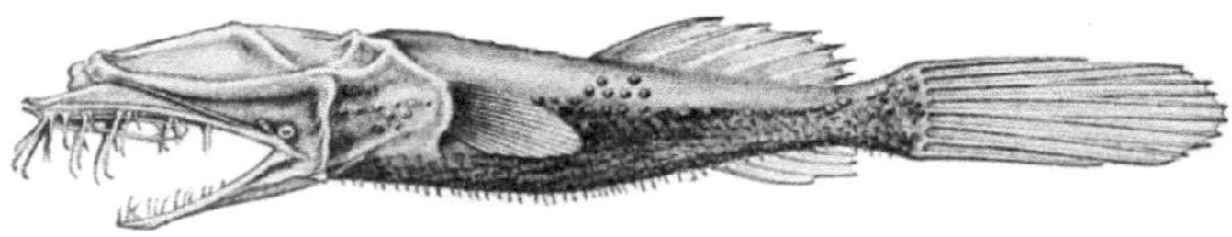

2

Credits: 1: Miya, et al. 2010. https://creativecommons.org/licenses/by/2.0/deed.en; 2: Public domain.

1. *Thaumatichthys binghami;* 2. *Thaumatichthys pagidostomus.*

Luminous organs and bioluminescence: Species of *Thaumatichthys* spp.

Unlike other anglerfishes, the bioluminescent lure ("esca") of the fishes of this genus is located inside its cavernous mouth (Anon., https://en.wikipedia.org/wiki/Thaumatichthys). In these fishes, the distinctive lure is achieved by an "upside-down" orientation of the illicium ("fishing rod"). Base of this lure is embedded in the skin fold and the short illicium projects down and back so that the escal tip hangs down from the roof of the mouth. Escal bulb ends in a pair of forked tendrils and bears several lateral lobes and a single curved denticle. Size of the transparent window in the bulb can be adjusted to alter the amount of light coming from the lure (Anon., https://en.wikipedia.org/wiki/Thaumatichthys). The luminous organs of the different species of *Thaumatichthys* are given below.

Thaumatichthys axeli: In this species, the pterygiophore (stalk) of illicium is short, and is completely hidden beneath the skin of the head. Illicium is also short, embedded within the esca which is hanging from the roof of the mouth and is bearing a single dermal denticle (Maia et al., 2021).

Thaumatichthys binghami: In this species, the esca bears 2–3 pairs of lateral lobes which are elongated into tentacle-like filaments, especially in the larger individuals. The uppermost medial appendage on the esca is finger-like and tapering (Anon., https://en.wikipedia.org/wiki/Thaumatichthys_ binghami).

Thaumatichthys pagidostomus: In this species, the esca is a simple small globe with the denticle protruding at the tip (Bertelsen and Struhsaker, 1977).

3.2.8 ORDER: TRACHICHTHYIFORMES

3.2.8.1 FLASHLIGHT FISHES (ANOMALOPIDAE)

The family Anomalopidae has a small circumtropical group of nocturnal neritic fishes. The most important feature of the group is the presence of a luminous organ beneath the eye in which symbiotic luminous bacteria are believed to be cultured. These luminous bacteria glow continuously, and the light organ is occluded mechanically, either by the erection of a black elastic membrane upward over it or by rotation of its luminous face downward (Rosenblatt and Johnson, 1991). This light is used in the wild by these fishes for communicating, luring prey and other activities.

1. Anomalops katoptron

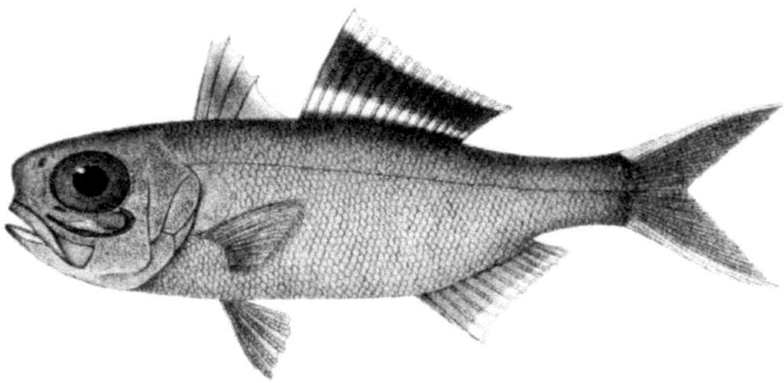

Source: Public domain.

Common name(s): Splitfin flashlightfish or two-fin flashlightfish.

Global distribution: It is found distributed in the warm waters of the central and western Pacific oceans.

Ecology: These fishes inhabit near shallow reefs at a depth range of 200–400 m. They may also be found in large schools during moonless nights.

Biology: This species is identified by its 6–7 Dorsal spines; 14–15 dorsal soft rays; 2 anal spines; and 9–11 anal soft rays. It feeds chiefly on zooplankton and attains a maximum length of 35 cm.

Luminous organs and bioluminescence: This species is characterized by its two half-moon shaped torch-like light organs under the eyes containing bioluminescent bacteria. These light organs are freely movable and they can be rolled inside. These light organs are embedded in suborbital cavities and are connected at the anterior edge through a cartilaginous rod like structures. The suborbital light organs are densely packed with luminous symbiotic bacteria which are cultured in tubular structures and are producing a constant bluish light (Anon., https://en.wikipedia.org/wiki/Splitfin_flashlightfish; Haneda, 1950). Tanet et al. (2020) reported that the light organ of this species grows the luminescent, symbiotic bacteria, *Candidatus Photodesmus katoptron.* Hellinger et al. (2017) reported that during the night time, this fish displays high blink frequencies of approximately 90 blinks/minute with equal on and off times. Widder et al. (1983) and Herring (1983) observed on the wavelengths of its emission maximum as 493 nm and 490 nm, respectively.

2. *Kryptophanaron alfredi*

Image courtesy: Mickey Charteris

Common name(s): Atlantic flashlight fish.

Global distribution: Western Central Atlantic Ocean.

Ecology: It is a reef-associated species with a depth range of 25–200 m. It is visible only on moonless nights and is swimming over the ocean floor. It has also been reported to move to deep water during the day and ascend to near top of drop-offs at night.

Biology: It is fairly a deep-bodied fish with a moderate-sized head. Mouth oblique and moderate-sized and snout is short and somewhat blunt. The eye is large and its diameter is longer than snout length. Caudal fin is deeply forked. Coloration of body, head, and pectoral fins is black. Lateral-line scales, ventral scutes, and enlarged scales at bases of second dorsal and anal fins are all white. It feeds on small shrimp and copepods and grows to a maximum length of 12.5 cm.

Luminous organs and bioluminescence: It has a very large light-producing, light yellow colored light organ below the eye. This light organ consists of bacteria-filled tubular invaginations of the epidermis which are connected to the surrounding seawater by ducts. This light organ emits blue-green light and a black membranous shutter can be raised to cover light organ (Anon., https://www.fishbase.se/summary/Kryptophanaron-alfredi.html; Anon., https://www.fao.org/3/y4161e/y4161e39.pdf). As in other related species, the light organ of this species harbors symbiotic, luminescent bacteria and the symbiosis of this species appears unique among the already known characterized light organ symbioses. Ribosomal RNA (rRNA) operon copy number and gene order have been determined for the luminous bacterial symbiont of this species (Wolfe and Haygood, 1993). Haygood et al. (1984) reported on the rate of release of bacteria into the surrounding seawater, bacterial population size in the light organ and the doubling times of bacteria in the light organ and the corresponding maximum values were 7×10^7 bacteria h^{-1}, 7.3×10^8 bacteria and 23 h, respectively.

2. Parmops coruscans

Common name(s): Flashlight fish.

Global distribution: It is native to the Pacific Ocean.

Ecology: This demersal species has a broad depth distribution and occurs up to a depth of 350 m.

Biology: Its body is compressed and back is somewhat elevated. Snout is blunt. Large eyes are set on either side. Profile is sloping forward without much curvature from occiput to before eye, then it is descending convexly to rostrum. The body coloration is black and fins and lower part of head are darkest. It grows to a maximum length of only 4.8 cm.

3. Parmops echinatus

Common name(s): Flashlight fish.

Global distribution: Western Central Pacific Ocean.

Ecology: This bathypelagic fish has a depth range of 440–550 m.

Biology: It is distinguished from *Parmops coruscans* in the following characteristics: absence of midventral scutes; an external tooth patch on the lateral face of the dentary, and the presence of papillose ridges on the gular isthmus. It attains a maximum length of 14 cm.

Luminous organs and bioluminescence: Parmops spp.: In the species of this genus, a mechanism of light organ occlusion is seen (https://en.wikipedia.org/wiki/Parmops_coruscans). Further, the light organ of these species is rotatable and there is an erectile shutter (Rosenblatt and Johnson, 1991).

4. Photoblepharon palpebratum

Common name(s): Eyelight fish or one-fin flashlightfish,

Global distribution: It is native to the western and central Pacific oceans.

Ecology: It is found along seaward reefs close to the ocean floor. It stays closer to coral and rocks, usually alone or in pairs.

Biology: It is a small fish, with a fairly stout body. It has a blunt snout and large eyes. There is a single dorsal fin and adipose fins are absent. Though most of its body is black, it is marked by a distinct white spot at the upper corner of its preopercle, and by a lateral line of white scales and reflective scales d in front of its pectoral fins. It attains a maximum length of 12 cm.

5. Photoblepharon steinitzi

Image credit: Peter Jäger

Common name(s): Flashlight fish.

Global distribution: It is native to the western Indian Ocean and the Red Sea.

Ecology: It is commonly found in caves around coral reefs up to a depth of 500 m during the day and is ascending to 3–20 m at night to feed. It is known for its characteristic of territorial like behavior, and during the night it dwells alone or in pairs near their cave entrance.

Biology: It has a short, stout and darkly colored body with a blunt snout and large eyes. Pectoral fins are preceded by a white splotch. It may have a spot of light discoloration at the top of its preopercle. Its lateral line is of reflective scales. It attains a maximum length of 11 cm.

Luminous organs and bioluminescence: Species of *Photoblepharon*

The species of this genus viz, *Photoblepharon palpebratus* and *Photoblepharon steinitzi* are known for their prominent subocular light organs which are known to harbor the symbiotic, luminescent bacteria viz. *Candidatus Photodesmus blepharus* (Tanet et al., 2020) and *Vbrio logei* (Dunlap et al. (2007). Shimomura (2012) reported that the light emitted by these light organs can be blocked by raising a black curtain. The luminous organs of these species are given below.

Photoblepharon palpebratum: Its most distinguishable feature is its cream-colored light organs underneath its eyes. These organs which are bean-shaped and are filled with bioluminescent, symbiotic bacteria are blinked on and off by the fish using a dark lid that slides up to cover them. Further, these organs contain bioluminescent bacteria which are fed by oxygen and nutrients from the fish's bloodstream. The constant blue-green light produced by these light organs are believed to be used by the fish to attract prey, communicate with other fish, and frighten and confuse them. Predators (Anon., https://en.wikipedia.org/wiki/Photoblepharon_palpebratum). Herring (1983) reported that the wavelength of its light emission maximum was at 490 nm. Like all flashlight fish, this species has been reported to control the amount of light it emits from its photophore at will. Haygood et al. (1984) reported that the light organ of this species consists of bacteria-filled tubular invaginations of the epidermis which are connected to the surrounding seawater by ducts. The maximum values relating to the rate of release of bacteria into the surrounding seawater, bacterial population size in the light organ and the doubling times of bacteria in the light organ of this species were found to be 2.2×108 bacteria h−1, 7.3×108 bacteria and 23 h, respectively.

Photoblepharon steinitzi: In this species also, the bioluminescent bacteria containing light organ is located under its eyes. The blue green light emitted by this fish may have various functions including intraspecific communication (Hellinger et al., 2020).

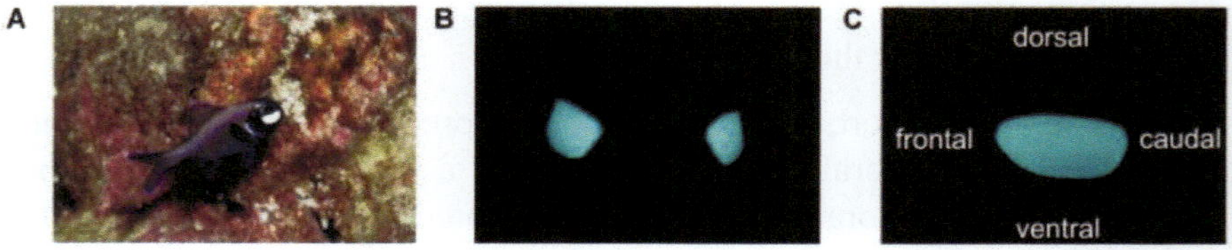

A: *Photoblepharon steinitzi* in its natural habitat, a reef cave entrance area on a reef slope

B, C: Frontal and side views of *Photoblepharon steinitzi* bean shaped light organs in the dark.

Images credit: Peter Jägers.

*6. **Protoblepharon mccoskeri***

Common name(s): Taiwanese flashlight fish.

Global distribution: Northwest Pacific Ocean.

Ecology: It is a demersal species living up to a depth of 300 m.

Biology: It has a thick and deeper body with a shorter snout and fairly small, bulging eyes. Head has very few scales. There are about 260 rows of scales along its body. Coloration of body is reddish black it can reach a lengths of up to 31 cm smaller light organ, and larger cup under the eye.

*7. **Protoblepharon rosenblatti***

Common name(s): Cook Islands flashlightfish.

Global distribution: Cook Islands of south Pacific Ocean.

Ecology: It is a shallow water species living up to a depth of 274 m.

Biology: The body of this species deep and laterally compressed. It is nearly twice as deep as it is wide. Profile is sloping gradually from occiput to snout. The eye is small. Mouth is oblique with jaws forced closed. The dorsal fin has 7 spines and 14 soft rays, and the anal fin is with 2 spines and 11 soft rays. It attains a maximum length of 22 cm.

Luminous organs and bioluminescence: Protoblepharon spp.

The species of this genus are characterized by the presence of rotatable light organs (Baldwin et al., 1997). The description of the light organs of two luminescent species of this genus is given below.

Protoblepharon mccoskeri: In this species, the light organ which is relatively narrow and small is free posteriorly. A fairly wide cup is supporting the anterior end of the light organ, and it is extending posteriorly along the entire ventral surface of the light organ as a thin ovoid plate. A strongly elevated ridge on the anterior portion of the cup is forming a deep groove between itself and the anterior portion of the light organ (Ho and Johnson, 2012).

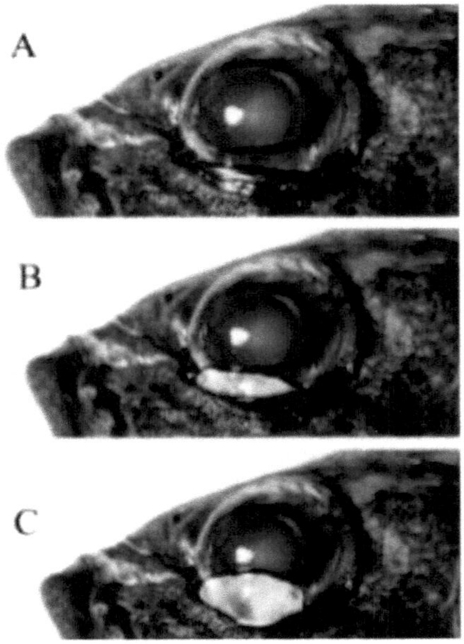

Protoblepharon mccoskeri (left side). A. Light organ fully occluded; B. Light organ half occluded; C. Light organ fully exposed.

Protoblepharon rosenblatti: In this species, the luminous organ which is located below the eye is small. This light organ which is free posteriorly is supported by fibrocartilaginous cup anteriorly. Its light organ is capable of rotating downward into a pocket below the eye and is medial to infraorbitals. When occluded, the dorsal margin of the light organ is found well below the infraorbital rim. A black elastic shutter membrane is attached along the outer margin of the suborbital pocket (Baldwin et al., 1997).

8. Phthanophaneron harveyi *(=Kryptophanaron harveyi)*

Image credit: Mike Kanzler, STRI, https://biogeodb.stri.si.edu/sftep/en/thefishes/species/937

Common name(s): Gulf flashlightfish or lanterneye fish.

Global distribution: It is endemic to the Gulf of California, the Pacific Ocean.

Ecology: It is often associated with coral reefs and can be found over soft and rocky substrates at a depth range of 0–100 m.

Biology: It has an oblong body and large head. Bones of the head and shoulder girdle are sculptured with several spine-bearing ridges. Broad ridges and skin are separating the sensory canals at the top of the head. Mouth is oblique and moderately sized; eye is large and is wider than the snout; and snout is short and blunt. Caudal fin is deeply forked. The body is blackish-brown with no reflective markings on it. It attains a maximum length of 8 cm.

Luminous organs and bioluminescence: It has a prominent pearly white luminous organ under its eye. This organ has a raisable shutter and it is rotatable (Anon., https://en.wikipedia.org/wiki/Phthanophaneron_harveyi). Nigel (2020) reported that the bacterial symbionts of this species have not so far been identified, but these bacteria are believed to be acquired anew with each generation through ducts that connect to the surrounding seawater.

3.2.8.2 PINEAPPLEFISHES (MONOCENTRIDAE)

The lower jaw of the pineapplefish is with two phosphorescent organs which appear as orange spots in daylight or blue-green ones at night. The light produced by luminescent bacteria is used by this fish to attract its prey viz. shrimps and other zooplankton at night (Anon., https://www.fishbase.se/summary/FamilySummary.php?ID=241).

1. Cleiodopus gloriomaris

Image credit: Richard Ling. https://creativecommons.org/licenses/by-sa/2.0/deed.en

Common name(s): Pineapplefish, knightfish or coat-of-mail fish.

Global distribution: It is native to Queensland, New South Wales, and Western Australia of the Eastern Indian Ocean.

Ecology: This reef-associated species is found at depths of 6–200 m. It is also commonly seen in caves, rocky habitats and muddy substrates.

Biology: It has a plump, rounded body that is almost completely covered with large, rough scales possessing sharp, backward-pointing spines. Head is large with mucous pits and is armored with a heavy bone. Snout is blunt and it overhangs the wide mouth. Scales are yellow to whitish with black rear margins and they form the striking pattern which gives this fish its name. The lips, chin, and parts of the jaw are black. A red stripe present on its lower jaw is running to the photophore. It grows up to a length 22 cm.

Luminous organs and bioluminescence: In this fish, there are two mucus pits (light organs) containing, symbiotic bioluminescent bacteria on its lower jaw near the corners of the mouth and these organs are often found concealed when the mouth is closed. (Anon., https://en.wikipedia.org/wiki/Pineapplefish; Anon., https://www.fishbase.se/summary/cleidopus-gloriamaris). These light organs are pale, resembling a ship's navigation lights. Tanet et al. (2020) and Dunlap et al. (2007) reported that the light organ of this species harbors the luminescent, symbiotic bacteria, *Aliivibrio fishcheri (= Vibrio fischeri*). The light emitted by these fishes are usually green, but occasionally red in older fish. (Anon., http://www.animalplanet.com/pets/pineapplefish/).

Widder et al. (1983) reported that this species is capable of emitting light of more than one color and its wavelengths of light emission maximum were 503 and 555 nm.

2. Monocentris japonica

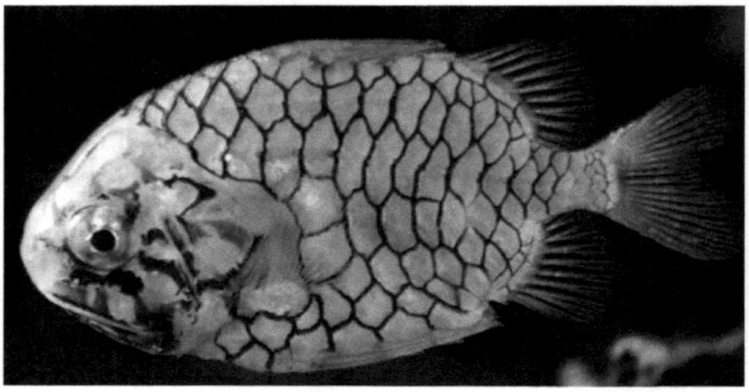

Image credit: Fernando Losada Rodríguez. https://creativecommons.org/licenses/by-sa/4.0/deed.en

Common name(s): Japanese pineapplefish.

Global distribution: It is found distributed in tropical Indo-West Pacific Oceans.

Ecology: This species can be found on both rocky and coral reefs at depths of 1–250 m. It is nocturnal and shelters in caves and under ledges during the day time. It has been reported to travel in schools near the ocean bottom.

Biology: The body of this species is yellow. Lower jaw is black. The body except for the caudal peduncle is armored with huge, immovable scales which are outlined in black. It grows to a maximum length of 17 cm.

Luminous organs and bioluminescence: It has a pair of luminous organs on each side of the lower jaw (Haneda, 1950). Tanet et al. (2020) and Dunlap et al. (2007) reported that the light organ of this species harbors the luminescent, symbiotic bacteria, *Aliivibrio fishcheri (= Vibrio fischeri).* The light organ of this species consists of bacteria-filled tubular invaginations of the epidermis which are connected to the surrounding seawater by ducts. Haygood et al. (1984) reported on the rate of release of bacteria into the surrounding seawater, bacterial population size in the light organ and the doubling times of bacteria in the light organ of this species and the corresponding maximum values were 2×10^7 bacteria h^{-1}, 1.5×10^8 bacteria and 23 h, respectively.

3. Monocentris neozelanicus

Common name(s): Pinecone fish.

Global distribution: Southwest Pacific Ocean.

Ecology: It is a demersal species found in deepwater marine habitats.

Biology: It has a characteristic oval body covered with enlarged, spiny plate-like scales which make this fish resembling a pinecone.

Luminous organs and bioluminescence: The light organ of this species has been reported to harbor the luminescent, symbiotic bacteria, *Vibrio fischeri* MJ101 (Guerrero-Ferreira and Nishiguchi, 2007).

3.2.8.3 SLIMEHEADS (TRACHICHTHYIDAE)

1. Aulotrachichthys spp.

Common name(s): Slimeheads or luminous roughies.

Global distribution: New Zealand, Australia, Namibia, west of Britain, and the mid-Atlantic Ridge.

Ecology: This meso- to bathypelagic species lives at depths of 75 to 5 000 m.

Biology: These are medium-sized species with a length range of 20–60 cm. The body of the fishes is oval and laterally compressed. Head is large; eyes are moderate to large in diameter; snout is rounded; and mouth is large and obliquely angled when closed. Large keeled scales are seen along the midventral belly between pelvic and anal fins. In most species, these scales form row of well-developed scutes. Coloration of the body is reddish orange, pinkish, or dusky silver. Adult fishes feed mainly on other fishes.

Image credit: Izuzuki. https://creativecommons.org/licenses/by-sa/3.0/deed.en

Aulotrachichthys prosthemius

Image credit: Museums Victoria, https://fishesofaustralia.net.au/home/species/3009; https://creativecommons.org/licenses/by/3.0/au/

Aulotrachichthys pulsator

Luminous organs and bioluminescence: Aulotrachichthys spp.

In the species of *Aulotrachichthys*, the light organ is more proximally placed within the body wall (with a smaller anal rim). The bacterial-lobule ducts unite and form fewer openings around the anal rim (Ghedotti, https://epublications.regis.edu/cgi/viewcontent.cgi?article=1023&context=csre_celebrationscholarship). The description of the light organs of the reported species of this genus is given below.

Aulotrachichthys prosthemius: In this species, a luminous organ is present in the body wall where it forms a ring around its anus located anteriorly between the bases of the ventral fins. An elongated cavity filled with gelatinous, luminous tissue extends posteriorly from the anus to the origin of the anal fin. Gelatinous tissue has also been reported beneath all striated skin of this species. Further, in this species, the ventral part of the hypaxial muscle mass is modified in order to reflect and diffuse light. (Cohen, 1964). Haneda (1980) and Dunlap et al. (2007) reported that its bioluminescent organ, a open type of gland is containing luminous, bacterial symbionts belonging to the species *Photobacterium kishitanii* and is used by these fishes for counter-illumination to be away from the predators

Aulotrachichthys pulsator: It has a broad silvery stripe as its luminescent organ which is found along the lower side of the fish from the pelvic-fin base to the caudal-fin base. (Anon., https://fishesofaustralia.net.au/home/species/ 3009).

2. *Paratrachichthys fernandezianus*

Common name(s): Chilean sandpaper fish.

Global distribution: Southeast Pacific Ocean.

Ecology: This demersal species is found between 0 m and 20 m from the surface of the ocean, i.e., the photic zone.

Biology: The body of this small fish is very deep to moderately deep and is laterally compressed. These fishes are known for their large, titular heads, large eyes, and bright colors. They have a length of 17 cm.

3. Paratrachichthys macleayi

Source: Australian National Fish Collection, CSIRO. https://creativecommons.org/licenses/by-nc-sa/3.0/

Common name(s): Sandpaper fish.

Global distribution: It is found distributed in the East Indian to Southwestern Pacific Ocean and is endemic to Australia.

Ecology: This reef associated and pelagic-neritic species is found in small schools in open rocky reefs at depths of 15–400 m.

Biology: This bronze to greenish-yellow roughy has silvery sides. Fins are pink to yellowish and black. It grows to a maximum length of 25 cm.

4. Paratrachichthys trailli

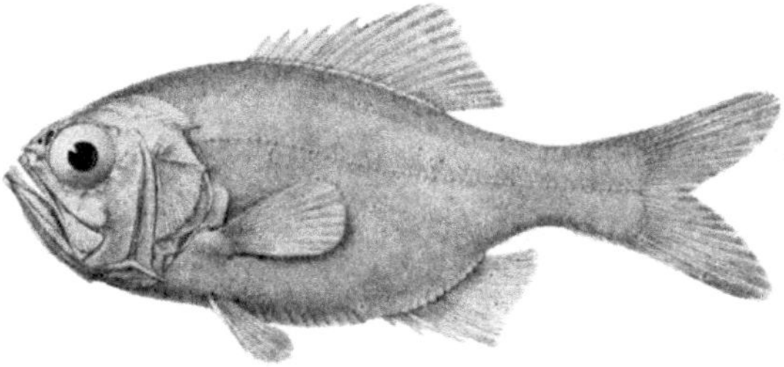

Source: Public domain

Common name(s): Common roughy or sandpaper fish.

Global distribution: Western Pacific Ocean.

Ecology: This bathypelagic species has a depth range of 20–500 m.

Biology: The body of this species is very deep to moderately deep and laterally compressed. These fishes are known for their large, titular heads, large eyes, and bright colors (in some species). They grow to a maximum length of 25 cm.

Luminous organs and bioluminescence: Paratrachichthys spp.

The fishes of the genus *Paratrachichthys,* like *Aulotrachichthys*, are bioluminescent with a bioluminescent organ containing the luminous symbiotic bacteria, *Photobacterium* sp. In these fishes, the light organ is seen immediately under the swollen anal rim on which the ducts of the bacterial lobules form openings (Ghedotti, https://epublications.regis.edu/cgi/viewcontent.cgi?article=1023&context=csre_celebrationscholarship).

3.3 MARINE FISHES WITH EXTRACELLULAR BIOLUMINESCENCE (LUMINOUS SECRETION)

The alepocephaloid fishes, or slickheads which are with two families viz. Alepocephalidae and Platytroctidae), are deep-sea fishes and are distributed in all three major oceans at depths of 100–5000 m. Among about 150 known species, all (about 40) platytroctid species have shown postcleithral luminous gland (postclavicular organ or shoulder organ) which has been reported to releases a luminescent secretion which presumably startles or blinds predators and allows the fish to escape (Sazonov, 1996). The fishes of the Platytroctid genera releasing luminous secretion are given in the following table.

Order	Family	Genera
Alepocephaliformes	Platytroctidae	*Platytroctes, Barbanius, Holtbyrnia, Mentodus, Searsia*

Source: Sazonov (1996).

3.3.1 ORDER: ALEPOCEPHALIFORMES

3.3.1.1 TUBESHOULDERS (PLATYTROCTIDAE)

The platytroctid fishes possess a pair of characteristic shoulder luminescent organs (postcleithral glands, supraclavicular organs or postclavicular organs). Each of these glands represents a sac-like reservoir imbedded in the skin just behind the cleithrum and supracleithrum, with a short extremal tube which is supported by a modified scale inside and is situated below the lateral line. The walls of the reservoir and tube are intense black. The internal walls of the reservoir consist of a thick layer of dense connective tissue and with a dense population of melanophores and numerous "violet cells" (photogenic bodies). The latter are the major component of the luminous secretion released by the shoulder gland. This luminescent secretion is used by these fishes for passive defense. It is believed that this bright flash of the luminous secretion may make sudden blinding of the attacking predator and this helps the fish to leave the site and to escape subsequent attacks (Sazonov, 1966). Matsui and Rosenblatt (1987) and Anon. (https://www.fishbase.se/summary/FamilySummary.php?ID=87) reported that these shoulder organs discharge a bright luminous fluid into the water which produces numerous bright points of blue-green light.

1. Barbantus curvifrons

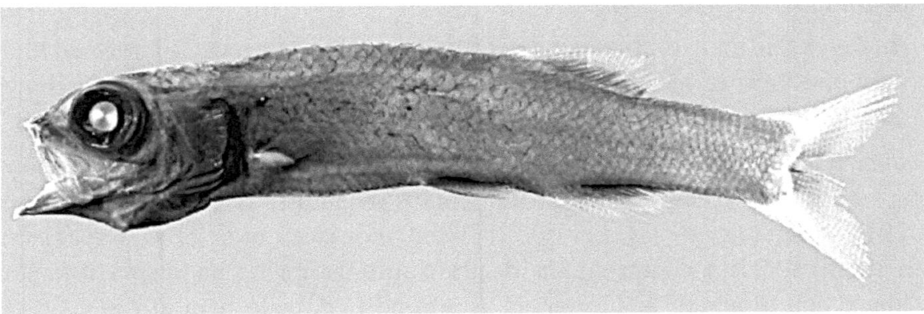

Credit: Reprinted from Moore et al., 2003

Common name(s): Palebelly searsid or tube shoulder.

Global distribution: Atlantic, Pacific and Indian Oceans.

Ecology: This mesopelagic species lives at depths of 600–4500 m. and is known to migrate every day, to shallower waters to feed on zooplankton, especially at night hours.

Biology: The body of this species is moderately elongate and laterally compressed. Upper and lower keels are shallow or absent. Predorsal margin is more than 1 scale wide and pelvic fin is present. It is uniformly dark colored except for a row of light-colored, scales along its belly. It grows to a maximum length of 13 cm.

Luminous organs and bioluminescence: This species has been reported to expel a blue to green luminous fluid from a small pore (the tubular papilla) which is found located below its lateral line. This brightly glowing fluid allows these fishes to escape from potential predators by distracting them (Matsui and Rosenblatt, 1987).

2. Holtbyrnia anomala

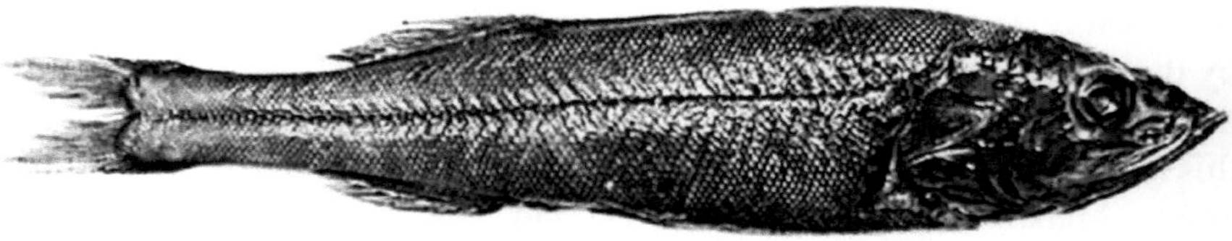

Image credit: Martin Nielson, University of Copenhagen

Common name(s): Bighead searsid.

Global distribution: Tropical and subtropical Atlantic ocean.

Ecology: This mesopelagic, bathypelagic and benthopelagic species has a depth range of 700–2,700 m.

Biology: The body of this species is compressed and its head is large and more than one-third of its body length. Snout is pointed. It grows to a maximum length of 25 cm.

Luminous organs and bioluminescence: In this species, the photophores are absent in young and are rudimentary in adults (Orrelland and Hartel, 2016). Herring (1972) reported that the shoulder organ of this species is a round black sac present inside the shoulder girdle and is visible as a black tubule which is supported by a modified scale behind the shoulder girdle, and just below the lateral line. Internally the afore said sac extends from the base of the pectoral fin to the dorsal tip of the cleithrum. This sac contains a blue-green luminous fluid which is expressed into the water. Matsui and Rosenblatt (1987) observed that the tubular opening of the luminescent shoulder organ is more conspicuous in the mesopelagic fishes than in the bathypelagic fishes of this family. Poulsen (2018) reported that this species discharges a green luminescent substance from its shoulder organ immediately after its capture and this discharge was continuous for about 30 min. Sazonov (1996) reported that when disturbed, this fish throws out its bright blue-green cloud of the luminous secretion from both of its postcleithral glands and this cloud shines for 4–5 s, then scatters into very numerous bright drops and disappears.

Image credit: Jan Y. Poulsen

The shoulder organ of *Holtbyrnia anomala* excreting bioluminescent substance.

3. Holtbyrnia laticauda

Common name(s): Tusked tube shoulder.

Global distribution: Eastern Atlantic Ocean.

Ecology: This mesopelagic and bathypelagic species has a depth range of 500–1630 m.

Biology: This species is distinguished by the following characteristics: absence of dorsal spines; shallow body; closely spaced vomerine teeth; gill filaments united at their base; scales are normally imbricated; 9–10 pelvic-fin rays; 19–21 dorsal soft rays; 15–17 anal soft rays; and 23 gill rakers in the first arch.

Luminous organs and bioluminescence: This species has been reported to possess both photophores (Matsui and Rosenblatt, 1987) and luminous shoulder organ, an external black tube on shoulder which connects to the internal sac producing the luminous fluid (Orrell and Hartel, 2016). Sazonov (1996) observed that this species emits flash when disturbed and repeated stimulations (up to more than 10 times) made the light emission. Further, it is also believed that one flash of this fish could illuminate a small room enough to distinguish printed texts.

4. *Mentodus facilis*

Source: Public domain

Common name(s): Tubeshoulder.

Global distribution: It is found distributed in all tropical and subtropical seas worldwide.

Ecology: This bathypelagic species has a depth range of 680–2000 m.

Biology: The body of this species is moderately compressed and its dorsal margin is rounded. Dorsal fin is ahead of anal-fin origin by about one-fourth to one-third of dorsal-fin base. It attains a maximum length of 12.5 cm.

Luminous organs and bioluminescence: In this species, the ventral photophores are absent (Priede, 2017) and it only possesses the luminous shoulder organ (Anon., https://www.sciencesource.com/archive/Dark-Searsid–Mentodus-facilis–SS2807313.html). It expels a blue to green luminous fluid from a pore (tubular papilla) which is located below the lateral line. As in other related species, the luminous fluid is present in a "shoulder sac apparatus" which is located under the shoulder girdle. The brightly glowing fluid released by these fishes allows them to escape from the potential predators by distracting them with this cloud.

5. *Platytroctes apus*

Source: Public domain

Common name(s): Legless searsid or legless tubeshoulder.

Global distribution: Atlantic and Indian oceans.

Ecology: This bathypelagic species has a depth range of 1000–2000 m.

Biology: The body of this species is deep and strongly compressed. Its upper and lower margins form a keel. Predorsal margin is sharp. The body coloration is dark brown and it attains a maximum length of 18 cm.

Luminous organs and bioluminescence: In this species there are no nodular light organs (photophores) (Anon., https://www.fishbase.se/summary/Platytroctes-apus). But it possesses shoulder luminous organ (postcleithral gland) on its side and caudal luminous glandules (luminous tissue) and both discharge luminous fluid which is capable of glowing brightly for several seconds (Sazonov, 1996; Anon., https://www.sciencesource.com/archive/Legless-Searsid-Fish-SS2713138.html). The 2–4 small, unpaired caudal glandules are located near the dorsal and ventral margin of the caudal peduncle. The bright flash of the luminous secretion of this fish makes sudden blinding of the attacking predator, and this helps the fish to leave the site and to escape the subsequent attacks of the predators. It is also reported that this species is apparently incapable of fast swimming and it may use the glow of its caudal glands to thwart repeated attack of the predators from behind (Sazonov, 1966).

6. Searsia koefoedi

Source: Public domain

Common name(s): Koefoed's searsid.

Global distribution: Tropical Indian and Pacific Ocean; and Eastern and NW Atlantic Ocean.

Ecology: This mesopelagic-bathypelagic species is found at depths from 450 to 1,500 m.

Biology: The body of this species is elongated, moderately deep, and is moderately compressed. Dorsal margin is rounded. Head is relatively short. Mouth is small. Eye is located below the head outline. Dorsal and anal fins are subequal and are nearly opposed. Pectorals are low on body. The body coloration is uniform dark. It attains a maximum length of 15 cm.

Luminous organs and bioluminescence: This species is known for the presence of regular photophore (Sazonov, 1996) and post-cleithral glands (shoulder organs) with luminous secretion (Saznov, 1996). Herring (1972) reported that the luminescent substance emitted from the post-cleithral organ of this species is in the form of granule-packed cells with the properties of photoprotein similar to that of the luminous polychaete, *Chaetopterus*. Saznov (1996) reported that when disturbed, these fishes have been reported to throw out a bright blue-green cloud of their luminous secretion from its postcleithral glands. This secretion seems to shine for 4–5 s, then it scatters into very numerous bright drops and disappears. This luminescent secretion is used by these fishes for passive defence. Nicol (1958) measured the light intensity of this species and it was about 2 x 10^6 mW cm^2 at a distance of 1 m in air.

3.4 MARINE FISHES WITH EXTRINSIC BIOLUMINESCENCE (LUMINOUS INGESTED CRUSTACEANS)

Recent studies have shown that there are three major groups of luminous bacteria. The first group contains *Photobacterium fischeri, Photobacterium leiognathi,* and *Photobacterium phosphoreum* comprise the second group of luminous bacteria; and the third group contains *Beneckea (Lucibacterium) harveyi, Beneckea (Photobacterium) splendida,* and *Vibrio cholerae biotype albensis*. While the first two Photobacterium groups contain members which are known to be symbiotically associated with higher organisms, the third group contains the three luminous *Beneckea* species which are not found as symbionts in luminous organs, but they are free-living and can be isolated directly from sources such as seawater, surfaces of decaying fish and squid, or surfaces of living marine animals. Luminous members of the genus *Beneckea,* as well as *Photobacterium* sp. which are shed into the seawater from luminous organs are termed "free-living" bioluminescent bacteria. The role of luminescence in symbiotically associated bacteria is well known. On the other hand, the advantage of light emission to free-living and saprophytic luminous bacteria is less clear. Free-living luminous bacteria viz. *Beneckea,* spp. have been reported to attach to the surface of marine crustaceans (e.g., penaeid shrimp) and if ingested by the proper fish, may establish residence in the gut of such fishes. The bioluminescent, free-living bacteria so entered into the gastrointestinal tract of a fish are responsible for the bioluminescence of that fish (O'Brien and Sizemore,

1979). The species of luminescent fish containing ingested crustaceans are given in the following table.

Order	Family	Species
Acropomatiformes	Pempheridae	*Parapriacanthus ransonneti; Pempheris klunzingeri*
Kurtiformes	Apogonidae	*Apogon striata; Epigonus macrops; Jaydia ellioti; Taeniamia fucata; Taeniamia lineolate; Verulux cypselurus;*

3.4.1 ORDER: ACROPOMATIFORMES

3.4.1.1 SWEEPERS OR BULLSEYES (PEMPHERIDAE)

1. Parapriacanthus ransonneti

Image credit: Wikipedia. https://creativecommons.org/licenses/by-sa/2.5/deed.en

Common name(s): Golden sweeper or golden bullseye.

Global distribution: Western Pacific Ocean.

Ecology: This coastal and reef-associated species has a depth range of 3–30 m. It is known to form large groups in caves under ledges or beneath overhangs in reef (under Acropora plate corals).

Biology: It is a semi-transparent fish with a greenish-gold head and eye. A horizontal blackish streak is seen beginning dorsally on the opercle and disappearing below the posterior end of the dorsal fin. These fishes forage at night mainly on zooplankton such as crustacean larvae and polychaetes. It attains a maximum length of 10 cm.

Luminous organs and bioluminescence: It has two types of light organs projecting from the digestive tract at the ventral side of the body. This fish uses its pyloric caeca -stored luciferin 'vargulin' obtained from the luminous crustacean *Vargula hilgendorfii* prey as a substrate for its bioluminescence. Interestingly this fish also obtains its luciferase enzyme from the bioluminescent ostracod prey. This enzyme which is purified from the fish's light organs is used in the bioluminescent reaction. was identical to the enzyme produced by the ostracods these fish eat. This autogenic luminescent phenomenon is known as "kleptoprotein bioluminescence" (Lau and Oakley, 2021; Anon., https://groups.oist.jp/mbu/event/kleptoprotein-bioluminescence-origin-luciferase-luminous-fish-parapriacanthus-ransonneti.

2. *Pempheris klunzingeri*

Image credit: © David Muirhead. Australian Museum, https://australian.museum/learn/animals/fishes/rough-bullseye-pempheris-klunzingeri/

Common name(s): Rough bullseye.

Global distribution: Eastern Indian Ocean.

Ecology: This demersal species has been reported to dwell up to a depth of 50 m.

Biology: It is a pale bronze to coppery-brown fish with the following characteristics: a darker bar on the rear of the gill cover above the pectoral-fin base; a clearcut yellow or brown lateral line; a black leading edge and tip on the dorsal fin; and dark margins on the anal and caudal fins. It attains a maximum length of 18 cm.

Luminous organs and bioluminescence: Parapriacanthus ransonneti and *Pempheris klunzingeri.*

The luminous organs of these species which contain ingested substances of luminous curstaceans are open to the outside (Haneda, 1980). Anon. (https://groups.oist.jp/mbu/event/kleptoprotein-bioluminescence-origin-luciferase-luminous-fish-parapriacanthus-ransonneti) reported that *Parapriacanthus ransonneti* has two types of light organs projecting from the digestive tract at the ventral side of its body. This fish uses luciferin 'vargulin' obtained from the luminous crustacean *Vargula hilgendorfii* prey as a substrate for its bioluminescence. Bessho-Uehara et al. (2020) reported that the luminous organ of *Parapriacanthus ransonneti* obtains not only its luciferin but also its luciferase enzyme from bioluminescent ostracod prey.

3.4.2 ORDER: KURTIFORMES

3.4.2.1 CARDINALFISHES (APOGONIDAE)

1. *Apogon striata*

Apogon sp.

Image credit: Williams et al., 2010. https://creativecommons.org/licenses/by/2.5/deed.en

Common name(s): Cardinal fish.

Global distribution: West Pacific Ocean.

Ecology: It inhabits coral and rocky reefs and other adjacent habitats including sand-rubble patches and seagrass beds. It is also seen in sand/rubble bottoms among rocks, gorgonians and sponges in the continental shelf.

Biology: The body of this species is ovate to elongate and is moderately compressed. The eyes are large and their diameter seems to exceed the snout length.

2. Epigonus macrops

Epigonus sp.

Source: Public domain

Common name(s): Luminous deepsea cardinalfish.

Global distribution: Indo-Pacific Ocean and Western Atlantic Ocean.

Ecology: This marine, brackish and bathydemersal species has a depth range of 550–1300 m.

Biology: Body of this species is elongated and laterally compressed; and is deepest at pectoral-fin base. Nape is not humped. Head is large and slightly compressed; snout is short and round; mouth is large, and terminal and gape is oblique. Caudal fin is deeply forked. In alcohol, body and all fins are uniformly light brown; opercular regions dark brown; and luminous window, tongue, and mouth cavity are all black. It attains a maximum length of 21 cm.

3. Jaydia ellioti *(= Apogon ellioti)*

Jaydia sp.

Source: Public domain

Common name(s): Flag-in cardinal-fish.

Global distribution: Indo-West Pacific Ocean.

Ecology: This reef-associated species has a depth range of 18–106 m. It is commonly found in sandy to muddy bottoms.

Biology: Body of this species is ovate to elongate and is moderately compressed. The eyes are large and their diameter may exceed snout length. Scales are usually ctenoid. Lateral line is complete and is not extending far onto caudal fin. It attains a maximum length of 16 cm.

4. Taeniamia fucata *(= Archamia fucata)*

Image credit: Museums Victoria, https://fishesofaustralia.net.au/home/species/4210; https://creativecommons.org/licenses/by/3.0/au/

Common name(s): Orangelined cardinalfish.

Global distribution: Indo-Pacific Ocean.

Ecology: It inhabits coral and rocky reefs in sheltered bays and lagoon; patch reefs, with muddy, sandy and silty bottoms or coral rubble; and mangroves of the continental and insular shelves. It is a nocturnal species with a depth range of 2–60 m.

Biology: It has the following characteristics: two short blue stripes from snout through eye; dark spot on tail base; and presence of dark bar beneath eye. Rear edges of scales of this species are iridescent yellow/orange, forming narrow wavy vertical lines. It is a mouthbrooder and is attaining a maximum length of 10 cm.

5. ***Taeniamia lineolata*** *(Archamia lineolata)*

Source: Public domain

Common name(s): Shimmering cardinal.

Global distribution: It is known only from the Red Sea of the Western Indian Ocean.

Ecology: It is a reef-associated, shallow-water species with a depth range of 1–20 m. It may form schools with other Archamia species especially in coral reef lagoons. It is also known to be cryptic, hiding in dark caves and crevices in coral reefs during the day.

Biology: It is distinguished by its posterior preopercular edge which is almost serrated. Ventral edge, however, is serrated on posterior half only. The scaly sheath along anal-fin base is well developed. It is a mouthbrooder attaining a maximum length of 10 cm only.

6. ***Verulux cypselurus*** *(= Rhabdamia cypselurus)*

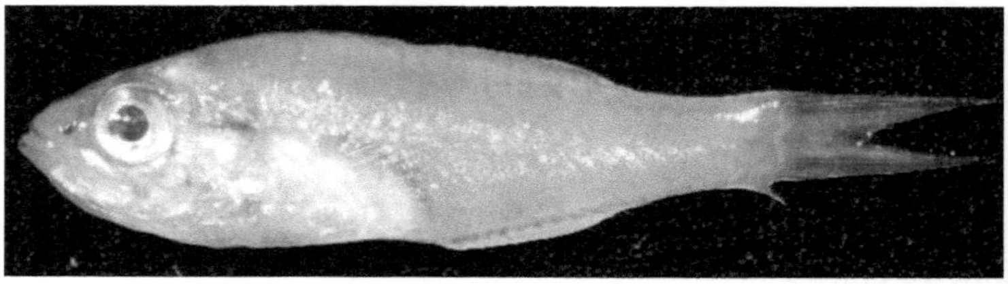

Source: Public domain

Common name(s): Swallowtail cardinalfish.

Global distribution: Indo-Pacific Ocean.

Ecology: This reef-associated species has a depth range of 2–15 m. It may form large shoals among rocks or coral reefs.

Biology: The body of this species is generally transparent with a few black pigments from snout to opercle. Upper and lower margins of the caudal fin are black and the caudal fin is forked. It attains a maximum length of 6 cm.

Luminous organs and bioluminescence: Luminescent apogonid fishes.

Haneda (1950) reported that certain species of the family Apogonidae possess unusually large luminous areas and they utilize half the muscles of their complicated body structure for this purpose (Haneda, 1950). Species of this family such as *Epigonus macrops, Jaydia ellioti (= Apogon ellioti), Apogon striatus, Taeniamia lineolata (= Archamia lineolate), Taeniamia fucata (= Archamia fucata)* and *Verulux cypselurus (= Rhabdamia cypselura*) have been reported to possess open glands containing ingested crustacean luminous substances (Haneda, 1980). Lau and Oakley (2021) also reported that the autogenic bioluminescence produced by these fishes depend largely on the dietary acquisition of luciferin, vargulin from their bioluminescent ostracod prey. Lau and Oakley(2021) reported that, instead of photophores, these fishes independently evolved visceral light organs that are connected to the gut and are less structurally complex. Okamoto and Nakayama (2016) reported that *Epigonus macrops* is a unique species of this genus in having a ventral luminous organ. In this species, one of the eight pyloric caeca has been modified into this light organ. The luminescent window of this species is located mid-ventrally between the anterior pelvic fins and is covered by a single large scale. Suntsov et al. (2008) observed that the species of the luminous genera of apogonids viz. *Apogon, Taeniamia (= Archamia)* and *Rhabdamia* possess endogenous (Cypridina-like) bioluminescence and show significant diversity in the morphology of their internal light organs which are varying from pouch-like protrusions of the intestine or rectum to even a modified pyloric caecum. Haneda et al. (1969) reported that the luminous organs of *Taenia* spp. (= *Archamia* spp.) viz. *Taenia lineolata, Taenia fucata* and *Taenia zosterophora; and Apogon striata* formed similarly and the luminous body in these species is located at the juncture of their pyloric caeca and the intestine. Further in these species, both the proximal part of the intestine and the pyloric caeca were found to emit light. Suntsov et al. (2008) reported that in the species of *Verulux (= Rhabdamia*), the light organs are located at the ends of two pyloric caeca. Haneda (1980) reported that the crude luciferin and luciferase extracts of the fishes of the genera *Taenia, Apogon* and *Verulux* produced light when these extracts were mixed. Fraser and Prokofiev (2016) found that in the species of Jaydia, the anterior light-emitting location is associated with a pouch located at the second bend of the intestine followed by a pair of small oval organs just anterior to the anus. Haneda (1969) reported that the luminescent organ system of *Jaidia ellioti (= Apogon ellioti)* consists of an oval-shaped, lemon yellow organ of 1–2 mm dia and this organ is connected by a duct to the second bend of the intestine. In this species, light is produced by the typical luciferase-luciferin type of reaction unlike other related species of *Apogon.*

7. *Siphamia* spp.

Common name(s): Cardinalfishes.

Global distribution: These fishes are native to the Indian and Pacific Oceans.

Ecology: All these species have been reported to be commensal with invertebrates such as sea urchins, crown-of-thorns starfish and corals (Gon and Allen, 2012).

Biology: The lower side of the body of these species possess a silvery, bioluminescent band. Caudal fin is emarginate or forked.

1. Siphamia cephalotes

2. Siphamia jebbi

3. Siphamia roseigaster

4. Siphamia tubifer (= Siphamia versicolor)

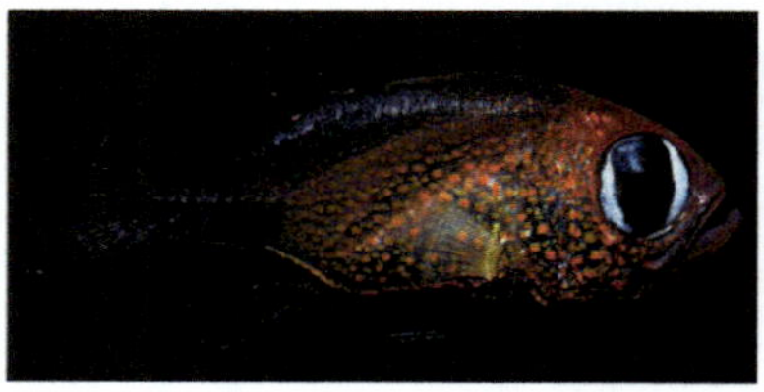

5. Siphamia arnazae

6. Siphamia cyanophthalma

7. Siphamia majimai

Image credit: 1-4,&6, 7: Museum of Victoria (CC); 5: Mark Erdmann, California Academy of Sciences

1. Siphamia cephalotes 2. Siphamia jebbi; 3. Siphamia roseigaster; 4. Siphamia tubifer (= Siphamia versicolor); 5. Siphamia arnazae; 6. Siphamia cyanophthal; 7. Siphamia jebbi; 8. Siphamia majimai

Luminous organs and bioluminescence: Siphamia spp.

Among the cardinal fishes of the family Apogonidae, fishes of the genus *Siphamia* are only bacterially luminescent. The species of this genus possess only a small luminous gland is situated on the ventral body. According to Fishelson et.al. (2005) an additional orobranchial luminous gland is also seen in the fishes of this genus. Gon and Allen (2012) observed that in these species, light producing bacteria are seen in a small pocket connected to the gut in the abdominal cavity and in a sac located on each side of the tip of the tongue (Gon and Allen, 2012). Haneda (1980) reported that the luminous system which is open to the outside consists of a luminous gland, a tube connecting the gland to the intestine, and an accessory structure made up of translucent abdominal muscle bundles.

Dunlap et al. (2011) and Wada et al. (2006) reported that most of the species of *Siphamia* possessed a ventral light organ with the luminous bacteria, *Photobacterium leiognathi* and it was located anterior to the pelvic fins of the fishes. The blue-green light produced by these luminous bacteria dispersed over the ventrum of the fish via translucent musculature. Further, it was reported that *Siphamia permutata (= Siphamia versicolor) Siphamia cephalotes,* and *Siphamia cuneiceps* were found to have a second light organ, at the anterior tip of the buccal cavity and this light organ functioned as a lure during feeding.

Based on the pattern of dark pigmentation on the light organ, the species of the genus *Siphamia* are divided into two main groups viz. *Siphamia tubifer* group (14 species) and *Siphamia tubulata* group (11 species). While the *Siphamia tuber* group species possess a striated light organ, the *Siphamia tubulata* group species have a dark-dotted light organ. The species associated with these groups are given below.

Siphamia tubifer group: The species of this group include *Siphamia arabica,, Siphamia argentea, Siphamia fraseri, Siphamia fuscolinea, Siphamia goreni, Siphamia guttulata, Siphamia jebbi, Siphamia. majimai, Siphamia mossambica, Siphamia papuensis, Siphamia randalli, Siphamia. spinicola, Siphamia stenotes,* and *Siphamia tubifer* (Gon and Allen, 2012; Gon et al., 2014).

Siphamia tubulata group: The species of this group include *Siphamia arnazae, Siphamia brevilux, Siphamia cephalotes, Siphamia corallicola, Siphamia cuneiceps, Siphamia cyanophthalma, Siphamia elongata, Siphamia fistulosa, Siphamia roseigaster Siphamia senoui,* and *Siphamia tubulata* (Gon and Allen, 2012; Allen and Erdmann, 2019).

The description of the light organs observed in certain species of *Siphamia* is given below.

Siphamia cephalotes: In this species, there are two luminescent sites with bioluminescent bacteria. In the first site a disc-like, bacteria-harboring "gland" and a paired ventral "bioluminescent reflectors" are seen. The second site which is situated more anteriorly possesses two sacs originating in the gular region and protruding into the oral cavity. The apical parts of these luminous sacs harbor the luminescent bacteria whose light diffuses within the oral cavity of this fish. This luminous oral cavity of this fish serves as a lure to attract the small prey as in other related species (Fishelson et al., 2005).

A buccal light organ containing bacteria and formed by tissues at the apical tips of the ventral diffuser was reported for this species. Apical tip tissues interpreted as a buccal light organ (Dunlap et al., 2011).

Siphamia cuneiceps; In this species, a buccal light organ containing luminous bacteria was found to develop by tissues at the apical tips of the ventral diffuser. These apical tip tissues are believed to constitute its buccal light organ (Dunlap et al., 2011).

Siphamia cyanophthalma: In this species, the luminous system consists of a pair of narrow blue stripes across the eye and a much longer posterior extension of the light organ, which reaches to about the level of the middle of the caudal peduncle (Allen and Erdmann, 2019).

Siphamia tubifer (= Siphamia versicolor; Siphamia permutata): The luminescent sites of this species are also very similar to that of *Siphamia cephalotes*. The luminescence system consists of a ventral light organ with luminous bacteria, reflector, lens, duct, and a ventral diffuser extending from the throat to the caudal peduncle. (Dunlap et al., 2011). In this species, the size of the light organs varies according to the standard length of the fishes. While it was 0.8 mm in diameter in the fishes of 1 cm standard length, it was found to be 2.9 mm in diameter in fishes of 3.6–3.7 cm length. Further, the light organ of this species has been reported to harbor the symbiotic luminous bacteria, *Photobacterium mandapamensis* with a population increasing from 7.0×10^6 to 8.7×10^7 cells in relation to the increase in the diameter of the light organ (Gould et al., 2016; Tanet et al., 2020). Suntsov et al. (2008) reported on the presence of *Photobacterium leiognathi* in the larval light organs of this species. The average maximum light emission from the luminescent bacterial symbionts of this species occurred at a wavelength of 487 nm. Dunlap et al. (2011) reported that this species possesses a buccal light organ formed by tissues at the apical tips of the ventral diffuser containing luminous bacteria. The ventral luminescence emitted at twilight is used by these fishes to attract and feed on zooplankton from the reef benthos; and as counterillumination to avoid from their predators.

Image courtesy: Paul Dunlap

Main components of the *Siphamia versicolor* luminescence system
lo: light organ; p: transparent primary diffuser; vd: ventral diffuser

BIBLIOGRAPHY

Academy of Natural Sciences, (1933). *Proceedings of The Academy of Natural Sciences* (Vol. LXXXV).

Allen, G. R., & Erdmann, M. V., (2019). *Siphamia arnazae*, a new species of cardinalfish (*Teleostei*: *Apogonidae*) from Papua New Guinea. *J. Ocean Sci. Found., 33*, 1–8.

Allen, L. G., & Horn, M. H., (2006). *The Ecology of Marine Fishes: California and Adjacent Waters.* University of California Press.

Anctil, M., & Case, J. F., (1977). The caudal luminous organs of lantern fishes: General innervations and ultrastrucutre. *Am. J. Anat., 149*, 1–22.

Anctil, M., (2018). *Luminous Creatures: The History and Science of Light Production in Living Organisms.* McGill-Queen's Press.

Anderson, M. E., & Leslie, R. W., (2001). Review of the deep-sea anglerfishes (*Lophiiformes*: *Cceratioidei*) of Southern Africa. *Ichthyol. Bull., 70*, 1–32.

Anon., (1933). *Proceedings of The Academy of Natural Sciences* (Vol. LXXXV).

Anon., (1972). Collected *Repri*nts. Southwest Fisheries Center (U.S.). National Oceanic and Atmospheric Administration, National Marine Fisheries Service, Southwest Fisheries Center, - Fisheries.

Anon., (1974). Osteology and relationships of ceratioid anglerfishes of the family *Oneirodidae*, with a review of the genus Oneirodes Lutken. *Natural History Museum of Los Angeles County, Science Bulletin, 18*, 1–113.

Anon., (1980). Fishery Bulletin (Vol. 78, No. 1). National Marine Fisheries Service, - Fisheries.

Anon., (1984). *NOAA Technical Report NMFS.* U.S. Department of Commerce, National Oceanic and Atmospheric Administration, National Marine Fisheries Service.

Anon., (1991). Fishery Bulletin (Vol. 89, No. 2). National Marine Fisheries Service. Fisheries.

Anon., (2020). *Identification Guide to the Mesopelagic Fishes of the Central and South East Atlantic Ocean* (p. 346). Food and Agriculture Organization of the United Nations. Food & Agriculture Org., Nature.

Anon., www.faoadriamed.org/html/Species/LophiusBudegassa.html (accessed on 9 January 2023).

Anon., *Fish Bioluminescence: Distinctive Flashing Patterns Might Facilitate Fish Mating.* American Museum of Natural History, https://www.sciencedaily.com/releases/2014/03/140304095107.htm (accessed on 9 January 2023).

Anon., http://lowlightintensity.weebly.com/bioluminescence-aiding-predation.html (accessed on 9 January 2023).

Anon., http://tolweb.org/Myctophidae/ (accessed on 9 January 2023).

Anon., http://www.animalplanet.com/pets/pineapplefish/ (accessed on 9 January 2023).

Anon., http://www.biodiversityexplorer.info/chondrichthyes/elasmobranchii/squalomorphii/heteroscymnoides_marleyi.htm (accessed on 9 January 2023).

Anon., http://www.fao.org/fishery/species/3163/en (accessed on 9 January 2023).

Anon., http://www.gma.org/fogm/Cyclothone_signata.htm (accessed on 9 January 2023).

Anon., http://www.marinespecies.org/aphia.php?p=taxdetails&id=126214 (accessed on 9 January 2023).

Anon., http://www.nbcnews.com/id/37093310/ns/technology_and_science-science/t/bioluminescence-lights-creatures-oceans/#.XsvBL2hKhPY (accessed on 9 January 2023).

Anon., https://www.sciencedaily.com/releases/2014/12/141208145138.htm (accessed on 9 January 2023).

Anon., https://animaldiversity.org/accounts/Idiacanthus_fasciola/ (accessed on 9 January 2023).

Anon., https://animaldiversity.org/accounts/Rhadinesthes_decimus/ (accessed on 9 January 2023).

Anon., https://australian.museum/learn/animals/fishes/silver-lightfish-phosichthys-argenteus/ (accessed on 9 January 2023).

Anon., https://bie.ala.org.au/species/37110003 (accessed on 9 January 2023).

Anon., https://bie.ala.org.au/species/urn:lsid:biodiversity.org.au:afd.taxon:a48dd410-bde2-42cd-9f38-2f5272e38625 (accessed on 9 January 2023).

Anon., https://biogeodb.stri.si.edu/sftep/en/thefishes/species/761 (accessed on 9 January 2023).

Anon., https://en.wikipedia.org/wiki/Acropoma (accessed on 9 January 2023).

Anon., https://en.wikipedia.org/wiki/Black_seadevil (accessed on 9 January 2023).

Anon., https://en.wikipedia.org/wiki/Cookiecutter_shark (accessed on 9 January 2023).

Anon., https://en.wikipedia.org/wiki/Cyclothone (accessed on 9 January 2023).

Anon., https://en.wikipedia.org/wiki/Footballfish (accessed on 9 January 2023).

Anon., https://en.wikipedia.org/wiki/Gigantactis (accessed on 9 January 2023).

Anon., https://en.wikipedia.org/wiki/Haplophryne (accessed on 9 January 2023).

Anon., https://en.wikipedia.org/wiki/Horned_lantern_fish (accessed on 9 January 2023).

Anon., https://en.wikipedia.org/wiki/Hymenocephalus_lethonemus (accessed on 9 January 2023).

Anon., https://en.wikipedia.org/wiki/Idiacanthus (accessed on 9 January 2023).

Anon., https://en.wikipedia.org/wiki/Lasiognathus_amphirhamphus (accessed on 9 January 2023).

Anon., https://en.wikipedia.org/wiki/Lasiognathus_intermedius (accessed on 9 January 2023).

Anon., https://en.wikipedia.org/wiki/Linophryne_indica (accessed on 9 January 2023).

Anon., https://en.wikipedia.org/wiki/Lophius (accessed on 9 January 2023).

Anon., https://en.wikipedia.org/wiki/Malacosteus_niger (accessed on 9 January 2023).

Anon., https://en.wikipedia.org/wiki/Neoscopelus_macrolepidotus (accessed on 9 January 2023).

Anon., https://en.wikipedia.org/wiki/Northern_lampfish (accessed on 9 January 2023).

Anon., https://en.wikipedia.org/wiki/Notoscopelus_bolini (accessed on 9 January 2023).

Anon., https://en.wikipedia.org/wiki/Notoscopelus_caudispinosus (accessed on 9 January 2023).

Anon., https://en.wikipedia.org/wiki/Notoscopelus_elongatus (accessed on 9 January 2023).

Anon., https://en.wikipedia.org/wiki/Notoscopelus_kroyeri (accessed on 9 January 2023).

Anon., https://en.wikipedia.org/wiki/Pacific_viperfish (accessed on 9 January 2023).

Anon., https://en.wikipedia.org/wiki/Pelican_eel (accessed on 9 January 2023).

Anon., https://en.wikipedia.org/wiki/Photocorynus_spiniceps (accessed on 9 January 2023).

Anon., https://en.wikipedia.org/wiki/Phthanophaneron_harveyi (accessed on 9 January 2023).

Anon., https://en.wikipedia.org/wiki/Phyllorhinichthys (accessed on 9 January 2023).

Anon., https://en.wikipedia.org/wiki/Pineapplefish (accessed on 9 January 2023).

Anon., https://en.wikipedia.org/wiki/Saccopharynx (accessed on 9 January 2023).

Anon., https://en.wikipedia.org/wiki/Sigmops_bathyphilus (accessed on 9 January 2023).

Anon., https://en.wikipedia.org/wiki/Spiniphryne_gladisfenae (accessed on 9 January 2023).

Anon., https://en.wikipedia.org/wiki/Splitfin_flashlightfish (accessed on 9 January 2023).

Anon., https://en.wikipedia.org/wiki/Thaumatichthys (accessed on 9 January 2023).

Anon., https://en.wikipedia.org/wiki/Thaumatichthys_binghami (accessed on 9 January 2023).

Anon., https://en.wikipedia.org/wiki/Triplewart_seadevil (accessed on 9 January 2023).

Anon., https://en.wikipedia.org/wiki/Velvet_dogfish (accessed on 9 January 2023).
Anon., https://eol.org/pages/46563342/articles (accessed on 9 January 2023).
Anon., https://fishbase.mnhn.fr/summary/9232 (accessed on 9 January 2023).
Anon., https://fishbase.mnhn.fr/summary/Gigantactis-meadi.html (accessed on 9 January 2023).
Anon., https://fishbase.mnhn.fr/summary/Leptacanthichthys-gracilispinis.html (accessed on 9 January 2023).
Anon., https://fishbase.mnhn.fr/summary/Maurolicus-mucronatus.html (accessed on 9 January 2023).
Anon., https://fishesofaustralia.net.au/home/family/99 (accessed on 9 January 2023).
Anon., https://fishesofaustralia.net.au/home/genus/1724 (accessed on 9 January 2023).
Anon., https://fishesofaustralia.net.au/home/genus/860 (accessed on 9 January 2023).
Anon., https://fishesofaustralia.net.au/home/species/1780 (accessed on 9 January 2023).
Anon., https://fishesofaustralia.net.au/home/species/1810 (accessed on 9 January 2023).
Anon., https://fishesofaustralia.net.au/home/species/2841 (accessed on 9 January 2023).
Anon., https://fishesofaustralia.net.au/home/species/3009 (accessed on 9 January 2023).
Anon., https://fishesofaustralia.net.au/home/species/3100 (accessed on 9 January 2023).
Anon., https://fishesofaustralia.net.au/home/species/3381 (accessed on 9 January 2023).
Anon., https://fishesofaustralia.net.au/home/species/3939 (accessed on 9 January 2023).
Anon., https://fishesofaustralia.net.au/home/species/5133 (accessed on 9 January 2023).
Anon., https://groups.oist.jp/mbu/event/kleptoprotein-bioluminescence-origin-luciferase-luminous-fish-parapriacanthus-ransonneti (accessed on 9 January 2023).
Anon., https://invertebrate.w.uib.no/2016/12/14/door-14-where-the-sun-doesnt-shine-lucifer-luciferin-and-luciferase/ (accessed on 9 January 2023).
Anon., https://mexican-fish.com/mexican-lampfish/ (accessed on 9 January 2023).
Anon., https://news.tulane.edu/pr/researchers-identify-new-species-pocket-shark (accessed on 9 January 2023).
Anon., https://noaateacheratsea.blog/tag/bioluminescence/ (accessed on 9 January 2023).
Anon., https://noaateacheratsea.blog/tag/photophores (accessed on 9 January 2023).
Anon., https://oceana.org/marine-life/ocean-fishes/pacific-blackdragon (accessed on 9 January 2023).
Anon., https://oceanexplorer.noaa.gov/explorations/19biolum/background/midnight-zone/midnight-zone.html (accessed on 9 January 2023).
Anon., https://sfmsi.wordpress.com/tag/bioluminescence/ (accessed on 9 January 2023).
Anon., https://shark-references.com/species/view/Aculeola-nigra (accessed on 9 January 2023).
Anon., https://sites.evergreen.edu/animatingthesea-s20/black-seadevil-melanocetidae/ (accessed on 9 January 2023).
Anon., https://sites.google.com/pingree.org/deepsea/species/vertebrates (accessed on 9 January 2023).
Anon., https://species.nbnatlas.org/species/NBNSYS0020420397 (accessed on 9 January 2023).
Anon., https://teara.govt.nz/en/photograph/5276/slender-lanternfish (accessed on 9 January 2023).
Anon., https://www.alamy.com/stock-photo-lepidophanes-guentheri-a-species-of-the-large-group-of-lanternfish-79235897.html (accessed on 9 January 2023).
Anon., https://www.amnh.org/explore/news-blogs/research-posts/bioluminescent-fish-flashing-patterns-might-facilitate-mating (accessed on 9 January 2023).
Anon., https://www.animalvogue.com/2015/02/bioluminescence-aggressive-mimicry-and-dating/ (accessed on 9 January 2023).

Anon., https://www.blakenz.org/person/myctophids/ (accessed on 9 January 2023).
Anon., https://www.britannica.com/animal/Cyclothone (accessed on 9 January 2023).
Anon., https://www.cancer.gov/publications/dictionaries/cancer-terms/def/car-t-cell-therapy (accessed on 9 January 2023).
Anon., https://www.cell.com/current-biology/pdf/S0960-9822(16)00046-4.pdf (accessed on 9 January 2023).
Anon., https://www.central.edu/writing-anthology/2019/04/30/bioluminescence-in-lanternsharks-and-its-impact-on-communication-via-identification/ (accessed on 9 January 2023).
Anon., https://www.discoverlife.org/20/q?search=Acropoma+boholensis&b=FB59419 (accessed on 9 January 2023).
Anon., https://www.etyfish.org/gadiformes1/ (accessed on 9 January 2023).
Anon., https://www.fao.org/3/t0243e/T0243E44.pdf (accessed on 9 January 2023).
Anon., https://www.fao.org/3/x2401e/x2401e43.pdf (accessed on 9 January 2023).
Anon., https://www.fao.org/3/y4161e/y4161e25.pdf (accessed on 9 January 2023).
Anon., https://www.fao.org/3/y4161e/y4161e30.pdf (accessed on 9 January 2023).
Anon., https://www.fao.org/3/y4161e/y4161e39.pdf (accessed on 9 January 2023).
Anon., https://www.fao.org/fishery/species/3380/en (accessed on 9 January 2023).
Anon., https://www.fishbase.de/summary/27427 (accessed on 9 January 2023).
Anon., https://www.fishbase.de/summary/28245 (accessed on 9 January 2023).
Anon., https://www.fishbase.de/summary/50780 (accessed on 9 January 2023).
Anon., https://www.fishbase.de/summary/Polymetme-corythaeola.html (accessed on 9 January 2023).
Anon., https://www.fishbase.de/summary/Polymetme-illustris.html (accessed on 9 January 2023).
Anon., https://www.fishbase.de/summary/Thaumatichthys-pagidostomus.html (accessed on 9 January 2023).
Anon., https://www.fishbase.in/Summary/SpeciesSummary.php?id=23260&lang=italian (accessed on 9 January 2023).
Anon., https://www.fishbase.se/summary/12545 (accessed on 9 January 2023).
Anon., https://www.fishbase.se/summary/16302 (accessed on 9 January 2023).
Anon., https://www.fishbase.se/summary/16888 (accessed on 9 January 2023).
Anon., https://www.fishbase.se/summary/16901 (accessed on 9 January 2023).
Anon., https://www.fishbase.se/summary/16980 (accessed on 9 January 2023).
Anon., https://www.fishbase.se/summary/16986 (accessed on 9 January 2023).
Anon., https://www.fishbase.se/summary/16990 (accessed on 9 January 2023).
Anon., https://www.fishbase.se/summary/16995 (accessed on 9 January 2023).
Anon., https://www.fishbase.se/summary/1726 (accessed on 9 January 2023).
Anon., https://www.fishbase.se/summary/1804 (accessed on 9 January 2023).
Anon., https://www.fishbase.se/summary/22719 (accessed on 9 January 2023).
Anon., https://www.fishbase.se/summary/27743 (accessed on 9 January 2023).
Anon., https://www.fishbase.se/summary/27745 (accessed on 9 January 2023).
Anon., https://www.fishbase.se/summary/3102 (accessed on 9 January 2023).
Anon., https://www.fishbase.se/summary/51609 (accessed on 9 January 2023).
Anon., https://www.fishbase.se/summary/51616 (accessed on 9 January 2023).
Anon., https://www.fishbase.se/summary/52849 (accessed on 9 January 2023).
Anon., https://www.fishbase.se/summary/52855 (accessed on 9 January 2023).
Anon., https://www.fishbase.se/summary/52856 (accessed on 9 January 2023).
Anon., https://www.fishbase.se/summary/5351 (accessed on 9 January 2023).

Anon., https://www.fishbase.se/summary/58267 (accessed on 9 January 2023).
Anon., https://www.fishbase.se/summary/58775 (accessed on 9 January 2023).
Anon., https://www.fishbase.se/summary/58781 (accessed on 9 January 2023).
Anon., https://www.fishbase.se/summary/60774 (accessed on 9 January 2023).
Anon., https://www.fishbase.se/summary/62981 (accessed on 9 January 2023).
Anon., https://www.fishbase.se/summary/7407 (accessed on 9 January 2023).
Anon., https://www.fishbase.se/summary/Argyropelecus-aculeatus.html (accessed on 9 January 2023).
Anon., https://www.fishbase.se/summary/Argyropelecus-gigas (accessed on 9 January 2023).
Anon., https://www.fishbase.se/summary/cleidopus-gloriamaris (accessed on 9 January 2023).
Anon., https://www.fishbase.se/summary/Diceratias-trilobus.html (accessed on 9 January 2023).
Anon., https://www.fishbase.se/summary/FamilySummary.php?ID=241 (accessed on 9 January 2023).
Anon., https://www.fishbase.se/summary/FamilySummary.php?ID=87 (accessed on 9 January 2023).
Anon., https://www.fishbase.se/summary/Gigantactis-gargantua (accessed on 9 January 2023).
Anon., https://www.fishbase.se/summary/Gigantactis-kreffti.html (accessed on 9 January 2023).
Anon., https://www.fishbase.se/summary/Gigantactis-paxtoni.html (accessed on 9 January 2023).
Anon., https://www.fishbase.se/summary/Gigantactis-sexfilis.html (accessed on 9 January 2023).
Anon., https://www.fishbase.se/summary/Gonostoma-elongatum.html (accessed on 9 January 2023).
Anon., https://www.fishbase.se/summary/Himantolophus-nigricornis.html (accessed on 9 January 2023).
Anon., https://www.fishbase.se/summary/Himantolophus-sagamius.html (accessed on 9 January 2023).
Anon., https://www.fishbase.se/summary/Kryptophanaron-alfredi.html (accessed on 9 January 2023).
Anon., https://www.fishbase.se/summary/Kumba-dentoni (accessed on 9 January 2023).
Anon., https://www.fishbase.se/summary/Melanocetus-eustalus.html (accessed on 9 January 2023).
Anon., https://www.fishbase.se/summary/Platytroctes-apus (accessed on 9 January 2023).
Anon., https://www.fishbase.se/summary/Tyrannophryne-pugnax.html (accessed on 9 January 2023).
Anon., https://www.fishbase.se/summary/Valenciennellus-tripunctulatus.html (accessed on 9 January 2023).
Anon., https://www.frogfish.ch/deepsea-anglerfish.html (accessed on 9 January 2023).
Anon., https://www.gbif.org/species/10051334/treatments (accessed on 9 January 2023).
Anon., https://www.livescience.com/17896-smallest-vertebrate-controversy-frog-angler-fish.html (accessed on 9 January 2023).
Anon., https://www.lkouniv.ac.in/site/writereaddata/siteContent/202003251324430605shelly_Sound_Producing.pdf (accessed on 9 January 2023).
Anon., https://www.nationalgeographic.com/animals/reference/bioluminescence-animals-ocean-glowing/ (accessed on 9 January 2023).
Anon., https://www.naturepl.com/search/preview/close-up-of-the-skin-of-a-bioluminescing-pygmy-shark-euprotomicrus/0_01534178.html (accessed on 9 January 2023).
Anon., https://www.pinterest.com/pin/33988172158016758 (accessed on 9 January 2023).
Anon., https://www.proquest.com/docview/1860272652 (accessed on 9 January 2023).

Anon., https://www.sciencesource.com/archive/Dark-Searsid–Mentodus-facilis–SS2807313.html (accessed on 9 January 2023).

Anon., https://www.sciencesource.com/archive/Stareye-Lightfish-SS2713156.html (accessed on 9 January 2023).

Anon., https://www.sciencesource.com/archive/Threadfin-Dragonfish--Echiostoma-barbatum-SS22345173.html (accessed on 9 January 2023).

Anon., https://www.scribd.com/document/93770802/pramodsingh (accessed on 9 January 2023).

Anon., https://www.sheppardsoftware.com/content/animals/animals/fish/flashlightfish.htm (accessed on 9 January 2023).

Anon., https://www.smore.com/qn0y0-flashlight-fish (accessed on 9 January 2023).

Anon., https://www.thoughtco.com/amazing-bioluminescent-organisms-373898 (accessed on 9 January 2023).

Anon., https://www.washington.edu/news/2006/01/27/flap-over-fishes-whos-the-smallest-of-them-all/ (accessed on 9 January 2023).

Anon., https://www.yourarticlelibrary.com/fish/anatomy-and-physiology/luminous-organs-or-photophore-of-the-fishes-with-diagram/88411 (accessed on 9 January 2023).

Anon., *In the Different Species of This Genus*. https://www.frogfish.ch/deepsea-anglerfish.html (accessed on 9 January 2023).

Anon., *Luciferin Details*. https://biolum.eemb.ucsb.edu/chem/detail2.html (accessed on 9 January 2023).

Anon., *Luring with Bioluminescence* https://www.frogfish.ch/deepsea-anglerfish.html (accessed on 9 January 2023).

Anon., *Vargulin*. https://en.wikipedia.org/wiki/Vargulin (accessed on 9 January 2023).

Antonio, L., (2017). *Periodic Tables Unifying Living Organisms at the Molecular Level: The Predictive Power of the Law of Periodicity*. World Sci.

Azuma, N., Furubayashi, C., Shichiri, T., Wada, M., Mizuno, N., & Suzuki, Y., (2005). Bioluminescence and mating behavior in pony fish, *Leiognathus nuchalis*. *J. Biolum. Chemilum.*, 3–6.

Baguet, F., & Marechal, G., (1976). Bioluminescence of bathypelagic fish from the strait of Messina. *Camp. Biochem. Physiol., 53*, 75–82.

Baker, H., (2020). *Seeing Red: The Incredible Vision of Dragonfish.* https://marinemadness.blog/2020/09/09/seeing-red-the-incredible-vision-of-dragonfish/ (accessed on 9 January 2023).

Baldwin, C. C., Johnson, G. D., & Paxton, J. R., (1997). *Protoblepharon rosenblatti*, a new genus and species of flashlight fish (*Beryciformes*: *Anomalopidae*) from the tropical South Pacific, with comments on anomalopid phylogeny. *Proc. Biol. Soc. Washington, 110*, 373–383.

Barnes, H., (2003). *Oceanography and Marine Biology*. CRC Press Boca Raton.

Barton, R., (2016). *Bioluminescent Fish, Bacteria, and Experiential Learning*. All Regis University Theses.

Bertelsen, E., & Struhsaker, P. J., (1977). The ceratioid fishes of the genus *Thaumatichthys* osteology, relationships, distribution and biology. *E. Galathea Rep., 14*, 7–40,

Bessho-Uehara, M., Yamamoto, N., Shigenobu, S., Mori, H., Kuwata, K., & Oba, Y., (2020). Kleptoprotein bioluminescence: *Parapriacanthus* fish obtain luciferase from ostracod prey. *Sci. Adv., 6*, eaax4942.

Bigelow, H. B., Cohen, D. M., Dick, M. M., Gibbs, Jr. R. H., Grey, M., Morrow, Jr. J. E., Schultz, L. P.,& Walters, V. (2018). Soft-Rayed Bony Fishes: Orders Isospondyli and Giganturoidei. Part 4. Series: Fishes of the Western North Atlantic, Yale University Press. https://doi.org/10.2307/j.ctvbcd0k1, p.620.

Causo, J. H., (1983). The systematics and distribution of the lophiid anglertfishes: II. Revision of the genera *Lophiomus* and *Lophius*. *Copeia*, 11–30.

Cavallaro, M., Battaglia, P., Guerrera, M. C., Abbate, F., Levanti, M. B., Ammendolia, G., Andaloro, F., et al., (2019). Structure and ultrastructure study on photophores of the Madeira lanternfish, *Ceratoscopelus maderensis* (Lowe, 1839), Pisces: *Myctophidae. Acta Zool., 100*, 89–95.

Chakrabarty, P., Davis, M. P., Smith, W. L., Berquist, R., Gledhill, K. M., Frank, L. R., & Sparks, J. S., (2011). Evolution of the light organ system in ponyfishes (*Teleostei*: *Leiognathidae). J. Morphol., 272*, 704–721.

Chakrabarty, P., Sparks, J. S., & Ho, H., (2010). Taxonomic review of the ponyfishes (*Perciformes: Leiognathidae)* of Taiwan. *Mar. Biodiv., 40*, 107–121.

Chiou, M., Shao, K., & Iwamoto, T., (2004). A new species, *Caelorinchus sheni*, and 19 new records of grenadiers (Pisces: *Gadiformes: Macrouridae*) from Taiwan. *Zool. Stud., 43*, 35–50.

Claes, J. M., & Mallefet, J., (2009). *Bioluminescence of Sharks: First Synthesis–Bioluminescence in Focus-A Collection of Illuminating Essays*, 51–65.

Claes, J. M., & Mallefet, J., (2011). Control of luminescence from lanternshark (*Etmopterus spinax*) photophores. *Commun. Integr. Biol., 4*, 251–253.

Claes, J. M., & Mallefet, J., (2015). Comparative control of luminescence in sharks: New insights from the slendertail lanternshark (*Etmopterus molleri*). *J. Exp. Mar. Biol. Ecol., 467*, 87–94.

Claes, J. M., Dean, M. N., Nilsson, D.,.Hart, N. S., & Mallefet, J., (2013). A Deepwater fish with 'lightsabers' – dorsal spine-associated luminescence in a counterilluminating lanternshark. *Sci. Rep., 3*, 1308.

Clarke, G. L., Conover, R. J., David, C. N., & Nicol, J. A. C., (1962). Comparative studies of luminescence in copepods and other pelagic marine animals. *J. Mar. Biol. Ass. U.K., 42*, 541–564.

Cohen, D. M., (1964). Bioluminescence in the Gulf of Mexico anacanthine fish *Steindachneria argentea. Copeia*, 406–409.

Cohen, D. M., lnada, T., lwamoto, T., & Scialabba, N., (1990). FAO species catalogue, Gadiform fishes of the world. An annotated and illustrated catalogue of cods, hakes, grenadiers, and others. *FAO Fish. Syn., 125*, 10.

Compagno, L. J. V., & Dando, M., (2005). Fowler, S. *Sharks of the World*. Princeton: Univ. Press.

Copeland, D. E., (1991). Fine structure of photophores in *Gonostoma elongatum*: Detail of a dual gland complex. *Biol Bull., 181*, 144–157.

Crane, J. M., (1968). Bioluminescence in the batfish *Dibranchus atlanticus. Copeia,* 410, 411.

Crescitelli, F., (2013). *The Visual System in Vertebrates.* Springer Science & Business Media.

Cruz-Agüero, J. D. L., (2016). Light organ anomalies in three species of myctophid fishes (*Actinopterygii*: *Myctophiformes*: *Myctophidae*) from the Pacific coast of Mexico. *Acta Ichthyol. Piscat., 46*, 247–250.

Davis, A. L., Sutton, T. T., Kier, W. M., & Johnsen, S., (2020). Evidence that eye-facing photophores serve as a reference for counterillumination in an order of deep-sea fishes. *Proc. Biol. Sci., 287*(1928), 20192918.

Davis, M. P., Sparks, J. S., & Smith, W. L., (2016). Repeated and widespread evolution of bioluminescence in marine fishes. *PLoS One, 11*, e0155154.

Deshmukh, A., (2017). Control of bioluminescence in myctophid fishes. *Indian J. Geo- Mar. Sci., 46*, 1436–1439.

Duchatelet, L., Delroisse, J., Flammang, P., Mahillon, J., & Mallefet, J., (2019). *Etmopterus spinax*, the velvet belly lanternshark, does not use bacterial luminescence. *Acta Histochem., 121*, 516–521.

Duchatelet, L., Hermans, C., Duhamel, G., Cherel, Y., Guinet, C., & Mallefet, J., (2019). Coelenterazine detection in five myctophid species from the Kerguelen Plateau. *Second Kerguelen Plateau Symposium: Marine Ecosystem and Fisheries*, 31–41.

Duchatelet, L., Marion, R., & Mallefet, J., (2021). A third luminous shark family: Confirmation of luminescence ability for *Zameus squamulosus* (*Squaliformes*; *Somniosidae*). *Photochem. Photobiol.* 10.1111/php.13393.

Dunlap, P. V., & Nakamura, M., (2011). Functional morphology of the luminescence system of *Siphamia versicolor* (*Perciformes: Apogonidae*), a bacterially luminous coral reef fish. *J. Morphol., 272*, 897–909.

Dunlap, P. V., (1984). Physiological and morphological state of the symbiotic bacteria from light organs of ponyfish. *Biol. Bull., 167,* 410–425.

Dunlap, P. V., Asta, J. C., Kimura, S., Fukui, A., Yoshino, T., & Endo, H., (2007). Phylogenetic analysis of host–symbiont specificity and codivergence in bioluminescent symbioses. *Cladistics, 23*, 507–532.

Dybas, C. L., (2019). Illuminating new biomedical discoveries: Bioluminescent, biofluorescent species glow with promise. *J. BioSci., 69*, 487–495.

Ebert, D. A., & Fowler, S., (2014). *An Illustrated Pocket Guide to the Sharks of the World.* Princeton University Press.

Ebert, D. A., Dando, M., & Fowler, S., (2021). *Sharks of the World: A Complete Guide.* Princeton University Press.

Farina, S. C., & Bemis, W. E., (2016). Functional morphology of gill ventilation of the goosefish, *Lophius americanus* (*Lophiiformes*: *Lophiidae*). *Zoology (Jena), 119*, 207–215.

Finn, R. N., (2020). *Fish Larval Physiology*. CRC Press: Boca Raton.

Fishelson, L., (2005). The oral cavity and bioluminescent organs of the cardinal fish species *Siphamia permutata* and *S. cephalotes* (*Perciformes, Apogonidae*). *Mar. Biol., 147*, 603–609.

Flynn, A. J., & Klepadlo, C., (2012). Two new species of *Photonectes* (*Teleostei*: *Stomiidae*) from the Indo-Pacific, and a re-examination of *P. achirus*. *Mem. Mus. Vic., 69*, 259–267.

Frank, T. M., Widder, E. A., Latz, M. I., & Case, J. F., (1984). Dietary maintenance of bioluminescence in a deep-sea mysid. *J. Exp. Biol., 109*, 385–389.

Fraser, T. H., & Prokofiev, A. M., (2016). A new genus and species of cardinalfish (*Percomorpha, Apogonidae*, Sphaeramiini) from the coastal waters of Vietnam: Luminescent or not?. *Zootaxa, 4144*, 227–242.

Freed, L. L., Easson, C., Baker, L. J., Fenolio, D., Sutton, T. T., Khan, Y., Blackwelder, P., et al., (2019). Characterization of the microbiome and bioluminescent symbionts across life stages of ceratioid anglerfishes of the Gulf of Mexico. *FEMS Microbiol. Ecol., 95*, fiz146.

Furness, B., *Bioluminescence in Lanternsharks and Its Impact on Communication Via Identification.* https://www.central.edu/writing-anthology/2019/04/30/bioluminescence-in-lanternsharks-and-its-impact-on-communication-via-identification/ (accessed on 9 January 2023).

Gago, F. J., & Lavenberg, R. J., (1992). Lanternfish Genus *Centrobranchus* (Pisces: Myctophidae). *Copeia,* 154–161

Ghedotti, https://epublications.regis.edu/cgi/viewcontent.cgi?article=1023&context=csre_celebrationscholarship (accessed on 9 January 2023).

Ghedotti, M. J., Gruber, J. N., Barton, R. W., Davis, M. P., & Smith, W. L., (2018). Morphology and evolution of bioluminescent organs in the glowbellies (*Percomorpha*: *Acropomatidae*) with comments on the taxonomy and phylogeny of Acropomatiformes. *J. Morphol., 279,* 1640–1653.

Gibbs, Jr. R. H., (1969). Taxonomy, sexual dimorphism, vertical distribution, and evolutionary zoogeography of the bathypelagic fish genus stomias (Stomiatidae). *Smithson. Contr. Zool., 31*, 25.

Gibbs, Jr. R. H., Clarke, T. A., & Gomon, J. R., (1983). Taxonomy and distribution of the stomioid fish genus *Eustomias* (Melanostomiidae), I subgenus *Nominostomias. Smithson. Contr. Zool.*, 380.

Gon, O., & Allen, G. R., (2012). Revision of the Indo-Pacific cardinalfish genus *Siphamia (Perciformes: Apogonidae). Zootaxa, 3294*, 1–84.

Gon, O., Allen, G. R., Erdmann, M. V., & Gouws, G., (2014). A new species of the cardinalfish genus *Siphamia* (*Perciformes, Apogonidae*) from West Papua, Indonesia. *Zootaxa,* 388.

Gonzalez, V. M. Jr., (2007). *Synthesis, Luminescence, and Applications of Coelenterazine and Its Analogs*. https://chemistry.illinois.edu/system/files/inline-files/03GonzalezFINALAbstract.pdf (accessed on 9 January 2023).

Gould, A. L., Dougan, K. E., Koenigbauer, S. T., & Dunlap, P. V., (2016). Life history of the symbiotically luminous cardinalfish *Siphamia tubifer* (*Perciformes: Apogonidae*). *J. Fish Biol., 89*, 1359–1377.

Green, A. A., & Mcelroy, W. D., (1956). Crystalline firefly luciferase. *Biochim Biophys Acta, 20*, 170–176.

Guerrero-Ferreira, R. C., & Nishiguchi, M. K., (2007). Biodiversity among luminescent symbionts from squid of the genera *Uroteuthis, Loliolus* and *Euprymna* (Mollusca: Cephalopoda). *Cladistics, 23*, 497–506.

Haddock, S. H. D., Moline, M. A., & Case, J. F., (2010). Bioluminescence in the sea. *Ann. Rev. Mar. Sci., 2*, 443–493.

Haffner, R. E., (1952). Zoogeography of the bathypelagic fish, *Chauliodus. Syst. Zool., 1*, 112–133.

Haneda, Y., (1950). Luminous organs of fish which emit light indirectly. *Pac. Sci., 4*, 214–227.

Haneda, Y., (1952). Some luminous fishes of the genera *Yarrella* and *Polyipnus. Pac. Sci.*, 13–16.

Haneda, Y., (1961). A preliminary report on two new luminous fish from Bombay and Hong Kong. *Sci. Rep. Yokosuka City Mus*., 45–50.

Haneda, Y., (1980). Luminescent fish with open- type glands containing luminous bacteria or luminous substances of ingested crustaceans. *Sci. Rep. Yokosuka City Mus., 27*, 5–12.

Haneda, Y., Tsuji, F. I., & Sugiyama, N., (1969). Luminescent systems in apogonid fishes from the Philippines. *Sci. New Ser., 165*, 188–190.

Hansen, K., & Herring, P. J., (1977). Dual bioluminescent systems in the anglerfish genus *Linophryne* (Pisces: Ceratioidea) *J. Zool., 182*, 103–124.

Haygood, M. G., Tebo, B. M., & Nealson, K. H., (1984). Luminous bacteria of a monocentrid fish (*Monocentris japonicus*) and two anomalopid fishes (*Photoblepharon palpebratus* and *Kryptophanaron alfredi*): Population sizes and growth within the light organs, and rates of release into the seawater. *Mar. Biol., 78*, 249–254.

Hellinger, J., Huhn, M., & Herlitze, S., (2017). Bioluminescence in fishes: Diversity and functions. *Fish & Ocean Opj.*, 2, OFOAJ.MS.ID.555587.

Hellinger, J., Jägers, P., Donner, M., Sutt, F., Mark, M. D., Senen, B., Tollrian, R., & Herlitze, S., (2017). The flashlight fish *Anomalops katoptron* uses bioluminescent light to detect prey in the dark. *PLoS One, 12*, e0170489.

Hellinger, J., Jägers, P., Spoida, K., Weiss, L. C., Mark, M. D., & Herlitze, S., (2020). Analysis of the territorial aggressive behavior of the bioluminescent flashlight fish *Photoblepharon steinitzi* in the red sea. *Front. Mar. Sci.* https://doi.org/10.3389/fmars.2020.00078.

Herring, P. J., & Cope, C., (2005). Red bioluminescence in fishes: On the suborbital photophores of *Malacosteus, Pachystomias* and *Aristostomias. Mar. Biol., 148*, 383–394.

Herring, P. J., & Munk, O., (1994). The escal light gland of the deep-sea anglerfish *Haplophryne mollis* (Pisces: Ceratioidei) with observations on luminescence control. *J. Mar. Biolog. Assoc. U.K., 74*, 747–763.

Herring, P. J., (1972). Bioluminescence of searsid fishes. *Mar. Biolog. Assoc. U. K., 52*, 879–887.

Herring, P. J., (1977). Bioluminescence in an evermannellid fish. *J. Zool., 181*, 297–307.

Herring, P. J., (1983). The spectral characteristics of luminous marine organisms. *Proc. R. Soc. B: Biol. Sci., 220*, 183–217. https://doi.org/10.1098/rspb.1983.0095.

Herring, P. J., (1992). Bioluminescence of the oceanic apogonid fishes *Howella brodiei* and *Florenciella lugubris. J. Mar. Biolog. Assoc. U.K., 72*, 139–148.

Ho, H., & Johnson, G. D., (2012). *Protoblepharon mccoskeri*, a new flashlight fish from eastern Taiwan (*Teleostei*: *Anomalopidae*). *Zootaxa, 3479*, 77–87.

Ho, H., Graham, K., & Russell, B., (2020). Three new species of the barracudina genus *Lestidium* (*Aulopiformes*: *Paralepididae*) from the Indo-West Pacific. *Zootaxa, 4767*, 71–88.

Ho, H., Kawai, T., & Amaoka, K., (2016). Records of deep-sea anglerfishes (*Lophiiformes*: *Ceratioidei*) from Indonesia, with descriptions of three new species. *Zootaxa, 4121*, 267–294.

Hubbs, C. L., & Wesner, R. L., (1964). *Parvilux*, a new genus of myctophid fishes from the Northeastern Pacific, with two new species. *Zool. Mededel., 39*, 445–463.

Hubbs, C. L., Tamotsu, I., & Matsubara, K., (1967). External and internal characters, horizontal and vertical distribution, luminescence, and food of the dwarf pelagic shark, *Euprotomicrus bispinatus. Bull. Scripps Inst. O'ceanogr., 10*, 83410818.

Iwamoto, T., & Arai, T., (1987). A new grenadier *Malacocephalus okamurai* (Pisces: Gadiformes: Macrouridae) from the Western Atlantic. *Copeia*, 204–208.

Iwamoto, T., Nakayama, N., Shao, K., & Ho, H., (2015). Synopsis of the grenadier fishes (*Gadiformes*; *Teleostei*) of Taiwan. *Proc. Calif. Acad. Sci., Series 4, 62*, 31–126.

Jha, S. *Luminous Organs or Photophore of the Fishes*. https://www.yourarticlelibrary.com/fish/anatomy-and-physiology/luminous-organs-or-photophore-of-the-fishes-with-diagram/88411 (accessed on 9 January 2023).

Johnsen, S. https://oceanexplorer.noaa.gov/explorations/15biolum/background/biolum/biolum.html (accessed on 9 January 2023).

Jørgensen, J. M., & Munk, O., (1979). Photophores and presumably luminous chin barbel and pectoral fin ray filaments of *Thysanactis dentex* (Pisces: *Stomiatoidea*). *Acta Zool.*, 33–42.

Kenaley, C. P., (2010). Comparative innervation of cephalic photophores of the loosejaw dragonfishes (*Teleostei*: *Stomiiformes*: *Stomiidae*): Evidence for parallel evolution of long-wave bioluminescence. *J. Morphol., 271*, 418–437.

Klepadlo, C., (2011). Three new species of the genus *Photonectes* (*Teleostei*: *Stomiiformes*: *Stomiidae*: *Melanostomiinae*) from the Pacific Ocean. *Copeia, 2*, 201–210.

Krönström, J., Holmgren, S., Baguet, F., Salpietro, L., & Mallefet, J., (2005). Nitric oxide in control of luminescence from hatchetfish (*Argyropelecus hemigymnus*) photophores. *J. Exp. Biol., 208*, 2951–2961.

Lau, E. S., & Oakley, T. H., (2021). Multi-level convergence of complex traits and the evolution of bioluminescence. *Biol. Rev. Camb. Philos. Soc., 96*, 673–691.

Lawry, J. V. Jr., (1973). Dioptric modifications of the scales overlying the photophores of the lantern fish, *Tarletonbeania crenularis* (*Myctophidae*). *J. Anat., 114*, 55–63.

Lee, H., Kim, J., Yu, H., & Kim, J., (2020). Ontogenetic comparison of larvae and juveniles of *Diaphus garmani* and *Benthosema pterotum* (Pisces: Myctophidae) collected from Korea. *Fish Aquatic Sci., 23*, 18.

Lee, J. Basic Bioluminescence. Photobiology. http://photobiology.info/LeeBasicBiolum.html (accessed on 28February 2023).

Liao, Y., Chen, L., & Shao, K., (2006). Review of the astronesthid fishes (*Stomiiformes*: *Stomiidae*: *Astronesthinae*) from Taiwan with a description of one new species. *Zool. Stud., 45*, 517–528.

Maia, M., Felix, M. G. V., Fabio, A., Nolé, D. D. L., Thierry, E., Flávia, F., Arnaud, L., & Pietsch, B. T. W., (2021). Deep-sea anglerfishes (Lophiiformes: Ceratioidei) from off Northeastern Brazil, with remarks on the ceratioids reported from the Brazilian exclusive economic zone. *Neotrop. Ichthyol., 19*(2), p. e200151 (28p.). doi: 10.1590/1982-0224-2020-0151.

Mallefet, J., Stevens, D. W., & Duchatelet, L., (2021). Bioluminescence of the largest luminous vertebrate, the kitefin shark, *Dalatias licha*: First insights and comparative aspects. *Front. Mar. Sci., 8*, 633582.

Martini, S., & Francis, W. R., (2020). *The Dark Ocean Is Full of Lights*. https://kids.frontiersin.org/articles/10.3389/frym.2020.00069 (accessed on 9 January 2023).

Matsui, T., & Rosenblatt, R. H., (1987). Review of the deep-sea fish family platytroctidae (Pisces: Salmoniformes). *Bull. Scripps Inst. Oceanogr., 26*. https://escholarship.org/content/qt35v4k0ks/qt35v4k0ks.pdf?t=lnq004 (accessed on 9 January 2023).

McEachran, J., (2021). *Fishes of the Gulf of Mexico: Myxiniformes to Gasterosteiformes* (Vol. 1). University of Texas Press.

McFall-Ngai, M. J., & Dunlap, P. V., (1983). Three new modes of luminescence in the leiognathid fish *Gazza minuta*: Discrete projected luminescence, ventral body flash, and buccal luminescence. *Mar. Biol., 73*, 227–237.

McMillan, P. J., et al., (2011). New Zealand fishes. A field guide to less common species caught by bottom and midwater fishing. *NZ Aquat. Envt. Biodiv., 2*, 78.

McMillan, P. J., et al., (2019). New Zealand fishes. A field guide to common species caught by bottom, midwater, and surface fishing- *NZ Aquat. Envt. Biodiv.*, 208.

Melo, M. R. S., (2009). Revision of the genus *Chiasmodon* (*Acanthomorpha*: *Chiasmodontidae*), with the description of two new species. *Copeia*, 583–608.

Melo, M. R. S., Walker, Jr. H. J., & Klepadlo, C., (2007). Two new species of *Pseudoscopelus* (*Teleostei*: *Chiasmodontidae*), with a new diagnosis for the genus. *Zootaxa, 1605*, 33–46.

Merrett, N. R., (1986). Macrouridae of the eastern North Atlantic. *Fich. Ident. Plancton, 173–175*, 14.

Merrett, N. R., (2003). Preliminary guide to the identification of the early life history stages of bathygadid & macrourid fishes of the Western Central North Atlantic. *NOAA Tech. Mem., NMFS-SEFSC-498*, 32.

Mincarone, M. M., Afonso, G. V. F., Dario, F. D., Eduardo, L. N., Frédou, T., Lucena-Frédou, F., Bertrand, A., & Pietsch, T. W., (2021). Deep-sea anglerfishes (*Lophiiformes*: *Ceratioidei*) from off Northeastern Brazil, with remarks on the ceratioids reported from the Brazilian exclusive economic zone. *Neotrop Ichthyol., 19*, e200151.

Miya, M., Pietsch, T.W., Orr, J.W. et al. (2010). Evolutionary history of anglerfishes (Teleostei: Lophiiformes): a mitogenomic perspective. BMC Evol Biol 10, 58. https://doi.org/10.1186/1471-2148-10-58.

Mochizuki, K., & Ohe, F., (1990). *Trigonognathus kabeyai*, a new genus and species of the squalid sharks from Japan. *Ichthyol. Res., 36*, 385–390.

Mooi, R. D., & Jubb, R. N., (1996). Descriptions of two new species of the genus *Pempheris* (Pisces: Pempherididae) from with a provisional key to Australian species. *Rec. Australian Mus., 48*, 117–130.

Moore, J.A., Hartel, K.E., James. (2003). An Annotated List of Deepwater Fishes from off the New England Region, with New Area Records - Scientific Figure on ResearchGate. https://www.researchgate.net/figure/Barbantus-curvifrons-MCZ-60746-76-mm-SL-38u56N-71u39W-0-1017-m_fig7_228692865 [accessed 24 Apr, 2023]

Moser, H. G., & Ahlstrom, E. H., (1972). Development of the lanternfish, *Scopelopszs multipunctatus* position in the family *Myctophidae* and its role in a proposed mechanism for the evolution of photophore patterns in lanternfishes. *Fish. Bull., 70*, 541–564.

Munk, O., (1992). Accessory escal gland (AEG) in some deep-sea anglerfishes. *Acta Zool., 73*, 33–37.

Munk, O., (1998). Light guides of the escal light organs in some deep-sea anglerfishes (Pisces; *Ceratioidei*). *Acta Zool., 79*, 175–186.

Nafpaktitis, B. G., & Paxton, J. R., (1978). *Idiolychnus*, a new genus of *Myctophidae Based* on *Diaphus urolampus*. *Copeia*, 492–497.

Nafpaktitus, B. G., Backus, R. H., Craddock, J. E., Haedrich, R. L., Robison, B. H., & Karnella, C., (2018). *Order Iniomi (Myctophiformes): Part 7*. Yale Univ. Press.

Nakayama, N., Endo, H., & Uchikawa, K., (2015). First record of the midwater grenadier, *Odontomacrurus murrayi* (*Actinopterygii*: *Gadiformes*: *Macrouridae*), from the Northwestern Pacific off Japan. *Species Div., 20*, 135–139.

Nealson, K. H., & Hastings, J. W., (1979). Bacterial bioluminescence: Its control and ecological significance. *Microbiol. Rev., 43*, 496.

Nicol, J. A. C., (1958). Observations on luminescence in pelagic animals. *J. Mar. Biol. Ass. U.K., 37*, 705–752.

Nigel, R., (2020). *Fish Larval Physiology*. CRC Press: Boca Raton.

O'Brien, C. H., & Sizemore, R. K., (1979). Distribution of the luminous bacterium *Beneckea harveyi* in a semitropical estuarine environment. *Appl. Environ. Microbiol*., 928–933.

O'Day, W. T., & Fernandez, H. R., (1974). *Aristostomias scintillans* (Malacosteidae): A deep-sea fish with visual pigments apparently adapted to its own bioluminescence. *Vision Res., 14*, 545–550.

Okamoto, M., & Golani, D., (2018). Three new species of the genus *Acropoma* (*Perciformes: Acropomatidae*) from the Indian Ocean. *Ichthyol. Res., 65*, 101–114.

Okamoto, M., & Ida, H., (2002). *Acropoma argentistigma*, a new species from the Andaman Sea, off southern Thailand (*Perciformes: Acropomatidae*). *Ichthyol. Res., 49*, 281–285.

Okamoto, M., & Nakayama, N., (2016). First records of the luminous Deepwater cardinalfish (*Perciformes: Epigonidae*), *Epigonus macrops* (Brauer, 1906), from the Southeastern Atlantic and Timor Sea. *Species Divers, 21*, 79–83.

Okamoto, M., (2014). *Acropoma profundum*, a new species of lantern belly (*Teleostei*: *Perciformes: Acropomatidae*) from the Solomon Islands. *Species Divers, 19*, 9–14.

Okamoto, M., Randall, J. E., & Motomura, H., (2021). *Acropoma musorstom*, a new lanternbelly (*Acropomatidae*) from the South Pacific and the first record of *Acropoma splendens* from the Andaman Sea off southwestern Thailand. *Ichthyol. Res., 68*, 517–528.

Okamoto, M., Williams, J. T., Carpenter, K., Santos, M. D., & Kimura, S., (2019). Description of a new species of the genus *Acropoma* (*Acropomatidae*) from the Arafura Sea with a redescription of *Acropoma leobergi* Prokofiev 2018 and the first record of *Acropoma boholensis* Yamanoue and Matsuura 2002 from the South China Sea. *Ichthyol. Res., 67*, 1–11.

Orlov, A. M., Orlova, S. Y., Volkov, A. A., & Pelenev, D. V. F., (2015). First record of humpback anglerfish (*Melanocetus johnsonii*) (*Melanocetidae*) in Antarctic waters. *Polar Res., 34*. https://doi.org/10.3402/polar.v34.25356.

Orrell, T. M., & Hartel, K. E., (2016). Platytroctidae. In: Carpenter, K. E., & Angelis, N. D., (eds.), *The Living Marine Resources of the Eastern Central Atlantic: Bony Fishes Part 1 (Elopiformes to Scorpaeniformes)* (Vol. 3, pp. 1771–1782). FAO Species Identification Guide for Fishery Purposes, FAO.

Paitio, J., Oba, Y., & Meyer-Rochow, V. B., (2016). Chapter 12. Bioluminescent fishes and their eyes. In: *Luminescence – An Outlook on the Phenomena and their Applications*. http://www.intechopen.com/books/luminescence-an-outlook-onthe-phenomena-and-their-applications (accessed on 9 January 2023).

Paitio, J., Yano, D., Muneyama, E., Takei, S., Asada, H., Iwasaka, M., & Oba, Y., (2020). Reflector of the body photophore in lanternfish is mechanistically tuned to project the biochemical emission in photocytes for counterillumination. *Biochem. Biophy. Res. Comm., 521*, 821–826.

Paulin, C. D., (1989). Review of the morid genera *Gadella, Physiculus* and *Salilota* (*Teleosta*: *Gadiformes*) with descriptions of seven new species. *New Zealand J. Zool., 16*, 93–133.

Pauly, D., (1977). The *Leiognathidae* (*Teleosts*): An hypothesis relating their mean depth occurrence to the intensity of their countershading bioluminescence. *Mar. Res. Indonesia, 19*, 137–146.

Paxton, J. R., (1963). A new lanternfish (Family *Myctophidae*) of the genus *Lampadena* from the Eastern Pacific Ocean. *Copeia*, 29–33.

Paxton, J. R., (1972). Osteology and relationships of the lanternfishes (Family Myctophidae). *Bull. Nat. Hist. Mus. Los Angeles Coun. Sci.*, 1–81.

Pietsch, T. W., & Sutton, T. T., (2015). A new species of the ceratioid anglerfish genus *Lasiognathus* Regan (*Lophiiformes*: *Oneirodidae*) from the Northern Gulf of Mexico. *Copeia, 103*, 429–432.

Pietsch, T. W., (2009). *Oceanic Anglerfishes: Extraordinary Diversity in the Deep Sea*. Ph.D. Thesis. University of California Press.

Pietsch, T. W., Balushkin, A. V., & Fedorov, V. V., (2006). New records of the rare deep-sea anglerfish *Diceratias trilobus* Balushkin and Fedorov (*Lophiiformes*: *Ceratioidei*: *Diceratiidae*) from the Western Pacific and Eastern Indian Oceans. *J. Ichthyol., 46*(Suppl. 1), S97–S100.

Poulsen, J. Y., (2018). New observations and ontogenetic transformation of photogenic tissues in the tubeshoulder *Sagamichthys schnakenbecki* (*Platytroctidae*, *Alepocephaliformes*). *J. Fish. Biol., 94*, 1–15.

Poulsen, J. Y., (2019). New records of *Leptacanthichthys gracilispinis* and *Microlophichthys microlophus* (*Actinopterygii*: *Lophiiformes*: *Oneirodidae*) from the Subarctic Atlantic Ocean, including new lophiiform barcoding data and a rare observation of a copepod parasite in ceratioid anglerfishes. *Acta Ichthyol. Piscat., 49*, 403–414.

Poulsen, J. Y., Sado, T., Hahn, C., Byrkjedal, I., Moku, M., & Miya, M., (2016). Preservation obscures pelagic deep-sea fish diversity: Doubling the number of sole-bearing opisthoproctids and resurrection of the genus *Monacoa* (*Opisthoproctidae*, *Argentiniformes*). *PLoS One*. https://doi.org/10.1371/journal.pone.0159762.

Priede, I. G., (2017). *Deep-Sea Fishes: Biology, Diversity, Ecology and Fisheries*. Cambridge University Press.

Pringgenies, D., Dewi, K., & Apriliyani, P., (2019). Isolation and identification of symbiont microorganisms from bioluminescent marine life. *Ann. Res. Rev. Biol., 33*, 1–16.

Pringgenies, D., Sastrodihardjo, S., Nganro, N. R., & Aryantha, N., (2001). Bacterial symbionts in the luminous organ of the squid, *Loligo duvauceli,* and cuttlefish, *Sepia* sp. *Phuket Mar. Biol. Cent. Res. Bull.*, 145–148.

Prokofiev, A. M., (2004). Two new species of *Protomyctophum* (Pisces, Teleostei, Myctophidae) from the Southern Ocean. *Beaufortia, 54*, 67–74.

Prokofiev, A. M., (2019). New species of luminous percoids of the *Acropoma* genus (Acropomatidae). *J. Ichthyol., 58*, 939–943.

Renwart, M., & Mallefet, J., (2013). First study of the chemistry of the luminous system in a deep-sea shark, *Etmopterus spinax* Linnaeus, 1758 (*Chondrichthyes*: *Etmopteridae*). *J. Exp.Mar.Biol. Ecol., 448*, 214–219.

Rosenblatt, R. H., & Johnson, G. D., (1991). *Parmops coruscans*, a new genus and species of flashlight fish (*Beryciformes*: *Anomalopidae*) from the South Pacific. *Proc. Biol. Soc. Wash., 104*, 328–334.

Ruby, E. G., & Morin, J. G., (1978). Specificity of symbiosis between deep-sea fishes and psychrotrophic luminous bacteria. *Deep Sea Res., 25*, 161–167.

Sasaki, A., Ikejima, K., Aoki, S., Azuma, N., Kashimura, N., & Wada, M., (2003). Field evidence for bioluminescent signaling in the pony fish, *Leiognathus elongatus. Env. Biol. Fish, 66*, 307–311.

Sazonov, Y. I., (1986). Morphology and classification of the fishes of the family *Platytroctidae* (*Salmoniformes, Alepocephaloidei*). *Tr. Inst Okeanol. Akad. Nauk SSSR, 121*, 51–96.

Sazonov, Y. I., Balanov, A. A., & Fedorov, V. V., (1993). Alepocephaloid fishes (*Alepocephaloidei*) from the Western North Pacific Ocean. *Tr. Inst. Okeanol. Akad. Nauk SSSR, 128*, 40–68.

Sazonov, Y. I., Uiblein, F., Ott, J., & Stachowitsch, M., (1996). Morphology and significance of the luminous organs in alepocephaloid fishes. *Biosys. Ecol., 11*, 151–163.

Schultz, L. P., (1961). Revision of the marine silver hatchetfishes (Family *Sternoptychidae*). *Proc. U. S. Natl. Mus., 112*, 3449.

Shimomura, O., (2012). *Bioluminescence: Chemical Principles and Methods*. World Scientific.

Smith, M. M., & Heemstra, P. C., (2012). *Smiths' Sea Fishes.* Springer Science & Business Media.

Somiya. H., (1977). Bacterial bioluminescence in chlorophthalmid deep-sea fish: A possible interrelationship between the light organ and the eyes. *Cell. Mol. Life Sci., 33*, 906–909.

Sparks, J. S., & Chakrabarty, P., (2015). Description of a new genus of ponyfishes (*Teleostei*: *Leiognathidae), with* a review of the current generic-level composition of the family. *Zootaxa, 3947*, 181–190.

Sparks, J. S., & Chakrabarty, P., (2019). Description of a New Species of Ponyfish (*Teleostei*: *Leiognathidae: Equulitini*: *Photolateralis*) from the Gulf of Oman. *Am. Mus. Novit., 3929*, 1–14.

Struhsakeri, P., (1973). *Argyripnus brocki*, a new species of stomiatoid fish from Hawaii, with observations on *A. ephippiatus* and *A. iridescens*. *Fish. Bull., 71*, 827–836.

Suntsov, A. V., Widder, E. A., & Sutton, T. T., (2008). Bioluminescence Ch.3. Bioluminescence. In: *Fish Larval Physiology* (pp. 51–88).

Syed, A. J., & Anderson, J. C., (2021). Applications of bioluminescence in biotechnology and beyond. *Chem. Soc. Rev., 50*, 5668–5705.

Tanet, L., Martini, S., Casalot, L., & Tamburini, C., (2020). Reviews and syntheses: Bacterial bioluminescence – ecology and impact in the biological carbon pump. *Biogeosciences, 17*, 3757–3778.

Tatsuta, N., Imamura, H., Nakaya, K., Kawai, T., Abe, T., Sakaoka, K., Takagi, S., & Yabe, M., (2014). Taxonomy of mesopelagic fishes collected around the Ogasawara Islands by the T/S oshoro- maru. *Mem. Facul. Fish. Sci. Hokkaido Univ., 56*, 1–64.

Vacquié-Garcia, J., Royer, F., Dragon, A. C., Viviant, M., Bailleul, F., & Guinet, C., (2012). Foraging in the darkness of the Southern Ocean: Influence of bioluminescence on a deep diving predator. *PLoS One, 7*, e43565. https://doi.org/10.1371/journal.pone.0043565.

Wada, M., Kamiya, A., Uchiyama, N., Yoshizawa, S., Kita-Tsukamoto, K., Ikejima, K., Yu, R., et al., (2006). LuxA gene of light organ symbionts of the bioluminescent fish *Acropoma japonicum* (*Acropomatidae*) and *Siphamia versicolor* (*Apogonidae*) forms a lineage closely related to that of *Photobacterium leiognathi* subsp. *mandapamensis*. *FEMS Microbiol Lett., 260*, 186–192.

Widder, E. A., Latz, M. I., & Case, J. F., (1983). Marine bioluminescence spectra measured with an optical multichannel detection system. *Biol. Bull., 165*, 791–810.

Williams, J. T.; Carpenter, K. E.; Van Tassell, J. L.; Hoetjes, P.; Toller, W.; Etnoyer, P.; Smith, M. (2010). Biodiversity Assessment of the Fishes of Saba Bank Atoll, Netherlands Antilles. PLoS ONE 5 (5): e10676. DOI:10.1371/journal.pone.0010676. PMID 20505760. PMC: 2873961.

Wilson, T. W., (2013). *Bioluminescence: Living Lights, Lights for Living*. Harvard University Press.

Wolfe, C. J., & Haygood, M. G., (1993). Bioluminescent symbionts of the Caribbean flashlight fish (*Kryptophanaron alfredi*) have a single rRNA operon. *Mol. Mar. Bio. Biotech., 2*, 189–197.

Yamanoue, Y., & Toda, M., (2008). Redescription of *Acropoma lecorneti* Fourmanoir 1988 (*Perciformes: Acropomatidae*) with the first record from the North Pacific. *Ichthyol. Res., 55*, 198–201.

INDEX

C

F

G

H

N

T